ENVIRONMENTAL and HUMAN SAFETY of MAJOR SURFACTANTS
Alcohol Ethoxylates and Alkylphenol Ethoxylates

Sylvia S. Talmage
Health Sciences Research Division
Oak Ridge National Laboratory
Oak Ridge, Tennessee

 The Soap and Detergent Association

Boca Raton   Ann Arbor   London   Tokyo

Library of Congress Cataloging-in-Publication Data

Talmage, Sylvia S.
　　Environmental and human safety of major surfactants : alcohol ethoxylates and alkylphenol ethoxylates / a report to the Soap and Detergent Association by Sylvia S. Talmage.
　　　　p.　cm.
　　Includes bibliographical references and index.
　　ISBN 1-56670-017-5
　　1. Alcohol ethoxylates--Toxicology.　2. Alkylphenol ethoxylates--Toxicology.　I. Soap and Detergent Association.　II. Title.
RA1242.E67T35　1994
574.2′4--dc20　　　　　　　　　　　　　　　　　　　　　　　　　　　　　　94-17354
　　CIP

　　The data and information contained in this book were compiled by the author and any views or opinions expressed are solely those of the author. Accordingly, neither the Soap and Detergent Association nor its member companies makes any warrant, express or implied, or assumes any responsibility or liability with respect to the use of the information contained in this publication. Further, nothing herein constitutes an endorsement of, or recommendation regarding any product or process by The Soap and Detergent Association.
　　All rights reserved. No part of this book may be reproduced or transmitted in any form or by any means, electronic or mechanical, including photocopying, recording, or by any information storage and retrieval system, without prior permission in writing from The Soap and Detergent Association, 475 Park Avenue South, NY, NY 10016.
　　Direct all inquiries to CRC Press, Inc., 2000 Corporate Blvd., N.W., Boca Raton, Florida 33431. Lewis Publishers is an imprint of CRC Press.

Copyright © 1994 The Soap and Detergent Association.

No claim to original U.S. Government works
International Standard Book Number 1-56670-017-5
Library of Congress Card Number 94-17354
Printed in the United States of America　　2 3 4 5 6 7 8 9 0
Printed on acid-free paper

About the Author

Sylvia S. Talmage is a toxicologist in the Health Sciences Research Division of Oak Ridge National Laboratory where she is involved in the assessment of risk to human health and the environment from chemical contaminants. She received a Ph.D. in Ecology/Environmental Toxicology from the University of Tennessee and has been certified as a Diplomate of the American Board of Toxicology. She is an active member of the Society of Environmental Toxicology and Chemistry, the Society of Toxicology, and Sigma Xi. Dr. Talmage has authored or coauthored several open literature publications and numerous technical manuscripts and reports. Her research experience has involved the transfer of chemical contaminants through the terrestrial ecosystem.

TABLE OF CONTENTS

PART I
ALCOHOL ETHOXYLATES

Chapter   Page

    Synopsis 3
    Nomenclature and Abbreviations 7

I. Introduction 13
    A. Background 13
    B. Chemistry and Manufacture 14
    C. Product CAS Identification 16
    References 18

II. Environmental Levels 19
    A. Analytical Methods 19
       1. Extraction/Separation Methods 20
       2. Physical Methods 21
       3. Chemical Methods 21
       4. Instrumental Analyses 24
    B. Water Quality Standards 27
       1. National Regulations 27
       2. State and Local Standards 27
    C. Nonionic Surfactants in Natural Bodies of Water 27
    References 29

III. Biodegradation 35
    A. Laboratory Investigations 35
       1. Test Systems 35
       2. Biodegradation Studies 36
    B. Field Studies 37
    C. Effect of Chemical Structure 40
       1. Ethoxylate Chain Length 41
       2. Alkyl Chain Structure 42
       3. Propylene Oxide Addition 45
    D. Effect of Environmental Variables 46
    E. Metabolic Pathways of Biodegradation 47
       1. Aerobic Biodegradation 48
       2. Anaerobic Biodegradation 50
    References 51

IV.	Environmental Toxicology	59
	A. Aquatic Toxicity	59
	1. Acute Toxicity to Fish	60
	2. Acute Toxicity to Aquatic Invertebrates	68
	3. Toxicity to Algae	68
	4. Sublethal Toxicity and Behavioral Responses	78
	5. Chronic Toxicity	78
	6. Structure-Activity Relationship	81
	7. Effects of Environmental Variables on Toxicity	83
	8. Bioaccumulation	84
	B. Effects on Microorganisms	84
	C. Effects on Higher Plants	85
	D. Mode of Action	86
	References	88
V.	Human Safety	95
	A. Animal Studies	95
	1. Acute Toxicity	95
	2. Subchronic Exposures	112
	3. Chronic Exposures	114
	4. Acute Irritation	115
	5. Skin Sensitization	120
	6. Carcinogenicity	124
	7. Genotoxicity	125
	8. Developmental/Reproductive Toxicity	125
	9. Absorption, Metabolism, and Disposition	130
	10. Anesthetic and Analgesic Effects	131
	B. Human Studies	132
	1. Dermal Irritation and Sensitization	132
	2. Absorption, Metabolism, and Disposition	133
	3. Therapeutic/Contraceptive Uses	133
	4. Cosmetic Uses	134
	5. Epidemiology	134
	References	135
Appendix		143
References		180

List of Tables

Table	Page
1-1. Nomenclature and CAS Numbers	16
3-1. Removal at Wastewater Treatment Facilities	38
3-2. Primary Biodegradation vs Chemical Structure in the OECD Screening Test	45
3-3. Primary Biodegradation vs Chemical Structure in the OECD Confirmatory Test	45
4-1. Acute Toxicity to Fish	61
4-2. Acute Toxicity to Aquatic Invertebrates	69
4-3. Effects on Algae	76
4-4. Chronic Toxicity Tests with Aquatic Species	79
5-1. Acute Oral Toxicity to Mammals	96
5-2. Acute Dermal Toxicity to Mammals	104
5-3. Acute Inhalation Toxicity to Rats	111
5-4. Rabbit Skin Irritation	116
5-5. Rabbit Eye Irritation	121
5-6. Genotoxicity Assays	126

List of Figures

Figure	Page
1-1. Typical distribution of EO units per mole of alcohol for a C_nAE_7 Product.	15
4-1. Relationship between Toxicity to Fish and Ethoxylate Chain Length	82
5-1. Acute Oral LD_{50} Values in the Rat	102

PART II
ALKYLPHENOL ETHOXYLATES

Chapter	Page
Synopsis	191
Nomenclature and Abbreviations	195

I. Introduction 199
   A. Chemistry and Manufacture 199
   B. Uses 200
   C. Product CAS Identification 201
   References 203

II. Environmental Levels 205
   A. Analytical Methods 205
      1. Physical Methods 205
      2. Chemical Methods 206
      3. Instrumental Analyses 208
   B. Standards and Regulations 217
      1. Water Quality Standards 217
      2. Air Emission Standards 217
   C. Nonionic Surfactants in Natural Water Bodies 217
      1. Water Column 217
      2. Sediments 223
      3. Groundwater 226
   References 227

III. Biodegradation 235
   A. Laboratory Investigations 235
      1. Test Methods 235
      2. Primary and Ultimate Biodegradation 237
      3. Other Methods of Removal 239
   B. Field Studies 240
      1. Waste Treatment Facilities 240
      2. Soil 250
      3. Pesticide Spraying Programs 251
   C. Effect of Chemical Structure 251
      1. Alkyl Chain Structure 251
      2. *o*-, *m*-, *p*-Substitution 252
      3. EO Chain Length 252
   D. Metabolic Pathways of Biodegradation 253
   References 256

IV. Environmental Toxicology 263
   A. Aquatic Toxicity 263
      1. Acute Toxicity to Fish 263
      2. Acute Toxicity to Aquatic Invertebrates 270
      3. Toxicity to Algae 276
      4. Sublethal Effects 280
      5. Chronic Toxicity 280
      6. Structure-Activity Relationship 281
      7. Effects of Environmental Variables on Toxicity 285
      8. Bioaccumulation 285
      9. Comparison of Toxicity Data with Environmental Concentrations 287
   B. Effects on Microorganisms 287
   C. Effects on Higher Plants 288
   D. Effects on Birds and Wildlife 291
   E. Mode of Action 291
   References 292

V. Human Safety 299
   A. Animal Studies 299
      1. Acute Exposures 299
      2. Subchronic Exposures 305
      3. Chronic Exposures 308
      4. Dermal and Ocular Irritation 308
      5. Skin Sensitization 311
      6. Carcinogenicity 311
      7. Genotoxicity 314
      8. Developmental/Reproductive Toxicity 314
      9. Other Effects 322
     10. Fate and Disposition 323
   B. Human Studies 324
      1. Dermal Irritation and Sensitization 325
      2. Pharmacokinetics 325
      3. Therapeutic/Contraceptive Uses 325
      4. Cosmetic Uses 326
      5. Epidemiology 326
   References 327

Appendix 335
References 358
Index 367

List of Tables

Table	Page
1-1. Nomenclature and CAS Numbers of Alkylphenol Ethoxylates and Metabolites	201
2-1. Analytical Methods for APE Surfactants and Metabolites	209
2-2. Concentrations of Alkylphenol Ethoxylates and Alkylphenols in Rivers	218
2-3. U.S. River Sampling Locations	221
2-4. Concentration of 4-Nonylphenol in Textile Waste and Receiving Waters	222
2-5. Concentrations of Nonylphenol and Nonylphenol Ethoxylates in Surface Waters and Sediments	224
3-1. Removal at Wastewater Treatment Facilities	241
3-2. Adsorption of Metabolites to Sludge	248
4-1. Acute Toxicity of Alkylphenol Ethoxylates and Alkylphenols to Fish	264
4-2. Acute Toxicity of Alkylphenol Ethoxylates and Alkylphenols to Aquatic Invertebrates	271
4-3. Effects on Growth of Algae	277
4-4. Chronic Effects on Aquatic Organisms	282
5-1. Acute Oral Toxicity of Alkylphenols and Alkylphenol Ethoxylates to Mammals	300
5-2. Acute Dermal Toxicity of Nonylphenol and Alkylphenol Ethoxylates to Rabbits	304
5-3. No-Observed Effect Level in 90-Day Feeding Studies	306
5-4. Cardiotoxicity in Dogs Fed APE of Various Chain Lengths	307
5-5. Rabbit Skin Irritation	309
5-6. Rabbit Eye Irritation	312
5-7. Genotoxicity Assays	315
5-8. ^{14}C in Tissues of the Rat 96 Hours after Oral Administration of Ethylene Oxide Labeled C_8APE_6	324

List of Figures

Figure	Page
3-1. Primary Biodegradation.	254
4-1. Relationship between Toxicity and Ethylene Oxide Chain Length.	284

PART I

ALCOHOL ETHOXYLATES

SYNOPSIS

Alcohol ethoxylates (AE) are presently the largest volume nonionic surfactants produced in the U.S., with linear primary AE the predominant type. Approximately 632 million pounds of linear AE were produced during 1988. Significant quantities of AE were converted to alcohol ethoxysulfates, with the remaining approximately 400 million pounds of AE used primarily in household laundry detergents. Linear primary AE, along with other AE, also find many smaller volume uses in household cleaners, institutional and industrial cleaners, cosmetics, agriculture, and in textile, paper, oil and other process industries. Growth in use of linear primary AE has been rapid over the past 20 years because of their many desirable qualities such as rapid biodegradation, low to moderate foaming ability, superior cleaning of man-made fibers, tolerance of water hardness, and ability to perform in cold water.

Alcohol ethoxylates are manufactured by the reaction of alcohols with ethylene oxide. The alcohols used are typically linear primary with a mixture of carbon chain lengths in the range of 8 to 18 carbons. Linear random secondary and branched alcohols are also used. Ethoxylate chains typically range from an average of 1 to 12 ethylene oxide units. Alcohol ethoxylates are described by the average number or range of carbon atoms in the alkyl chain and the average number of ethylene oxide units, e.g., $C_{12-15}AE_6$.

Sensitive analytical methods specific to the component chemical moieties of AE have been developed. These methods include (1) separation by gas chromatography and by normal- and reversed-phase high performance liquid chromatography coupled with UV or fluorescent detection and (2) mass spectrometry methods. These methods have primarily been applied to biodegradation studies in the laboratory. These methods have also been applied to environmental samples of river water and the influent, effluent, and sludge from wastewater treatment plants. In the future, detection systems based on evaporative light scattering will likely be increasingly used. Chemical and physical methods used in the past for environmental samples did not distinguish between AE, alkylphenol ethoxylates (APE, another major class of nonionic surfactants), and other interferences or between intact surfactants and degradation intermediates or products and thus tended to overestimate concentrations. The most commonly used of these chemical methods that detect nonionic surfactants are measurement of cobalt thiocyanate active substances and measurement of bismuth iodide active substances. The two principal physical procedures are measurement of foaming and measurement of surface tension.

Alcohol ethoxylates with predominantly linear alkyl chains exhibit a high degree of biodegradability under most test procedures. Both rapid primary and ultimate biodegradation have been demonstrated under a variety of laboratory and field conditions. The factors that affect the rate and extent of biodegradation are the type and extent of branching in the alkyl chain, the extent of incorporation of propylene oxide into the ethoxylate chain, and the length of the ethoxylate chain. Biodegradation is retarded by a high degree of alkyl chain branching, secondary alkyl structure, the addition of more than about three equivalents of propylene oxide to the ethoxylate moiety, and by an ethoxylate chain length of >20 units. These characteristics that retard biodegradation are not present in the type of AE most commonly used in consumer detergent formulations.

Biodegradation is due to bacterial action. The primary mechanism of degradation for linear primary AE appears to be hydrolysis of the ether linkage next to the alkyl chain followed by rapid oxidation of the alkyl chain with liberation of two-carbon units as acetyl groups (β-oxidation) and a slower oxidation of the polyoxyethylene chain, also by two-carbon units. Both chains are subsequently biodegraded to carbon dioxide and water. The terminal ends of the fatty alcohol (ω-oxidation followed by β-oxidation) and the polyoxyethylene chains are also initial points of bacterial attack. For highly branched alkyl chains, oxidation has been reported to start at the terminal end of the polyoxyethylene chain with little or no oxidation of the branched fatty alcohol.

Limited sampling in U.S. rivers showed concentrations of total nonionic surfactants, as measured by chemical non-specific methods, of 0.03 to 0.08 mg/L; near wastewater treatment plant outfalls, concentrations ranged up to 0.42 mg/L. In surface waters worldwide, nonionic surfactants, as measured by chemical non-specific methods, ranged up to 2.0 mg/L, with most concentrations ≤ 0.5 mg/L. Few data on environmental concentrations of AE using AE-specific analytical methods are available. The only measured concentration is that of the commonly used $C_{14-15}AE_7$ which was present in a single U.S. riverwater sample at a concentration of 0.5 μg/L. Below a wastewater treatment plant outfall the concentration was 1.1 μg/L.

At wastewater treatment plants in the U.S., nonionic surfactants are generally >90% removed during secondary treatment. Measurement of AE in the effluent following dosing of intact linear AE into the influent also indicated extensive removals regardless of seasonal changes in water temperature. Concentrations of nonionic surfactants in the influent ranged up to 9 mg/L whereas effluent concentrations ranged up to 2.3 mg/L. Concentrations in receiving waters are usually much lower due to rapid dilution.

Although the toxicity of AE surfactants to aquatic organisms is compound and species-specific, several generalizations concerning chemical structure can be made. Toxicity increases with an increase in the alkyl chain length, decreases with an increase in the ethoxylate chain length, and decreases with methyl branching of the alkyl chain. Surfactant liposolubility appears to be the major factor in determining toxicity. The uptake, distribution, and elimination of radiolabeled compounds in fish is rapid. Whole-body bioaccumulation factors range up to 800; half-lives range from one to several days.

In laboratory studies, linear primary AE surfactants are acutely toxic to freshwater fish and crustaceans at median lethal (LC_{50}) concentrations ranging from 0.4 to 10 mg/L and 0.29-0.4 to 20 mg/L, respectively. Most LC_{50} values are <5 mg/L. The lowest reported values were for the brown trout (*Salmo trutta*), 0.4 mg/L of $C_{16,18}AE_{14}$, and the cladoceran (*Daphnia magna*), 0.29-0.4 mg/L of $C_{14-15}AE_7$. Concentrations that decreased growth of algae by 50% (EC_{50}) ranged from 0.09 to >10 mg/L. In chronic studies, survival and reproduction of fish (*Pimephales promelas*) and invertebrates (*Daphnia magna*) were not affected at concentrations of $C_{14-15}AE_7$ of 0.18 mg/L and 0.24 mg/L, respectively. Alcohol ethoxylates with highly branched alkyl groups or with secondary attachment of the alkyl group are less acutely toxic than linear primary AE but are slower to degrade.

Sublethal effects and behavioral responses to AE may make organisms more susceptible to disease or predation. For several intact surfactants, the respiratory rate of bluegill sunfish was affected at 1.2 mg/L and the swimming activity of cod was affected at 0.5 mg/L.

The effect of AE surfactants on soil microorganisms depends on the species present, their acclimation history, and the type and concentration of surfactant. While a concentration of 1.5 mg/L of n-pri-$C_{12-15}AE_9$ was inhibitory to one species of bacteria, concentrations of 80-100 mg/L of $C_{12-15}AE_7$ did not inhibit nitrifying microbes present in sewage. Watering soil with a solution of 1000 mg of n-sec-$C_{11-15}AE_9$/L reduced microfungi biomass. Intact higher plants watered with or grown in surfactant solutions were not affected at concentrations ≤ 100 mg/L; in some cases, low concentrations were growth promoting. Growth of cell suspensions of soybeans was inhibited at concentrations ranging from 2 to 75 mg/L depending on the individual surfactant.

An evaluation of the data dealing with various measures of mammalian toxicity as indicators of potential human toxicity as well as actual human exposures indicate that AE do not represent a hazard to human health. In laboratory animals, AE exhibit a low order of acute toxicity by the oral, dermal, and inhalation routes of exposure. In the rat, oral LD_{50} values range from 544 to >25,000 mg/kg, with toxicity increasing with an increase in ethoxylate units from 1 to 11. Dermal LD_{50} values in rabbits ranged from 2000 to >5000 mg/kg. At toxic dermal doses, a pattern of delayed deaths and visual evidence of lung injury associated with the presence of aspirated food particles has been reported in rabbits but not in rats. Inhalation studies indicate that AE are not acutely toxic at concentrations less than or equal to their saturated concentration in air. Undiluted AE are moderately to severely irritating to the skin and eye of rabbits. Concentrations of 1.0% are slightly irritating and concentrations of 0.1% are generally non-irritating. Repeated dermal applications do not cause a sensitization reaction.

Alcohol ethoxylates test negative in *in vitro* genotoxicity assays. In longer-term studies with laboratory animals, AE did not cause reproductive, genotoxic, or carcinogenic effects. Orally-administered AE are rapidly absorbed from the gastrointestinal tract and rapidly eliminated. Metabolism and route of elimination depend on chemical structure, with the urine, feces, and expired CO_2 being routes of elimination.

The use of AE in cosmetic and contraceptive preparations and as analgesics and anesthetics in human therapy have not produced any reported unfavorable reactions.

NOMENCLATURE AND ABBREVIATIONS

Throughout this review the abbreviation AE has been used to designate alcohol ethoxylate(s). Alcohol ethoxylates are nonionic surfactants composed of a long-chain fatty alcohol (hydrophobe moiety) combined, via an ether linkage, with one or more ethylene oxide (EO) or (poly)oxyethylene units (hydrophile moiety). Some AE may contain propylene oxide (PO) units as part of the hydrophile moiety. The average number or range of carbon atoms in the alkyl chain (C) and the average number or average range of ethoxylate units (AE or EO) are designated by subscripts ($C_{12-15}AE_{7-9}$ or EO_7). Alcohols derived from natural sources contain an even number of carbon atoms (e.g., $C_{12,14}$). Alcohols derived from petrochemical sources may be even carbon numbered (e.g., $C_{12,14}$), mixtures of even and odd (e.g., C_{12-15}, i.e., C_{12}, C_{13}, C_{14}, C_{15}) or all odd carbon chains (e.g., $C_{13,15}$).

If the information was available, the alcohol portion of the AE was designated as (1) linear or branched and (2) primary or secondary. Most of the alcohols used in the manufacture of AE surfactants are saturated primary alcohols with a high degree of linearity. Examples of the representative structures of the alcohols and an AE are given:

$$CH_3-CH_2-CH_2-CH_2-CH_2-CH_2-CH_2-CH_2-CH_2-CH_2-CH_2-CH_2-OH$$

C_{12} linear primary alcohol, n-pri-C_{12}-OH

$$CH_3-CH_2-CH_2-CH_2-CH_2-CH_2-CH_2-CH_2-CH_2-\underset{\underset{CH_3}{|}}{CH}-CH_2-OH$$

C_{12} high linearity primary alcohol, branched isomer

$$CH_3-CH_2-CH_2-CH_2-CH_2-\underset{\underset{OH}{|}}{CH}-CH_2-CH_2-CH_2-CH_2-CH_2-CH_3$$

C_{12} linear random secondary alcohol, example isomer

$$H-(CH_2-\underset{\underset{CH_3}{|}}{\overset{}{CH}})_4-CH_2-OH$$

iso-C_{13} branched primary alcohol, example isomer

$$CH_3-(CH_2)_{11}-O-(CH_2-CH_2-O-)_3H$$

Linear primary alcohol ethoxylate (n-pri-$C_{12}AE_3$)

If provided, surfactant tradenames for alcohol ethoxylates and other nonionic surface active agents, enclosed in parentheses, follow the chemical names. In-text references to the products are not capitalized and do not carry the registered trademark symbol. Product tradenames/suppliers for U.S. and U.S.-affiliated manufactured products and the manufacturers/suppliers follow.[1]

Product	Company
ADSEE®	Witco Corporation
ALFONIC®	Vista Chemical Company
ALLWET	W.A. Cleary Corporation
AMEROXOL®	Amerchol Corporation
ANATAROX	Rhone-Poulenc
ARLASOLVE	ICI Americas Inc.
ARMIX, ARMUL	Witco Corporation
AROSURF	Sherex Chemical Company, Inc.
ATRANONIC POLYMER 20	Atramax Inc.
BIOSOFT	Stepan Company
BRIJ®	ICI Americas Inc.
	Atlas Chemical Industries, London, UK
CERFAK 1400	E.F. Houghton & Company
CETOMACROGOL 1000 BP	Croda Inc.
CHEMAL	Chemax Inc.
COREXIT	Exxon Chemical Company
DECERESOL	American Cyanamid Company
DEHYDOL	Henkel Canada Ltd.
DESONIC®	Witco Corporation
DIONIL	Huls America Inc.
DISPONIL	Henkel Corporation/Functional Products Group
DOBANOL®a	Shell International Chemical Company, London, UK
ECCOSCOUR, ECCOTERGE	Eastern Color & Chemical Company
EMERY, EMTHOX	Henkel Corporation/Emery Group

[1]*McCutcheon's Emulsifiers and Detergents.* 1992. North American Ed., MC Publishing Company, McCutcheon Division, 175 Rock Pond, Glen Rock, New Jersey.

Product	Company
ETHAL	Ethox Chemicals, Inc.
ETHOSPERSE®	Lonza Inc.
EMULGIN	Henkel Corporation
EXAMIDE	Soluol Chemical Company, Inc.
FLO MO®	Witco Corporation
FLUORAD	3M Company
FORLAN	R.I.T.A. Corporation
G-1120	ICI Americas
GENAPOL®	Hoechst Celanese Corporation
GENEROL	Henkel Canada Ltd.
GRADONIC	Graden Chemical
HARTOPOL, HARTOWET	Hart Chemical Ltd.
HETOXIDE, HETOXOL	Heterene Chemical Company
HIPOCHEM, HIPOSCOUR	High Point Chemical Corp.
HOSTACERIN	Hoechst Celanese Corporation
HYONIC	Henkel Corporation
IBERTERG	A. Harrison & Company, Inc.
ICONOL	BASF Corporation
INCROPOL	Croda Inc.
INDUSTROL®	BASF Corporation
INTRATEX	Crompton & Knowles
KIERALON	BASF Corporation/Fibers Division
LEVELENE	Ciba-Geigy Corporation
LEXEMUL	Inolex Chemical Company
LIPOCOL	Lipo Chemicals, Inc.
MACOL®	PPG Industries
MAKON	Stepan Company
MARLIPAL, MARLOWET	Huls America Inc.
MAZAWET®	PPG Industries
MERPOL	E.I. DuPont De Nemours & Company
NEODOL®[a]	Shell Chemical Company
NOVEL®	Vista Chemical Company
NORFOX	Norman, Fox & Co.
OCENOL	Henkel Corporation/Emery Group
PLURAFAC®, PLURONIC®	BASF Corporation
POLYCHOL, PROCETYL	Croda Inc.
POLYLUBE	Hart Chemical Ltd.

Product	Company
POLYTERGENT®	Olin Chemical Corporation
PROMULGEN®	Amerchol Corporation
RENEX	ICI Americas Inc.
REXOL, REXONIC, REXOPAL	Hart Chemical Ltd.
RHODASURF	Rhone-Poulenc
RITACHOL, RITAPRO, RITAWAX, RITOLETH	R.I.T.A. Corporation
Ross Emulsifier	Ross Chemical Company
SANDOXYLATE	Sandoz Chemicals Corporation
SOULAN®	Amerchol Corporation
STEPANOL	Stepan Company
SULFOTEX	Henkel Canada Ltd.
SURFLO	Exxon Chemical Company
SURFONIC®	Texaco Chemical Company
SYN FAC	Milliken Chemical Company
SYNTHRAPOL	ICI Americas Inc.
T-DET	Harcros Organics
TERGITOL®	Union Carbide Corporation
TEX-WET	Intex Chemical Company
TINEGAL®	Ciba-Geigy Corporation
TRITON®	Union Carbide Corporation
TRYCOL, TRYLON	Henkel Corporation
UNIPEROL	BASF Corporation
VALDET	Valchem Chemical Company
VARONIC	Sherex Chemical Company, Inc.
VOLPO	Croda Inc.
WITCONOL	Witco Corporation

[a]The NEODOL® and DOBANOL® tradenames are for equivalent Shell products. The Neodol name is used in North America and the Dobanol name in the rest of the world.

NOMENCLATURE AND ABBREVIATIONS

The following nomenclature and abbreviations have been used throughout the text:

AE	Alcohol ethoxylate
AE_n or EO_n	Ethylene oxide moiety with an average of n units or moles
AES	Alcohol ethoxysulfates
APE	Alkylphenol ethoxylates
APHA	American Public Health Association
ASTM	American Society for Testing and Materials
BCF	Bioconcentration factor
BIAS	Bismuth iodide active substances
BOD	Biochemical oxygen demand
br	Branched
Cetyl alcohol	C_{16} alcohol
C_n	Alkyl group with n carbons
CMC	Critical micelle concentration
Coconut alcohol	$C_{12,14,16}$ alcohol
COD	Chemical oxygen demand
CTAS	Cobalt thiocyanate active substances
DOC	Dissolved organic carbon
EC_{50}	Concentration effective to 50% of organisms
EO	Ethylene oxide
FOEC	First-observed-effect concentration
g	Gram
GC	Gas chromatography
HPLC	High performance liquid chromatography
K_{ow}	Octanol/water partition coefficient
L	Liter
LAS	Linear alkylbenzenesulfonate
Lauryl alcohol	C_{12} fatty alcohol
LC_0	Concentration lethal to 0% of organisms
LC_{50}	Concentration lethal to 50% of organisms
LD_{50}	Dose lethal to 50% of organisms
LOEC	Lowest-observed-effects concentration
mg	Milligram
mL	Milliliter
n	Normal, linear
NOEC	No-observed-effects concentration
Oleyl alcohol	C_{18} natural fatty alcohol with double bond at C_9
oxo-Alcohol	Primary alcohols produced from an olefin by reaction with carbon monoxide and hydrogen
PEG	Polyethylene glycols

PO	Propylene oxide
POE	Polyoxyethylene
PPAS	Potassium picrate active substances
ppm	Parts per million (=mg/L)
pri	Primary
SCAS	Semi-continuous activated sludge
sec	Secondary
SDA	Soap and Detergent Association
Stearyl alcohol	C_{18} fatty alcohol
Tallow alcohol	$C_{16,18}$ fatty alcohol (derived from beef tallow)
TCO_2	Theoretical carbon dioxide
TLC	Thin layer chromatography
TOC	Total organic carbon
tp-	Tetrapropylene
μg	Microgram
UV	Ultraviolet

I. INTRODUCTION

A. BACKGROUND

Alcohol ethoxylates (AE) are nonionic surfactants that have been used in significant amounts in industrial products since the 1930's. Usage grew after World War II in household and institutional cleaners, and from the mid-1960's in laundry products. Growth through the 1970's and 1980's has been rapid, especially in consumer products (Cahn and Lynn, 1983).

Alcohol ethoxylates are the largest volume nonionic surfactants produced, with ethoxylates based on high linearity primary alcohols the predominant type. Household laundry detergents are the largest single end-use for AE. Linear primary AE are the preferred nonionic surfactants for this use due to their rapid biodegradability. Linear primary AE, along with other AE, also find many smaller volume and more specialized uses in household cleaners, institutional and industrial cleaners, cosmetics, agriculture, and in textile, paper, oil and other process industries (Schick, 1967; 1987; Richtler and Knaut, 1988).

The rapid growth in usage of AE in household laundry products over the past 20 years has been driven by trends to low- and no-phosphate formulations, more man-made fibers in the wash load, and cooler washing temperatures. Alcohol ethoxylates are less sensitive to water hardness than competing materials. They are also particularly effective at removing oily soils and are relatively low foamers. Alcohol ethoxylates are used in powdered and particularly in liquid laundry detergents which now account for 35-40% of the U.S. market (Chemical Week, 1990; Hepworth, 1990).

The U.S. International Trade Commission (1988) reported that of 1745 million pounds of nonionic surfactants produced in the USA during 1988, 721 million pounds were AE and other non-benzenoid ethers. About 632 million pounds were linear AE.

Schirber (1989) reported that approximately 300 million pounds of AE were consumed in 1988 in the U.S. in household detergents. An additional 100 million pounds were utilized in institutional and industrial applications. Significant quantities of AE, particularly those with 1 to 3 oxyethylene units on average, are further reacted to produce alcohol ethoxysulfates (AES). Other small volume specialized surfactant derivatives of AE include phosphate esters, carboxymethylates, sulfosuccinates, etc.

This report reviews available safety information on AE with respect to

- Product chemistry and analysis
- Biodegradation
- Environmental levels, including fate and distribution
- Aquatic toxicity
- Human safety

The report covers information in the public domain along with unpublished data submitted by Soap and Detergent Association (SDA) member companies. It provides a summary and critical

B. CHEMISTRY AND MANUFACTURE

AE may be derived from linear or branched primary alcohols, and from linear random secondary alcohols. The alcohols typically contain alkyl chains with 8 to 18 carbon atoms, with C_{12} to C_{16} being the most common range. The ethoxylate chain typically averages from 1 to 12 oxyethylene units.

Commercially, primary AE are manufactured by the reaction of detergent-range primary alcohols with ethylene oxide (EO). The parent primary alcohols typically have linear or essentially linear alkyl chains. Linear alcohols include those derived from vegetable or animal sources and those derived from ethylene via Ziegler chemistry. Also included in this category are alcohols derived from linear olefins and synthesis gas (CO/H_2) via oxo-type chemistry. The olefins may be ethylene or paraffin derived. Such oxo-alcohols are predominantly linear with, typically in the U.S., around 20-25% of the alkyl chains containing a branch at the 2-position, mainly 2-methyl. Commercial grades of linear alcohols typically contain a range of carbon chain lengths, e.g., $C_{12,14,16}$, C_{12-15}, etc. (Gautreaux et al., 1978; Peters, 1978).

Ethoxylation of primary alcohols is typically carried out in a batch reactor under mild pressure and temperature using a basic catalyst, usually potassium or sodium hydroxide.

$$ROH + nCH_2\underset{O}{\diagdown\diagup}CH_2 \xrightarrow[\text{2) HAc}]{\text{1) KOH}} RO(CH_2CH_2O)_nH$$

The reaction product is a mixture of polyoxyethylene adducts with a broad distribution of EO-adduct chain lengths (Figure 1-1) and some unreacted alcohol (Geissler, 1989; Johnson et al., 1990). Commercial materials are typically described by the average moles of EO added per mole of alcohol (EO_n or AE_{m-n}), or by the percent EO in the product.

After addition of EO, the reaction mixture is reacted down or "soaked" to complete reaction of the final traces of ethylene oxide, then usually neutralized with a trace of acid such as acetic or phosphoric acid. The product is essentially 100% active and is shipped as is. Alcohol ethoxylates range in physical form from clear liquids to pastes to solids depending on the molecular weight of the starting alcohol and degree of ethoxylation.

Relatively small quantities of AE are produced for specialty uses from other alcohols. These include primary oxo-alcohols derived from highly branched propylene and butylene oligomers, unsaturated linear primary alcohols such as oleyl alcohol, and random linear secondary alcohols:

$$CH_3(CH_2)_x\underset{OH}{\overset{|}{C}H}(CH_2)_yCH_3$$

INTRODUCTION 15

The latter, diagrammed above, are produced by the borate-modified oxidation of normal paraffins (Kurata and Koshida, 1978). They are much less reactive than primary alcohols and require a specialized two-step procedure for ethoxylation. A seed ethoxylate is first produced using an acid-catalyzed reaction followed by product purification steps. Additional EO is then added using conventional base catalysis. Secondary alcohol ethoxylates were more widely used in the USA in the 1960's and 1970's. Usage is now limited to specialized applications for economic reasons.

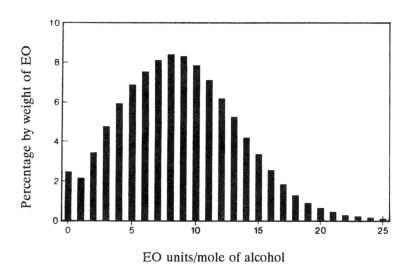

Figure 1-1. **Typical distribution of EO units per mole of alcohol in a C_nAE_7 product.**
Source: Shell Chemical Company

The properties of AE surfactants may be modified by various means. The oldest technique involves the use of propylene oxide (PO) in addition to EO during alkoxylation. The PO is either incorporated randomly by alkoxylation with an epoxide mixture, or in a block by sequencing the addition of EO and PO to the reactor. The uses of such alkoxylates are relatively specialized. Structures will be noted where data are reported.

Alcohol ethoxylate products with modified EO-adduct distributions have also been introduced. These include AE where the more volatile components have been removed by evaporation (Smith et al., 1972). More recently, peaked distribution ethoxylates have been commercialized using a variety of catalytic systems (Dillan, 1985; Matheson et al., 1986).

In addition to the principal components (ethoxylated alcohol adducts and some unethoxylated starting alcohol) a conventional commercial grade of AE typically contains on the order of 0.5 to 2.0% weight polyethylene glycols and 0.1 to 0.2% weight neutralized catalyst, commonly potassium acetate.

Trace levels of certain regulated chemicals may also be present. The following levels are representative of those reported on supplier's material safety data sheets as an "upper bound concentration" or "typical maximum" (MSDS, 1990).

	Level (ppm)
Ethylene oxide	<10
1,4-Dioxane	<15
Formaldehyde	<4
Acetaldehyde	<6

Actual levels will typically be lower, dependent on grade and intended end use. Stafford (1980), for example, has reported no detectable 1,4-dioxane in a series of base catalyzed linear primary AE using an analytical method with a detection limit of 0.5 ppm.

C. PRODUCT CAS IDENTIFICATION

The CAS (Chemical Abstract Service) numbers of commercially available AE surfactants are listed in Table 1-1 (TSCA, 1979). Most of the CAS numbers refer to a generic class of ethoxylated alcohols with average EO units >1.

Table 1-1. Nomenclature and CAS Numbers

Chemical	CAS Number
C_{12} Alcohol ethoxylates	9002-92-0
C_4 Alcohol ethoxylates	9004-77-7
C_{16} Alcohol ethoxylates	9004-95-9
Oleyl alcohol ethoxylates	9004-98-2
C_{18} Alcohol ethoxylates; stearyl alcohol, ethoxylated	9005-00-9
Mono(trimethylnonyl)polyethylene glycol	9008-57-5
Monobutyl methylpolyethoxylate, polypropoxylate	9038-95-3
C_{13} Alcohol ethoxylates	24938-91-8
C_{10} Alcohol ethoxylates; Poly(oxy-1,2-ethanediyl), α-decyl-w-hydroxyl	26183-52-8
C_8 Alcohol ethoxylates	27252-75-1
C_{14} Alcohol ethoxylates	27306-79-2
C_6 Alcohol ethoxylates	31729-34-8
C_{15} Alcohol ethoxylates	34398-05-5
Polyoxyethylene isolauryl ether	60828-78-6
br-C_{12} Alcohol ethoxylates	61702-78-1
Tallow alcohol, ethoxylated	61791-28-4

INTRODUCTION

Table 1-1 (cont.)

Chemical	CAS Number
C_{12-13} Alcohol ethoxylates	66455-14-9
C_{10-14} Alcohol ethoxylates	66455-15-0
C_{16-18} Alcohols, ethoxylated, propoxylated	68002-96-0
C_{10-16} Alcohol ethoxylates	68002-97-1
C_{12-15} Alcohol ethoxylates	68131-39-5
C_{11-15} sec-Alcohol ethoxylates	68131-40-8
C_{14-18} Alcohol ethoxylates	68154-96-1
C_{10-12} Alcohols, ethoxylated, propoxylated	68154-97-2
C_{14-18} Alcohols, ethoxylated, propoxylated	68154-98-3
C_{12-18} Alcohol ethoxylates	68213-23-0
C_{12-16} Alcohols, ethoxylated, propoxylated	68213-24-1
C_{6-12} Alcohol ethoxylates	68439-45-2
C_{9-11} Alcohol ethoxylates	68439-46-3
C_{16-18} Alcohol ethoxylates	68439-49-6
C_{12-14} Alcohol ethoxylates	68439-50-9
C_{12-14} Alcohols, ethoxylated, propoxylated	68439-51-0
C_{11-13} sec-Alcohol ethoxylates	68439-54-3
C_{12-20} Alcohol ethoxylates	68526-94-3
C_{12-20} Alcohols, ethoxylated, propoxylated	68526-95-4
C_{12-16} Alcohol ethoxylates	68551-12-2
C_{12-15} Alcohols, ethoxylated, propoxylated	68551-13-3
C_{11-15} sec-Alcohols, ethoxylated, propoxylated	68551-14-4
C_{12-19} Alcohol ethoxylates	68603-20-3
C_{8-10} Alcohols, ethoxylated, propoxylated	68603-25-8
C_{6-12} Alcohols, ethoxylated, propoxylated	68937-66-6
C_{6-10} Alkyl and C_6 branched aliphatic alcohol ethoxylates	68987-81-5
C_{7-21} Alcohol ethoxylates	68991-48-0
C_{1-2} Alcohol ethoxylates	69012-85-7
C_{8-18} Alcohols, ethoxylated, propoxylated	69013-18-9
C_{8-22} Alcohol ethoxylates	69013-19-0
C_{16-22} Alcohol ethoxylates	69227-20-9
C_{12-18} Alkyl alcohols, ethoxylated, propoxylated	69227-21-0
C_{6-10} Alkyl alcohols, ethoxylated, propoxylated	69227-22-1
C_{6-10} Alcohol ethoxylates	70879-83-3
C_{14-26} Alcohol ethoxylates	71011-10-4
C_{8-10} Alcohol ethoxylates	71060-57-6
C_{8-16} Alcohol ethoxylates	71243-46-4
C_{13-18} Alcohol ethoxylates	73905-87-4

REFERENCES

Cahn, A. and J.L. Lynn. 1983. Surfactants and detersive systems. In: Encyclopedia of Chemical Technology, 3rd ed., Vol. 22. John Wiley & Sons, New York, pp. 360-363.
Chemical Week. Special report. January 31, 1990, p. 60.
Dillan, K.W. 1985. Effects of the ethylene oxide distribution on nonionic surfactant properties. J. Am. Oil Chem. Soc. 62:1144-1151.
Gautreaux, M.F., W.T. Davis and E.D. Travis. 1978. Higher aliphatic alcohols (synthetic). In: Encyclopedia of Chemical Technology, Vol. 1. John Wiley & Sons, New York.
Geissler, P.R. 1989. Quantitative analysis of ethoxylated alcohols by supercritical fluid chromatography. J. Am. Oil Chem. Soc. 66:685-689.
Hepworth, P. 1990. Heavy duty laundry liquids. Chemistry and Industry, March 19th, p. 166.
Johnson, A.E., Jr., P.R. Geissler and L.D. Talley. 1990. Determination of relative ethoxylation rate constants from supercritical fluid chromatographic analysis of ethoxylated alcohols. J. Am. Oil. Chem. Soc. 67:123-131.
Kurata, N. and K. Koshida. 1978. Oxidize n-paraffins for sec-alcohols. Hydrocarbon Processing (January), p. 145-151.
Matheson, K.L., T.P. Matson and K. Yang. 1986. Peaked distribution ethoxylates - their preparation, characterization and performance evaluation. J. Am. Oil. Chem. Soc. 63:365-370.
MSDS (Material Safety Data Sheets). 1990. Suppliers material safety data sheets on AE.
Peters, R.A. 1978. Higher aliphatic alcohols (natural). In: Encyclopedia of Chemical Technology, Vol. 1. John Wiley & Sons, New York.
Richtler, H.J. and J. Knaut. 1988. World prospects for surfactants. Proceedings of the 2nd CESIO World Surfactants Congress, Paris, France, May 24-27, 1988, 1:3-58.
Schick, M.J. (Ed.). 1967. *Nonionic Surfactants*, Vol. 1, Surfactant Science Series. Marcel Dekker, New York.
Schick, M.J. (Ed.). 1987. *Nonionic Surfactants: Physical Chemistry*, Vol. 23, Surfactant Science Series. Marcel Dekker, New York.
Schirber, C.A. 1989. Linear alkylbenzene sulfonate and alcohol ethoxylates -the workhorse surfactants. Presented at the Chemical Marketing Research Association meeting February 1989, Houston, Texas.
Smith, G., W.M. Sawyer, Jr. and R.C. Morris. 1972. U.S. Patent 3,682,849.
Stafford, M.L., K.F. Guin, G.A. Johnson, L.A. Sanders and S.L. Rocky. 1980. Analysis of 1,4-dioxane in ethoxylated surfactants. J. Soc. Cosmet. Chem. 31:281-287.
TSCA (Toxic Substances Control Act) Chemical Substance Inventory. 1979. Volume I. U.S. Environmental Protection Agency, Washington, DC.
U.S. International Trade Commission. 1988. Synthetic organic chemicals, United States production and sales, September 1989.

II. ENVIRONMENTAL LEVELS

Before the development of chemical-specific methods for the measurement of AE in biodegradation studies and environmental samples, AE surfactants were determined by methods that measured the loss of a chemical or physical property of the parent molecule. These chemical non-specific techniques such as measurement of foaming potential, changes in surface tension, and detection of cobalt thiocyanate, bismuth iodide, and potassium picrate active substances do not differentiate between AE and APE surfactants and do not detect the lower EO oligomers of AE. These methods, thus, have limitations when applied to environmental samples.

Although AE surfactants are not presently being monitored in the environment, concern over the possible persistence of some classes of surfactants and their degradation products has led to the development of sensitive and specific measurement techniques. These analytical techniques, which are specific to the component chemical moieties of AE and other surfactants, have recently been described in the literature. The use of normal-phase HPLC coupled with UV or the more sensitive fluorescence detection, used for the separation and identification of EO chains, has been applied to AE surfactants. Reversed-phase HPLC/UV detection has been used to identify the structure of alkyl chains. Alcohol ethoxylate surfactants must be derivatized for UV and fluorescent detection. Gas chromatography coupled with various detection methods has been applied to environmental samples and is applicable to determination of AE. In the future, various mass spectrometry (MS) techniques including evaporative light scattering and fast atom bombardment will likely be increasingly used.

Few data on environmental concentrations of AE are available. In surface waters worldwide, nonionic surfactants, as measured by chemical non-specific methods, ranged up to 2 mg/L, with most concentrations ≤ 0.5 mg/L. Concentrations in U.S. rivers, as measured by CTAS, generally ranged from 0.03-0.08 mg/L, but ranged up to 0.42 mg/L below a wastewater treatment plant outfall. Only one study documented a commonly used AE in a U.S. river; C_{14-15} AE_7 was present at a concentration of 0.5 μg/L above a wastewater treatment plant outfall and 1.1 μg/L below the outfall.

A. ANALYTICAL METHODS

The first comprehensive review of analytical procedures for the determination of nonionic surfactants in commercial products and in environmental samples was *Nonionic Surfactants*, edited by M.J. Schick (1967). Although several chemical-specific methods are described, most of the methods are for nonionic-type compounds in general. Advances in analytical techniques since that time, particularly in separation techniques and chemical-specific analyses, have resulted in several updates: *Nonionic Surfactants: Chemical Analysis*, edited by J. Cross (1987), *Surfactant Biodegradation*, edited by R.D. Swisher (1987), and *Detergent Analysis* edited by Midwidsky and Gabriel (1982). These books discuss in considerable detail the available analytical procedures and the problems involved in analysis of environmental samples.

One of the major problems in the analysis of environmental samples is the lack of a reliable AE-specific method able to distinguish between AE and other nonionic surfactants and between AE and their degradation products. Based on the literature reviewed for this report, the most commonly used chemical non-specific and chemical specific analyses methods are discussed here. The available methods may be divided into physical, chemical, physicochemical, and instrumental. In most cases, clean-up methods including extraction, concentration, and separation must be employed prior to analysis of environmental samples.

1. Extraction/Separation Methods

Because of the low concentrations of nonionic surfactants in the environment as well as the presence of chemically similar degradation products and naturally-occurring materials which interfere with these measurements, concentration and cleanup steps are usually necessary when making chemical and physicochemical determinations. These concentration/purification steps include solvent or solid-phase extraction, sublation of samples, and ion-exchange (Crisp, 1987; Swisher, 1987).

In solvent extraction, nonionic surfactants are extracted by (1) multiple or continuous solvent extraction, or (2) salting out of the surfactants which involves reduction of the solubility of the surfactant in water by addition of an inorganic salt. In 1966 Patterson et al. developed a solvent extraction method for sewage and river water samples which is the basis for many methods used today. Magnesium sulfate (75 g $MgSO_4 \cdot 7H_2O/L$) was added to the sample followed by three extractions with chloroform. The extracts were washed successively with strongly acidic and strongly basic solutions to which sodium chloride had been added, and then evaporated to dryness. In addition to chloroform, many other extraction solvents, such as ethylene dichloride and methylene chloride have given satisfactory results (Crisp, 1987).

In solvent sublation (also called foam stripping or gas stripping), nonionic surfactants are isolated from a dilute aqueous sample to yield a residue relatively free of nonsurfactant substances (APHA, 1989). Nitrogen is bubbled through a column containing the sample and an overlying layer of ethyl acetate. The surfactant is adsorbed at the water-gas interfaces and carried into the ethyl acetate. The ethyl acetate is separated, dehydrated, and evaporated, leaving the surfactant as a residue suitable for analysis. The method was developed by Wickbold (1971, 1972) and is used by the Organization for Economic Cooperation and Development (OECD) (1976).

Jones and Nickless (1978a; 1978b) used Amberlite XAD-4 resin to adsorb polyethoxylated material from river water and sewage systems. They achieved 80-100% adsorption/desorption efficiencies. Exhaustive purification of the resin was necessary to remove contaminants prior to use.

Anion and cation exchange resins are useful for removal of anionic and cationic surfactants which interfere with sample quantification. Prior to chemical determinations, Waters et al. (1986) suggested concentration and cleanup involving four sublation steps and a cation/anion exchange resin to minimize interferences. For environmental samples, Wee (1981) used a three-step procedure to remove interferences. Cation and anion exchange resins were used to remove the cationic and anionic organic compounds that interfere with colorimetric and GC analysis. A toluene extraction procedure was used to remove biodegradation products such as polyethylene glycols (PEG), and silica gel was used to remove other interferences.

2. Physical Methods

The two principal physical procedures used to monitor the degradation of nonionic surfactants in the laboratory and in environmental samples are foaming and surface tension. The usefulness of the foam method is limited because the foaming phenomenon is complex and transient, and foaming is not a linear function of surfactant concentration. Furthermore, other substances present in environmental samples such as protein, partial degradation products, and other surfactants make quantification impossible. Measurement of nonionics with the foam method is further confounded by the fact that they are generally low in foaming potential compared to anionic surfactants (Swisher, 1987).

The surface tension of water is lowered significantly at a concentration of a few ppm surfactant. It decreases further with increasing concentration of surfactant up to a critical micelle concentration (CMC), but the relationship is not linear. Above the CMC there is little change in surface tension with increase in concentration. As in the case of foaming properties, surface tension lacks specificity and is subject to interferences from other substances (Swisher, 1987). The accuracy and repeatability of these two physical methods were reviewed by the SDA (Mausner et al., 1969). Greater consistency among and within testing laboratories was achieved through the use of a foam loss procedure.

3. Chemical Methods

Chemical methods for determination of nonionic surfactants are based on the formation of a complex between the polyoxyethylene chain and an inorganic cation followed by reaction of the complex ion with an anion to produce a salt. The resulting salt either forms a precipitate which can be determined gravimetrically or can be extracted into a polar solvent and determined colorimetrically. The two chemical procedures most widely used in biodegradation studies and for the analysis of environmental samples are the cobalt thiocyanate spectrophotometric and iodobismuthate titration methods. Both of these methods tend to overestimate the AE present because of interferences by many organic materials such as cationic and anionic surfactants, biodegradation intermediates such as PEG, lipids, and other substances. In addition, these analytical methods respond to the entire class of nonionic surfactants and thus do not distinguish between types of nonionic surfactants and their degradation products and do not provide information on specific chemical homologs (Cross, 1987; Swisher, 1987).

Cobalt Thiocyanate Active Substances (CTAS). Cobalt thiocyanate active substances such as nonionics form a blue-colored complex with the ammonium cobaltothiocyanate reagent. The complex can be extracted into an organic solvent and the nonionic content determined spectrophotometrically by comparing the absorbance of the complex with standard curves for the individual ethoxylates (Crabb and Persinger, 1964; APHA, 1989). Quantities as low as 1 mg/L (1 ppm) can be determined. Sensitivity of the method is dependent on the number of moles of EO in the polyether and on the extraction efficiency of the complex. When applied to environmental samples, Wee (1981) found a detection limit of 0.2 mg/L.

The procedure is most sensitive for surfactants having 6 to 25 ethylene oxide groups per mole of hydrophobe. Molar absorbance is not linear below 6 ethylene oxide units and for higher molecular weight complexes. The absorption coefficients of the latter may be 10-15 times greater than lower molecular weight complexes. Without use of a sublation technique, the method is subject to interferences from strongly acidic or basic solutions, cationic and anionic surfactants, PEG, and naturally occurring substances (Swisher, 1987; APHA, 1989).

In 1977 the SDA (Boyer et al., 1977) developed an improved CTAS procedure for determination of nonionic surfactants in biodegradation and environmental studies. Their procedure involves an improved sublation step, and an ion-exchange step prior to reaction with cobalt thiocyanate; methylene chloride was the extraction solvent. The method is applicable to both linear and branched AE. Recovery of radiotagged AE in a variety of biodegradation tests was consistently above 90%. Without the ion-exchange step, CTAS values for environmental samples more than doubled, due to anionic surfactants. A round robin series of tests among six laboratories using both the CTAS and bismuth iodide active substances (BIAS) methods and using the same environmental samples indicated good intralaboratory reproducibility, but significant variation among laboratories. This method was applied to influent and effluent samples from a wastewater treatment plant (Sedlak and Booman, 1986; Gledhill et al., 1989).

Bismuth Iodide Active Substances (BIAS). In the BIAS method, Dragendorff reagent, a preparation containing barium tetraiodobismuthate, forms a precipitate with polyoxyethylene compounds through interaction with the ethoxylate oxygen atoms. These complexes are formed with polyethers higher than EO_5. Polyethylene glycols, the major degradation products of nonionic surfactants, which also give a positive reaction are removed during an initial butanone extraction step. The PEG can be isolated and determined spectrophotometrically following treatment with Dragendorff reagent (Bürger, 1963; Swisher, 1987).

Wickbold (1966, 1972, 1973) modified the BIAS method into what is now a widely accepted procedure for nonionic surfactants. The present method involves isolation of the nonionics by sublation, precipitation with Dragendorff's reagent, and potentiometric titration of the bismuth in the precipitate. The procedure, as adopted by the OECD (1976), involves these steps: removal of particulates by filtration, sublation into ethyl acetate using a stream of nitrogen bubbles presaturated with ethyl acetate vapor, removal of cationic surfactants with cation exchange resin, precipitation of nonionics by barium bismuth iodide reagent, and measurement of bismuth content of the precipitate by potentiometric titration with pyrrolidine dithiocarbamate solution. This method has a detection limit of approximately 0.01 mg/L. In Europe the BIAS procedure of Wickbold has been officially adopted for determination of nonionic surfactants in biodegradation test studies (EEC, 1982). Block et al. (1987) suggested a modification of the German biodegradation test which results in improved accuracy, particularly for nonionics with more than 30 EO units. They replaced the sublation enrichment step using ethyl acetate with an evaporation step.

Although the BIAS method is widely used in laboratory biodegradation tests, it usually overestimates concentrations of AE in environmental samples in the presence of PEG, cationic surfactants, biodegradation intermediates and other substances (Swisher, 1987). For environmental samples, an optimized BIAS method is suggested: four 10-minute sublation steps and a cation/anion exchange cleanup of sublation extracts (Waters et al., 1986).

For environmental samples the BIAS method is preferable over the CTAS method in that it provides more accurate results for samples that require an arbitrary calibration factor. The BIAS precipitates show a fairly constant ratio of one barium atom to about 9.8 ether oxygens whereas the CTAS response depends on both the number of ether oxygens and the extraction efficiency of the cobalt complex. Thus, less variation is expected with the BIAS method (Swisher, 1987).

Invernizzi and Gafà (1973) compared five different analytical methods. They concluded that surface tension and the Wickbold sublation/BIAS method were the best followed by foaming potential and the Patterson thin-layer chromatography method. CTAS was the least preferred method. In river-water tests on nonionics, however, Ruiz Cruz and Dobarganes Garcia (1976) found that the results from foam, surface tension, and CTAS were generally in agreement.

Potassium Picrate Active Substances (PPAS). Favretto and co-workers (Favretto and Tunis, 1976; Favretto et al., 1980a, 1983) developed and improved a paired ion extraction method for nonionic surfactants in water and wastewaters. In the improved PPAS method, the sample is purified and/or concentrated and extracted into ethylene dichloride; the ethoxylate chain is then complexed with aqueous K^+ and paired with picrate anion; the ion pair is extracted into an organic solvent (methylene chloride) and the picrate content of the organic phase is determined by its spectrophotometric absorbance. The lower limit of detection is 0.1 mg/L. This method can be used for ethoxylate chains ranging from 7-28 units. Of several potentially interfering substances tested (sodium chloride, anionic surfactants, cationic surfactants, sodium soaps, and crude oil) only cationic surfactants gave positive results at low concentrations (Favretto et al., 1982).

Other Methods. In addition to the above methods, polyoxyethylene or polyoxypropylene units form water-insoluble complexes with mercuric and cadmium chlorides and with the calcium and barium salts of phosphosilicic, phosphomolybdic, and silico- and phosphotungstic acids. Analyses may be gravimetric or spectrometric. These methods are not commonly used because they are time consuming and subject to many interferences (Cross, 1987; Swisher, 1987). Han (1967) developed a method which involved extraction of polyethoxylates into chloroform and sodium chloride, sulfation of the polyethoxylates, and determination by methylene blue activity.

Thin-layer chromatography (TLC) has been a useful physicochemical method in the determination of the extent of nonionic surfactant biodegradation. The procedure involves concentration of a sample into a solvent, deposition near the edge of the chromatographic medium, drying of the solvent, separation of the sample components by addition and advancement of a suitable solvent along the chromatographic medium, development with a suitable reagent, and identification of component spots by comparison with standards. The choice of solvent and color reagent influences mobility and resolution. The ethoxylates separate according to their degree of ethoxylation. Polyethylene glycols appear as streaks tailing behind the nonionic spots (Swisher, 1987).

The system developed by Patterson et al. (1966, 1967) involves adding magnesium sulfate to the aqueous sample, extracting the nonionics and polyglycols into chloroform, washing the chloroform successively with acidic and basic solutions, placement on activated silica gel plates, and development with ethyl acetate:water:acetic acid (40:30:30). The spots can be detected following spraying with barium bismuth iodide reagent. The method was applied to the detection of 1-5 μg quantities of nonionics in sewage and river water samples. Neither anionic surfactants nor PEG interfere with the results. Variations of the TLC method are discussed in Swisher (1987).

4. Instrumental Analyses

Several types of chromatographic separation techniques interfaced with ultraviolet (UV), infrared (IR), or mass spectroscopy (MS) detection have been used to identify APE and their degradation intermediates during primary degradation and can presumably be applied to AE. Because these compounds have no near-UV absorption, they must be derivatized before UV or fluorescence detection. Generally a combination of methods, preceded by specific extraction or concentration techniques, is required for complete identification of APE in environmental samples. These methods provide greater specificity and sensitivity than physical and chemical methods.

Gas Chromatography. Gas chromatography coupled with various detection methods has been applied to APE in environmental samples (Giger et al., 1981; Schaffner et al., 1982; Stephanou and Giger, 1982; Stephanou, 1984; 1986; Wahlberg et al., 1990) and is applicable to determination of AE.

Gas chromatography is not commonly applied to the analysis of nonionic surfactants for several reasons: nonionic surfactants above EO_7 are generally not volatile enough for GC analysis, they tend to decompose at the temperatures used, and they are too polar. In addition, derivatization of AE is required for UV or florescence detection. Two techniques that extend the detection range of ethoxylate chain length are (1) derivatization with trimethylsilane or acetate and (2) fission of the surfactants into identifiable products (Cross, 1987). Gas chromatography has more often been applied to analyses of neat surfactants in the laboratory. Gas chromatography was applied to the analysis of alkyl carbon distributions of surfactants in detergent formulations (Sones et al., 1979) and to the alkyl bromide cleavage products of a commercial formulation ($C_{12-15}AE_9$) in order to identify biodegradation products (Tobin et al., 1976a; 1976b). Gas chromatography/MS was used to determine carbon number distributions and GC/FI (flame ionization) was used to determine polyoxyethylene chain lengths from 1 to 6 for trimethylsilyl ether derivatives of highly branched alcohol ethoxylates (Kravetz et al., 1991).

Wee (1981) described analysis of linear AE surfactants present in wastewater. Alcohol ethoxylate surfactants were converted to hydrogen bromide cleavage products and analyzed by GC. Interfering substances were removed by a cleanup procedure of mixed-bed ion exchange, toluene extraction, and silica gel adsorption. AE were determined indirectly (FI detection) by quantification of their reaction products, the alkyl bromides. The cleavage products from the branched APE present in the sample eluted prior to the AE. The method was validated using $C_{14-15}AE_7$ (Neodol 45-7). The detection limit was 10 ug of AE/sample. Stephanou (1984) applied a GC/CI-MS (chemical ionization-mass spectroscopy) method to the qualitative analysis of linear alcohol ethoxylates in the effluent of a wastewater treatment plant. The results provided information on molecular weight as well as structural characteristics of the compounds present and appears superior to GC/EI (electron impact) analysis.

High Performance Liquid Chromatography (HPLC). HPLC is a useful technique for separating and identifying the ethoxy components of mixtures of nonionic surfactants and their PEG metabolites as well as the alkyl and ethoxy chain lengths of individual AE. Detection may be by any of several methods; UV absorption and fluorescence are commonly used, but MS and FI detectors have been tested (Swisher, 1987).

This technique was first utilized to determine the structure of intact surfactants. In 1975 Nakamura and Matsumoto showed that the ethoxylate units of commercially available, single-carbon number alcohols could be separated by adsorption HPLC of the acetate derivatives and detected with a FI detector. Retention time on the column was directly proportional to the number of moles of ethylene oxide present. This method can also be applied to PEG. Allen and Linder (1981a; 1981b) resolved linear primary AE with ethoxy chains of 3 to 20 units into well defined peaks. Resolution of secondary alcohols and linear oxo-alcohol ethoxylates was less distinct, with minor peaks adjacent to some components and broadening of peaks indicating the presence of several components in each peak. Derivatization by reaction with phenyl isocyanate, allowed detection by UV absorbance. McClure (1982) got similar results for acetylated ethoxylate oligomers ranging from 0 to 30 units and detected by a rotating disc-flame ionization detector. Double bonds, as in the case of oleyl alcohol ethoxylates, made derivatization unnecessary for UV detection and identification of the peaks (Aserin et al., 1984). Kravetz et al. (1991) used HPLC coupled with

UV detection to determine polyoxyethylene chain lengths greater than 5 EO units in several types of alcohol ethoxylates including highly branched alcohols. Alcohol ethoxylates were injected as their phenylurethane derivatives.

Recently, the SDA outlined a comprehensive procedure that can be applied to the chemically complex wastewaters at municipal wastewater treatment plants (Schmitt et al., 1990). The procedure involved sample concentration and cleanup with XAD-2 resin, liquid-liquid extraction with ethyl acetate, removal of interfering ionic surfactants by anion and cation exchange resins and cobalt thiocyanate extraction, derivatization with phenyl isocyanate, separation by both normal-phase and reversed-phase HPLC, and detection of the phenyl isocyanate derivatives by UV absorbance. The alkyl chain length distribution was determined using reversed-phase HPLC; the ethoxy chain length was determined by normal-phase HPLC. The limit of quantification was 0.1 ppm. The method was applied to samples from a sewage treatment plant. The method needs further refinement to remove APE surfactants which appeared in some of the chromatogram peaks.

Several new analytical techniques for surfactants have been reported recently. These focus on more precise detection methods following HPLC. In one case, Kubeck and Naylor (1990) used fluorescence detection to determine APE in environmental samples in order to achieve sensitivity at the 1-2 $\mu g/L$ level. Bear (1988) has investigated the use of evaporative light scattering for "universal" detection of both anionic and nonionic surfactants. It is likely that both of these techniques are applicable to determination of AE in environmental samples.

Fast Atom Bombardment/Mass Spectrometry (FAB/MS). Ventura et al. (1988, 1991) used GC/MS, FAB/MS, and MS/MS to identify linear AE in raw and drinking water samples from Barcelona, Spain. In the second study, acidic and neutral plus base sample extracts were passed through Amberlite XAD-2 resin to concentrate organics. The samples were analyzed without further separation by FAB using thioglycerol saturated with NaCl as a matrix. The acidic components were derivatized to methyl esters. APE were used as internal standards.

Nuclear Magnetic Resonance (NMR). Specific nonionic surfactants and PEG can be identified in environmental samples by NMR. Proton (^{1}H)-NMR has been used to determine the degree of ethoxylation, the hydrophile-lipophile balance and the solution of water by the polyethoxylate chain (Llenado and Neubacker, 1983). Leenheer et al. (1991) identified PEG residues in surface water samples from the Mississippi River and its tributaries. Nuclear magnetic resonance has been used to identify the presence of polyethoxylates in humic substances extracted from sewage sludges (Bayer et al., 1984), APE in groundwater (Thurman et al., 1987), and AE and APE in river water and sewage (Jones and Nickless, 1978b). Following selective extraction, methylation, and chromatographic separation procedures, Leenheer et al. (1991) quantitatively determined acidic and neutral PEG and polypropylene glycols in the river water samples.

B. WATER QUALITY STANDARDS

1. National Regulations

National criteria for regulation of nonionic surfactants in waters of the United States were not located.

2. State and Local Standards

Recent information on state and local standards was not located. The sale of detergents containing nonionic and methylene blue active surfactants is prohibited in Suffolk, New York and in New Shoreham, Rhode Island. The sale of detergents that do not biodegrade in a secondary sewage treatment plant is prohibited in Oregon. The sale, possession, or use of non-degradable detergents is unlawful in Dade County, Florida (SDA, 1991).

C. NONIONIC SURFACTANTS IN NATURAL BODIES OF WATER

Alcohol ethoxylates are not currently being monitored in the United States and few data on environmental concentrations of AE are available. Where data are available, analyses were usually by chemical non-specific methods and a distinction between AE and other types of nonionic surfactants was not made. In addition, as stated earlier, non-specific measurement techniques such as BIAS and CTAS are prone to interferences from polyethylene glycols, lipids, cationic and anionic surfactants, and biodegradation intermediates and therefore, overestimate environmental concentrations (Swisher, 1987). There is a need for a specific analytical method for AE in environmental samples.

Swisher (1987), reviewing data on surface waters and groundwaters world wide, found a range for nonionics of undetectable up to 2 mg/L, with most values ≤ 0.5 mg/L. Concentrations of 0.01 to 1.0 mg/L nonionic surfactant have been reported for United Kingdom (U.K.) rivers and 0.01 to 0.11 mg/L in West German rivers.

Analytical data on the surfactant content of many U.K. rivers were published annually by the Standing Committee on Synthetic Detergents (STCSD, 1980), a group that has since disbanded. Fischer (1980) has published data collected since 1960 on the Rhine River, Germany. Concentrations of nonionic surfactants in river waters in Europe were reviewed by Arthur D. Little (1977). During the period 1967-1974, concentrations in several rivers in Great Britain ranged from 0.01-0.8 mg/L. Higher values, up to 1.0 mg/L were found in the Aire and Calder Rivers in the wool-treatment district of Yorkshire. Concentrations in the Lippe and Rhine Rivers, Germany, were 0.014-0.11 mg/L. Analysis was by the Patterson TLC (Great Britain) and Wickbold BIAS (Germany) methods. A recent study documented that the load of nonionics in the Rhine River decreased by 67% between 1974 and 1987; the average concentration in 1987 was 0.01 mg/L as determined by the BIAS method without ion exchange cleanup (Gerike et al., 1989). Concentrations of nonionics in earlier studies were probably overestimated because of the lack of ion exchange cleanup of samples; additional factors contributing to this observation, such as a decrease in the use of the "hard" anionic detergent, tetrapropylenebenzene sulfonate, and the construction and improvement of sewage treatment plants were noted by Gerike et al. The concentration of nonionics (secondary AE + APE) in the River Avon below a sewage plant was 8.0 µg/L (Jones and Nickless, 1978b). XAD resin was used to concentrate the surfactants and TLC, UV, IR and NMR techniques were used to characterize the components. Favretto et al. (1978, 1980b), using the PPAS method, measured concentrations up to 3.0 µg/L (3 µg/kg) in seawater at Trieste, Italy. Using GC/MS and FAB/MS, Ventura et al. (1988) identified AE, APE, alkylphenol carboxylic acids and their brominated derivatives in the Llobregat River and tap water of Barcelona, Spain. Quantitative determinations were not made.

Some studies reported concentrations of total nonionics in U.S. rivers; one study presented concentrations of a commonly used AE. The concentration of nonionic surfactants above a sewage treatment plant on Blacklick Creek, Ohio, as measured by CTAS, was 0.03-0.08 mg/L (Maki et al., 1979). Concentrations 100 meters and 3.2 kilometers downstream of the plant were 0.42 and 0.12 mg/L, respectively. The concentrations of nonionic surfactants, as measured by CTAS, in grab samples from an unidentified U.S. river were 0.03 mg/L above a sewage outfall and 0.24 mg/L below the outfall (Wee, 1981). The concentrations of $C_{14-15}AE_7$ as detected by GC analysis in the corresponding samples were 0.5 µg/L and 1.1 µg/L. A river water sample obtained from a raw-water tap at a water treatment plant contained 4.2 µg/L of $C_{14-15}AE_7$.

Two studies documented the presence of PEG, intermediates in the degradation of surfactants, in natural waters. Patterson (1966) found 0.1-9 mg/L in British waters. Polyethylene glycol residues were measured in surface water of the lower Mississippi River and its tributaries in summer and fall of 1987 (Leenheer et al., 1991). Concentrations based on CTAS analyses ranged from undetectable to 28 µg/L and concentrations based on ^{1}H-NMR spectrometry ranged from undetectable (2.9 µg/L) to 145 µg/L. Analyses by ^{13}C-NMR spectrometry revealed the presence of neutral and acidic residues and polypropylene glycol moieties.

REFERENCES

Allen, M.C. and D.E. Linder. 1981a. Ethylene oxide oligomer distribution in nonionic surfactants via high performance liquid chromatography (HPLC). J. Am. Oil Chem. Soc. 58:950-957.

Allen, M.C. and D.E. Linder. 1981b. An improved HPLC method for the determination of ethylene oxide distribution in nonionic surfactants. Presented at the 72nd annual meeting of the American Oil Chemists' Society, New Orleans, May 19, 1981, Paper No. 160, p.23.

APHA. (American Public Health Association). 1989. Standard Methods for the Examination of Water and Wastewater, 17th Ed., L.S. Clesceri, Eds. American Public Health Association, Washington, DC.

Arthur D. Little. 1977. Human Safety and Environmental Aspects of Major Surfactants. A Report to The Soap and Detergent Association. Arthur D. Little, Inc., Cambridge, MA.

Aserin, A.M., M. Frenkel and N. Garti. 1984. HPLC analysis of nonionic surfactants. IV. Polyoxyethylene fatty alcohols. J. Am. Oil Chem. Soc. 61:805-809.

Bayer, E., K. Albert, W. Bergmann, K. Jahns, W. Eisener and H.K. Peters. 1984. Angew Chem. 96:151-153. (Cited in Leenheer et al., 1991).

Bear, G.R. 1988. Universal detection and quantitation of surfactants by high-performance liquid chromatography by means of the evaporative light-scattering detector. J. Chromatog. 459:91-107.

Block, H., W. Gniewkowski and W. Baltes. 1987. Accumulation of nonionic surfactants during the degradation test. Tenside 24:160-163.

Boyer, S.L., K.F. Guin, R.M. Kelley, M.L. Mausner, et al. 1977. Analytical method for nonionic surfactants in laboratory biodegradation and environmental studies. Environ. Sci. Technol. 11:1167-1171.

Bürger, K. 1963. Methods for the micro determination and trace detection of surfactants. III. Trace detection and determination of surface active poly EO compounds and polyethylene glycols. Z. Anal. Chem. 196:251-259.

Crabb, N.T. and H.E. Persinger. 1964. The determination of poly EO nonionic surfactants in water at the parts per million level. J. Oil Chem. Soc. 41:752-755.

Crisp, P.T. 1987. Trace Analysis of Nonionic Surfactants. In: J. Cross (Ed.), *Nonionic Surfactants: Chemical Analysis*. Marcel Dekker, Inc., New York, pp. 77-116.

Cross, J. 1987. *Nonionic Surfactants: Chemical Analysis*. Surfactant Science Series, Vol. 19, Marcel Dekker, Inc., New York.

EEC (European Economic Community). 1982. Directive 82/242/EEC amending 73/404/EEC, Brussels, January, 1982.

Favretto, L. and F. Tunis. 1976. Determination of polyoxyalkylene ether nonionic surfactants in waters. Analyst 101:198-202.

Favretto, L., B. Stancher and F. Tunis. 1978. Extraction and determination of polyoxyethylene alkyl ether nonionic surfactants in water at trace levels. Analyst 103:955-962.

Favretto, L., B. Stancher and F. Tunis. 1980a. Determination of polyoxyethylene alkyl ether nonionic surfactants in waters at trace levels as PPAS. Analyst 105:833-840.

Favretto, L., B. Stancher and F. Tunis. 1980b. Pollution of coastal sea-water from the harbour of Trieste by non-ionic surfactants. Proc. 1st Internat. Symp. Technol. Environ. Econ. Trends Detergent, Rome, October 22-24.

Favretto, L., B. Stancher and F. Tunis. 1982. Investigations on possible interferences in trace determination of polyoxyethylene non-ionic surfactants in waters as potassium picrate active substances. La Rivista Ital. Delle Sos. Grasse 59:23-27.

Favretto, L., B. Stancher and F. Tunis. 1983. An improved method for the spectrophotometric determination of polyoxyethylene nonionic surfactants in water as PPAS in presence of cationic surfactants. Intl. J. Environ. Anal. Chem. 14:201-214.

Fischer, W.K. 1980. Course of the surfactant concentrations in German waters 1960-1980. Tenside 17:250-261.

Gerike, P., K. Winkler, W. Schneider and W. Jacob. 1989. Detergent components in surface waters in the Federal Republic of Germany. Presented at the Seminar on the Role of the Chemistry Industry in Environmental Protection, Geneva, Switzerland, November 13-17, 1989.

Giger, W., E. Stephanou and C. Schaffner. 1981. Persistent organic chemicals in sewage effluents: 1. Identification of nonylphenols and nonylphenolethoxylates by glass capillary gas chromatography/mass spectrometry. Chemosphere 10:1253-1263.

Gledhill, W.E., R.L. Huddleston, L. Kravetz, A.M. Nielsen, R.I. Sedlak and R.D. Vashon. 1989. Treatability of surfactants at a wastewater treatment plant. Tenside 26:276-281.

Han, K.W. 1967. Determination of biodegradability of nonionic surfactants by sulfation and methylene blue extraction. Tenside 4:43-45

Invernizzi, F., and S. Gafà. 1973. Biodegradation of nonionic surfactants. I. Study of some analytical methods for nonionic surfactants applied in biodegradation tests. Riv. Ital. Sost. Grasse 50:365-372.

Jones, P. and G. Nickless. 1978a. Characterization of non-ionic detergents of the polyethoxylated type from water systems. I. Evaluation of Amberlite XAD-4 resin as an extractant for polyethoxylated material. J. Chromatog. 156:87-97.

Jones, P. and G. Nickless. 1978b. Characterization of non-ionic detergents of the polyethoxylated type from water systems. II. Isolation and examination of polyethoxylated material before and after passage through a sewage plant. J. Chromatog. 156:99-110.

Kravetz, L., J.P. Salanitro, P.B.Dorn and K.F. Guin. 1991. Influence of hydrophobe type and extent of branching on environmental response factors of nonionic surfactants. J. Am. Oil Chem. Soc. 68:610-618.

Kubeck, E. and C.G. Naylor. 1990. Trace analysis of alkylphenol ethoxylates. J. Am. Oil Chem. Soc. 67:400-405.

Leenheer, J.A., R.L. Wershaw, P.A. Brown and T.I. Noyes. 1991. Detection of poly(ethylene glycol) residues from nonionic surfactants in surface water by ^{1}H and ^{13}C nuclear magnetic resonance spectrometry. Environ. Sci. Technol. 25:161-168.

Llenado, R.A. and T.A. Neubecker. 1983. Surfactants. Anal. Chem. 55:93R-102R.

Maki, A.W., A.J. Rubin, R.M. Sykes, R.L. Shanks. 1979. Reduction of nonionic surfactant toxicity following secondary treatment. J. Water Pollut. Control Fed. 51:2301-2313.

Mausner, M., J.H. Benedict, K.A. Booman, et al. 1969. The status of biodegradability testing on nonionic surfactants. J. Am. Oil Chem. Soc. 46:432-444.

McClure, J.D. 1982. Determination of ethylene oxide oligomer distributions in alcohol ethoxylates by HPLC using a rotating disc-flame ionization detector. J. Am. Oil Chem. Soc. 59:364-373.

Midwidsky, B. and D.M. Gabriel. 1982. *Detergent Analysis*. Halstead Press, New York.

Nakamura, K. and I. Matsumoto. 1975. Analysis of nonionic surfactants. I. Ethylene oxide adducts by HPLC. J. Chem. Soc. Japan 8:1342-1347.

OECD (Organization for Economic Cooperation and Development). 1976. Environment Directorate. Proposed method for the determination of the biodegradability of surfactants used in synthetic detergents, OECD, Paris.

Patterson, S.J., E.C. Hunt and K.B.E. Tucker. 1966. The determination of commonly used nonionic detergents in sewage effluents by a TLC method. J. Proc. Inst. Sew. Purif., pp. 190-198.

Patterson, S.J., C.C. Scott and K.B.E. Tucker. 1967. Nonionic detergent degradation. I. Thin-layer chromatography and foaming properties of alcohol polyethoxylates. J. Am. Oil Chem. Soc. 44:407-412.

Ruiz Cruz, J. and M.C. Dobarganes Garcia. 1976. Pollution of natural waters by synthetic detergents. X. Biodegradation of nonionic surfactants in river water. Grasas Aceites 27:309-322.

Schaffner, C., E. Stephanou and W. Giger. 1982. Determination of nonylphenols and nonylphenolethoxylates in secondary sewage effluents. Comm. Eur. Communities Rep.

Schick, M.J. (Ed.) 1967. *Nonionic Surfactants*. Volume I, Surfactant Science Series. Marcel Dekker, Inc., New York.

Schmitt, T.M., M.C. Allen, D.K. Brain, K.F. Guin, D.E. Lemmel and Q.W. Osburn. 1990. HPLC determination of ethoxylated alcohol surfactants in wastewater. J. Am. Oil Chem. Soc. 67:103-109.

SDA. (Soap and Detergent Association). 1991. Enacted Detergent Legislation: State and Local, Summary No. 110.

Sedlak, R.I. and K.A. Booman. 1986. LAS and alcohol ethoxylate: a study of their removal at a municipal wastewater treatment plant. Soap Cosmetics Chem. Spec. (April).

Sones, E.L., J.L. Hoyt and A.J. Sooter. 1979. The determination of alcohol and ether sulfates and their respective alkyl carbon distributions in detergent formulations by gas chromatography. J. Am. Oil Chem. Soc. 56:689-700.

STCSD (Standing Technical Committee on Synthetic Detergents). 1980. Twentieth Progress Report, HMSO, London.

Stephanou, E. 1984. Identification of nonionic detergents by GC/CI-MS: 1. A complementary method or an attractive alternative to GC/EI-MS and other methods? Chemosphere 13:43-51.

Stephanou, E. 1986. Determination of acidic and neutral residues of alkylphenol polyethoxylate surfactants using GC/MS analysis of their TMS derivatives. Comm. Eur. Communities Rep.

Stephanou, E. and W. Giger. 1982. Persistent organic chemicals in sewage effluents. 2. Quantitative determinations of nonylphenols and nonylphenol ethoxylates by glass capillary gas chromatography. Environ. Sci. Technol. 16:800-805.

Swisher, R.D. 1987. *Surfactant Biodegradation*, 2nd. Ed. Surfactant Science Series, Vol. 18. Marcel Dekker, Inc., New York. pp. 47-146.

Thurman, E.M., T. Willoughby, L.B. Barber Jr., and K.A. Thorn. 1987. Surfactants. Anal. Chem. 59:1798-1802. (Cited in Leenheer et al., 1991)

Tobin, R.S., F.I. Onuska, D.H.J. Anthony and M.E. Comba. 1976a. Nonionic surfactants: conventional biodegradation test methods do not detect persistent polyglycol products. Ambio 5:30-31.

Tobin, R.S., F.I. Onuska, B.G. Brownlee, D.H.J. Anthony and M.E. Comba. 1976b. The application of an ether cleavage technique to a study of the biodegradation of a linear alcohol ethoxylate nonionic surfactant. Water Res. 10:529-535.

Ventura, F., A. Figueras, J. Caixach, I. Espadaler, J. Romero, J. Guardiola and J. Rivera. 1988. Characterization of polyethoxylated surfactants and their brominated derivatives formed at the water treatment plant of Barcelona by GC/MS and FAB mass spectrometry. Water Res. 10:1211-1217.

Ventura, F., D. Fraisse, J. Caixach and J. Rivera. 1991. Identification of [(alkyloxy)polyethoxy] carboxylates in raw and drinking water by mass spectrometry/mass spectrometry and mass determination using fast atom bombardment and nonionic surfactants as internal standards. Anal. Chem. 63:2095-2099.

Wahlberg, C., L. Renberg and U. Wideqvist. 1990. Determination of nonylphenol and nonylphenol ethoxylates as their pentafluorobenzoates in water, sewage sludge and biota. Chemosphere 20:179-195.

Waters, J., J.T. Garrigan and A.M. Paulson. 1986. Investigations into the scope and limitations of the bismuth active substances procedure (Wickbold) for the determination of nonionic surfactants in environmental samples. Water Res. 20:247-253.

Wee. V.T. 1981. Determination of linear alcohol ethoxylates in waste- and surface water. In: L.H. Keith, Ed., Advances in the Identification & Analysis of Organic Pollutants in Water, Vol. 1, Ann Arbor Science.

Wickbold, R. 1966. Analysis for nonionic surfactants in water and wastewater. Vom Wasser 33:229-241.

Wickbold. 1971. Enrichment and separation of surfactants from surface waters by transport at the gas/water interface. Tenside 8:61-63.

Wickbold, R. 1972. On the determination of nonionic surfactants in river-and wastewaters. Tenside 9:173-177.

Wickbold, R. 1973. Analytical determination of small amounts of nonionic surfactants. Tenside 10:179-182.

III. BIODEGRADATION

Biodegradation, the breakdown of chemical compounds via biotic systems to less complex structures, is primarily the result of bacterial action (Swisher, 1987). Biodegradation can be subdivided into primary, the oxidation or alteration of a molecule resulting in the loss of measurable physical or chemical properties, and ultimate, the complete conversion or mineralization of a compound to carbon dioxide, water, and other inorganic compounds. Because of adsorption of surfactants to sludge and the incorporation of metabolites into bacterial structure, the change in surfactant concentration, particularly in systems containing sludge, is often referred to as removal.

As a class, AE undergo rapid primary and ultimate biodegradation under both laboratory and field conditions. Linear AE degrade by hydrolysis of the ether linkage at the hydrophile-hydrophobe bond followed by oxidation of the alkyl and polyoxyethylene moieties. Highly branched AE appear to degrade by stepwise shortening of the polyoxyethylene chain starting at the omega (ω) end with little or no metabolism of the alkyl chain.

A. LABORATORY INVESTIGATIONS

1. Test Systems

In addition to the analytical methods outlined in Section II.A, the ultimate biodegradation of surfactants may be followed in the laboratory by a number of chemically non-specific methods. These methods are directed at the organic content of the medium — chemical oxygen demand (COD), biological oxygen demand (BOD), total organic carbon (TOC), and carbon dioxide (CO_2) formation. Standard methods for these analyses are published by the American Public Health Association (Part 5000: Determination of Organic Constituents) (APHA, 1989).

Three widely used test systems for the measurement of biodegradability are the river-water dieaway, the shake-flask, and the activated-sludge test which may be semi-continuous (SCAS) or continuous (CAS). The first test involves adding a specific quantity of surfactant to river water in a glass jar and allowing the solution to incubate at room temperature. Degradation is determined by measurement of the final substrate concentration using a suitable method. Although the natural microbial populations and solids of river water are usually low, this test imitates conditions in natural bodies of water. In the shake-flask test, surfactant is added to a medium containing adapted microbial cultures. Again, the final concentration of the substrate is analyzed at the end of the test. In the CAS test, a specific concentration of surfactant and feed sewage containing nutrients are continuously introduced into an activated-sludge unit and the overflow effluent is collected and analyzed for undegraded surfactant (Schick, 1967). These units are designed to simulate municipal activated sludge sewage treatment plants. In SCAS units the nutrients and test surfactant are fed once a day. CAS and SCAS tests measure overall removal via both biodegradation and adsorption to sludge.

Several of the test methods have become standard tests or guidelines. The shake-flask test is standard E/279-89 in the ASTM Standards (1990). Using the shake-flask culture as a presumptive step and the SCAS system as a confirming test, U.S. manufacturers of detergents have agreed upon a lower limit of 90% degradation in test studies of branched sulfonates and linear alkylbenzene sulfonates (SDA, 1965). This two-step procedure has been adopted by the ASTM (D2667).

In 1973 Sturm published a screening test for predicting the rate and ultimate biodegradation of nonionic surfactants based on CO_2 production and BOD. This procedure allowed a rapid screening of surfactant biodegradability without the need to develop specific analytical methods for each surfactant under investigation. The Sturm test is widely used and a modification has been adopted as an OECD (1981) guideline.

The European Economic Community (EEC) has issued a directive recommending standard tests for the determination of biodegradability (Birch, 1984). These tests, adopted by the OECD, are a dieaway screening test and, for nonionics failing the screening test, a continuous-flow activated sludge test using synthetic wastewater. Primary degradation is measured by the Wickbold BIAS method.

The U.S. Environmental Protection Agency (U.S. EPA) provides guidelines for the determination of the rate and extent of aerobic biodegradation that might occur when chemicals are released to aquatic environments (Federal Register, 1985). Their shake-flask system for CO_2 evolution is based on the method of Gledhill (1975). Carbon dioxide values should be in the range of 80-100% of the theoretical CO_2. Flasks should also be monitored for dissolved organic carbon (DOC) removal. The total organic carbon is analyzed in order to calculate the percent of theoretical yield of CO_2 and percent of DOC loss.

2. Biodegradation Studies

A variety of these test methods and microbial test systems have been employed in laboratory studies of biodegradation or removal. Because of the large number of studies, they have been summarized in tabular form (see Appendix). Test systems listed in Column 2 of the Appendix include uninoculated natural media such as river water and synthetic media inoculated with acclimated or unacclimated bacteria, as well as sludge-inoculated and CAS and SCAS model sewage systems. Initial concentrations of AE fed into the flasks or systems ranged from <1 to 100 mg/L, with most concentrations between 1 and 10 mg/L. Even at high concentrations, AE showed no deleterious effects on acclimated bacteria. In one study, bench scale biotreaters fed 100 mg/L influent dosages of linear AE ($C_{12-15}AE_7$) showed no deleterious effects of the linear AE on foaming, nitrification, feed BOD removal, or biosolids growth (Salanitro et al., 1988). Because of the variety of test systems and conditions, data are not necessarily comparable from study to study. In addition, accurate measurement of biodegradation may be complicated by the adsorption of surfactants onto organic matter in some systems and the metabolic uptake of degradation products into the biomass present (Birch, 1984; Steber and Wierich, 1987; Swisher, 1987). Nevertheless, results of these experimental studies do indicate that as a class, essentially linear AE undergo rapid primary and ultimate biodegradation.

Birch (1991a) developed a mathematical model for assessing the biodegradability of nonionic surfactants in continuous-flow activated sludge plants. The model indicates the importance of sludge retention time (SRT) and temperature on the treatment efficiency. In the case of linear primary AE, the effects of temperature and SRT are marginal. The validity of the model was tested in laboratory-scale porous-pot units using a series of C_nAE with 7, 11, 15, and 20 EO units. At an SRT of six days and a temperature of 15°C, the effluent concentration remained unchanged and similar for the four AE despite a five-fold increase in influent concentration (5 and 25 mg/L).

In addition to river water and activated sludge units, biodegradation of AE in soil is of concern. Howells et al. (1984) determined the extent of biodegradation of $C_{12-15}AE_7$ (Neodol 25-7) in laboratory-scale soil percolation columns. Approximately 50% of the surfactant biodegraded to CO_2 and water within two days; the undegraded surfactant was located within the top 6.4 millimeters of soil. Within two weeks, over 90% of the surfactant had biodegraded to CO_2 and water. Knaebel et al. (1990) studied the biodegradation of $C_{12}AE_{8-9}$ at a concentration of 50 ng/g in 11 different natural soils without historical exposure to surfactants. Mineralization exhibited first-order kinetics and occurred without a lag period. Yields of CO_2 ranged between 25 and 69%.

B. FIELD STUDIES

Although early studies of AE removal at wastewater treatment plants used chemical non-specific measurement methods, the available data indicated extensive removals (Arthur D. Little, 1977; Goyer et al., 1981), presumably through both degradation and adsorption to sludge. At sewage treatment plants with either activated sludge or trickling filter systems, greater than 80% of surfactants (90% for activated sludge plants) normally present or dosed into the plants were removed as determined by CTAS, BIAS, TLC, or BOD (Table 3-1). Most of the plants employed both primary and secondary (trickling filter or activated sludge) treatment processes; secondary processes degrade most classes of surfactants relatively effectively (Maki et al., 1979).

In most cases, lower temperatures in winter did not greatly influence the extent of removal. In a field test, plant influent at an activated sludge treatment plant in Ohio was dosed with $C_{14-15}AE_7$ (Neodol 45-7) at 10 mg/L under both summer and winter conditions (Maki et al., 1979; Sykes et al., 1979). Plant performance and surfactant removal at various sampling locations were judged by several methods including BOD and CTAS, respectively. Results indicated 90% surfactant removal and were consistent through summer and winter periods.

Influent concentrations for all classes of nonionic surfactants at U.S. wastewater treatment plants may range up to 9 mg/L (Maki et al., 1979); however, maximum values are generally lower at most plants. Highest influent concentrations at more than a dozen sewage treatment plants, as determined by CTAS were: 1.6, 3.4, 4.4, 4.7, and 5.4 mg/L (Sedlak and Booman, 1986; Gledhill et al., 1989; Tabak and Bunch, 1981; Kravetz et al., 1984; Boyer et al.,

Table 3-1. Removal at Wastewater Treatment Facilities

Type of Plant/ Location	Surfactant	Influent Concentration (mg/L)	Effluent Concentration (mg/L)	Removal (%)	Measurement Method	Reference
Activated sludge/ United States	n-sec-$C_{11-15}AE_9$	—	—	94	CTAS	Conway et al., 1965
Trickling filter/ England	$C_{12-15}AE_9$ $C_{9-11}AE_8$	4.2-12.0 (7.2) 9.0-12.5 (10.3)	0.5-1.0 (0.8) 1.0-2.5 (1.7)	89 (86-93) (winter) 84 (80-90) (winter)	TLC TLC	Mann and Reid, 1971
Trickling filter/ Germany	Nonionics	2.3	0.5	—	BIAS	Bock, 1973
Trickling filter/ England	$C_{14-15}AE_7$ $C_{14-15}AE_{11}$	10, 25[a] 10, 25[a]	— —	96-98 (winter) 96-98 (winter)	CTAS CTAS	Abram et al., 1977
Activated sludge/ Germany	Nonionics	—	—	91	BIAS	Wagner, 1978
Not given/ England	Nonionics	0.7	0.07	90	TLC	Jones and Nickless, 1978
Activated sludge/ United States	Nonionics $C_{14-15}AE_7$	6.0 10[b]	0.92 —	(summer) 90 (all seasons)	CTAS CTAS	Maki et al., 1979; Sykes et al., 1979
Not given/ United States	Nonionics	0.71-3.35	0.18-0.61	—	CTAS	Tabak and Bunch, 1981
Not given/ United States	Nonionics $C_{14-15}AE_7$	3.1-5.4 0.19-0.47	0.4-0.5 0.006-0.12	— —	CTAS[c] GC	Wee, 1981

Table 3-1. (cont.)

Type of Plant/ Location	Surfactant	Influent Concentration (mg/L)	Effluent Concentration (mg/L)	Removal (%)	Measurement Method	Reference
Activated sludge, aeration treaters/ United States	Nonionics C_nAE_n	0.22-4.4; 2 —; 1.7	— —	— —	CTAS CTAS	Kravetz et al., 1982; 1984
Activated sludge/ United States	Nonionics C_nAE_n	1.62 0.86	0.13-0.18 0.01	90 99	CTAS[d] HPLC	Sedlak and Booman, 1986; Gledhill et al., 1989
Activated sludge/ Germany	Nonionics	—	—	96	BIAS[e]	Brown et al, 1986
Trickling filter/ Germany	Nonionics	4.0 4.1	0.7 0.5	81 (March) 88 (March)	BIAS[e] BIAS[e]	Brown et al, 1987
Not given (primary treatment)/ United States	Nonionics C_nAE_n	— —	2.5[f]; 1.6 2.1; —	— —	CTAS HPLC	Ankley and Burkhard, 1992

[a] Dosed into the plant.
[b] Added above normal influent concentration of 6.7 mg/L; normal range, 3.0-9.0 mg/L.
[c] $C_{14-15}AE_7$ was 6.2-11% of the influent and 1.2-3.2% of the effluent measurement.
[d] C_nAE_n was 50% of the influent and 7% of the effluent CTAS measurement.
[e] APE was 18-20% of the influent and 33-37% of the effluent measurement.
[f] APE was 13.6% of the nonionic measurement; LAS was additionally present at 3.8 mg/L.
A dash (—) indicates no data.

1977; Wee, 1981). At one plant where the influent CTAS measurement ranged from 0.22 to 4.4 mg/L, 85% of the CTAS measurement was attributed to AE (CTAS, 2 mg/L; AE, 1.7 mg/L) (Kravetz et al., 1982; 1984). At another plant AE, as determined by an HPLC method, was 53% of the influent CTAS measurement (Sedlak and Booman, 1986). Of the 3.1-5.4 mg/L measured in the influent of a third plant, only 0.19-0.47 mg/L, as determined by GC, was C_{14-15} AE_7, a commonly used AE (Wee, 1981).

Effluent concentrations of total nonionic surfactants at these U.S. plants ranged up to 2.3 mg/L. Chemical-specific methods showed that 1 to 3% of the influent AE was present in the effluent (Sedlak and Booman, 1986; Wee, 1981). Effluent concentrations are further diluted in the receiving waters. Nonionic surfactant concentration, as measured by CTAS, fell from 0.4-0.5 in the effluent to 0.24 mg/L below the sewage outfall (Wee, 1981). Maki et al. (1979) found concentrations of 0.92, 0.42, and 0.12 mg/L in the effluent, 100 meters downstream of the effluent outfall, and a recovery zone approximately 3.2 kilometers below the plant, respectively. The average concentration at a distance of 100 meters above the plant was 0.08 mg/L. Polyethylene glycols, which might have arisen from microbially-initiated central cleavage of AE (See Section E, Metabolic Pathways of Biodegradation) have also been measured in sewage effluent samples. Jones and Nickless (1978) measured 0.02 mg/L in both sewage influent and effluent (silica gel TLC) at a sewage treatment plant in the U.K.

In field tests at sites other than wastewater treatment plants, removals were also greater than 90% for surfactants in brackish pond water as measured by the CTAS method (Mann and Schöberl, 1976) and ambient bay water during the fall months as measured by the BIAS method (Tobin et al., 1976b).

Mineralization of ^{14}C-labeled $C_{12}AE_{8-9}$ in sediments taken from a pond receiving laundromat wastewater and a reference pond was rapid. Rate constants were first order with half-lives of 2.8-8.6 and 3.5-137 days in the laundromat and reference sediments, respectively. Below a depth of 2.5 meters in the laundromat sediment and 1.8 m in the reference pond sediment, mineralization was preceded by a short lag period which correlated with bacterial number. The short lag periods in the upper sediment suggest rapid removal of surfactant resulting in little opportunity for bacterial adaption in the lower depths. Bacterial cells/g sediment in the laundromat sediment averaged an order of magnitude greater than that in the reference sediment (Federle and Pastwa, 1988).

C. EFFECT OF CHEMICAL STRUCTURE

Based on laboratory-scale experiments, the primary factors affecting rate of biodegradation of nonionic surfactants are the length of the ethoxylate chain and the linearity of the alkyl chain. For linear oxo-alcohol ethoxylates, branching of the alkyl chain has little effect on primary biodegradation whereas ethoxylates derived from highly branched oxo alcohols biodegrade more slowly. The length of the alkyl chain appears to influence the biodegradation rate only slightly. The introduction of propylene alkyl or higher oxide groups in the hydrophilic moiety hinders the speed of biodegradation, probably by increasing hydrophobicity and methyl branching, but only if the block size is greater than four units (Birch, 1984; Swisher, 1987; Naylor et al., 1988). The effects of these variables have not been validated by field monitoring conditions.

1. Ethoxylate Chain Length

The effect of added ethylene oxide units on biodegradability of alcohol ethoxylates has been extensively studied (Arthur D. Little, 1977; Swisher, 1987). Within the range generally utilized in detergent formulations, ethoxylate chain length has little effect on the rate and extent of degradation. Rates generally decrease slightly with increased chain length, and above 20 ethylene oxide units the rate of degradation is retarded.

Many of the studies cited in the Appendix reported on removal of nonionics with varying EO chain length. Most of these studies are on linear primary AE. Sturm (1973) examined the role of ethoxylate chain length using both a series of C_{17} (average) nonionics with EO chain lengths ranging from 3 to 30. Chain length above 20 EO units (corresponding to a molecular weight of ~900) exhibited some resistance to degradation as measured by CO_2 production.

In tests lasting five days, degradation (BOD) of n-pri-$C_{12-13}AE_n$ with 10, 20, 30, or 40 EO units was approximately 35, 20, 15, and 12% complete, respectively (Dai Ichi Kogyo Pharmaceutical Co., 1974). Studies on ultimate biodegradation of linear primary AE reported by Huyser (1960), Blankenship and Piccolini (1963), Ruschenberg (1963), Borstlap and Kortland (1967), Heinz and Fischer (1967), Patterson et al. (1967), Pitter (1968a, 1968b), Fischer (1971), Möller (1972), Gerike and Schmid (1973), and Stühler and Wellens (1976) show similar results.

Kravetz and co-workers (Kravetz et al., 1979; Scharer et al., 1979) compared the ultimate biodegradation of 75% n-$C_{12-15}AE$ (Neodol 25 series) containing 7, 18, 30, and 100 EO units. At a surfactant concentration of 15 mg/L, degradation, as measured by CO_2 evolution, was $\geq 80\%$ in 32 days for the three shorter polyoxyethylene chains. $C_{12-15}AE_{100}$ (Neodol 25-100) was poorly biodegradable (<20%). Other investigations on primary and ultimate biodegradation of AE of oxo alcohols derived from linear olefins showed similar results, i.e., a decreased rate of biodegradation at >20 EO units (Borstlap and Kortland 1967; Brown, 1976; Laboureur et al., 1976; Birch, 1982, 1984).

Booman et al. (1967) compared the primary biodegradation, as measured by foam loss, of three linear secondary AE. In 21-day tests, extent of degradation was 95% for $C_{11-15}AE_9$, 70% for $C_{11-15}AE_{12.5}$, and 40% for $C_{11-15}AE_{16.5}$. Differences were also apparent for primary degradation of the linear secondary series $C_{14.7}$ with 8 to 36 EO units. Time periods for greater than 95% degradation ranged from two to six days, with slower rates above 20 EO units (Gerbil and Naim, 1969).

Birch (1982, 1984; 1991b) compared the primary biodegradation of linear primary, oxo-alcohol (50 and 75% linear) and linear secondary AE all containing 10, 20, 30, 40, or 50 EO units. While degradation of the linear primary alcohol ethoxylates were all 98-99%, as measured by BOD, regardless of ethoxylate chain length, there was a decrease in primary degradation with increasing ethoxylate chain length for the oxo-alcohol and linear secondary AE.

Sturm (1973) tested the biodegradability of unsubstituted PEG ranging in molecular weight from 300-4000. Using acclimated sewage bacteria and the CO_2 evolution test, PEG 300, 400, and 600 (EO_7, EO_9, and EO_{13}) degraded to 87-90% of the theoretical CO_2 in 24 days while PEG 1000, 1540, and 4000 (EO_{22}, EO_{35}, and EO_{90}) degraded to only about 10-20%. These data correlated with his findings for AE with comparable ethylene oxide chain lengths. Patterson et al. (1967) and Pitter (1972) similarly reported a decreased rate of degradation with increased molecular weight. PEG with molecular weights less than 600 degraded completely within 15 days while PEG 1000 degraded only 15% and PEG 1500 was unchanged (Pitter, 1972).

2. Alkyl Chain Structure

Chain length. Variations in the alkyl chain length do not appear to greatly affect the rate and extent of linear AE biodegradation. However, data reviewed by Swisher (1987) are meager and conflicting and are often obscured by accompanying variations in ethoxylate chain length.

Huyser (1960) found that linear primary C_8AE_4 degraded faster than $C_{18}AE_4$ in river water, with primary degradation complete in 7 and 27 days, respectively, as measured by the phosphomolybdate method. The data of Pitter (1968a; 1968b) show a reverse pattern: degradation of linear primary $C_{10-16}AE_n$ was slower than $C_{16-18}AE_n$. Sturm (1973) studied a series of linear primary AE, C_8AE_3-$C_{20}AE_3$, in which the length of the alkyl group changed by increments of two carbons. Although the rate of biodegradation as indicated by CO_2 production varied, 66-84% in 28 days, there was no consistent pattern, and it was concluded that variations in alkyl chain length did not affect biodegradability of linear AE. Gledhill (1975) reported that alkyl chain length (C_8-C_{20}) had little effect on the rate and extent of CO_2 evolution. Larson and Games (1981) looked at $C_{12}AE_9$ and $C_{16}AE_3$ (uniformly labeled with ^{14}C in the ethoxylate chain or at the α carbon in the alkyl chain) and found no effect of alkyl chain length or ethoxylate chain length on $^{14}CO_2$ evolution. Blankenship and Piccolini (1963) found that linear primary ethoxylates containing equal fractional weights of EO ($C_{10}AE_{6.7}$, $C_{12}AE_8$, $C_{14}AE_{9.5}$, $C_{16}AE_{10.4}$) underwent primary degradation (surface tension measurements) in river water to ~97% in 5-7 days.

Chain branching. The degree of branching or nonlinearity of the alkyl chain may affect the rate of biodegradation. Some studies show that the highly branched AE synthesized from oxo alcohols derived from polypropylenes, such as tetrapropylene, are only slowly and incompletely broken down. The slight branching of presently-used linear-oxo derivatives (around 20-50% alkyl-branched) reported below, however, pose little hindrance to biodegradation (Swisher, 1987).

Borstlap and Kortland (1967) compared the biodegradation, as measured by the weight of soluble organics, of linear primary (C_{12-14}) and linear oxo (C_{12-15}) alcohol ethoxylates with ethoxylate groups of 6-9, 10-12, 15-19, and 30-50 and found the differences between the two negligible. Likewise, Albanese and Capuci (1974), Arpino et al. (1974), and Bosari et al. (1974), using surface tension tests, showed that primary AE (derived from oxo-alcohols) with ~ 50% methyl branching biodegrade at rates which are similar to primary AE with virtually no branching. Sturm (1973) found that slight methyl branching (~ 12%) in a primary AE had no significant effect on biodegradation. Using the 24-hr SCAS test, Han (1967) found extents of 100 and 86% for the n-pri-$C_{14-16}AE_8$ and $C_{14-16}AE_{16}$ (sulfation/methylene blue method) and extents of 100 and 92% for the linear oxo-alcohol derived $C_{11-14}AE_3$ and $C_{11-14}AE_{13}$. Using the same test, both the n-pri-$C_{12-14}AE_{10}$ and the linear oxo-alcohol derived $C_{11-15}AE_9$ were degraded approximately 99% in 17 hours (Mausner et al., 1969).

Based on SCAS tests and shake flask tests, Huddleston and Allred (1964a) and Allred and Huddleston (1967) ranked the biodegradability of several types of surfactants. Linear AE ($C_8AE_{5.5}$, $C_{10}AE_{6.7}$, and $C_{12}AE_{7.9}$) were more degradable as measured by CTAS than the branched chain tp-$C_{13}AE_{14.4}$, with complete degradation within 3 days for the linear AE compared to 30% in 8 days for the branched-chain AE. Birch (1982, 1984) confirmed that increased branching retarded the rate of biodegradation, but the effect was greatest at ≥ 30 EO units.

Wickbold (1974) isolated and separated (GC analysis of alkyl iodide products) a series of linear oxo-alcohol derived ethoxylates with alkyl chains ranging from C_{11} to C_{14} and branching at the 1 to 6 positions. The extent of biodegradation increased with alkyl chain length and decreased with the more internal attachment of the methylol group.

Kravetz et al. (1978) compared the biodegradability of a 45% and a 75% normal alcohol ethoxylate, $C_{12-15}AE_9$. CTAS was used to measure primary degradation and CO_2 evolution, DOC, and BOD were used to measure ultimate degradation. Both AE biodegraded at about the same rate, indicating little effect of this range of branching on biodegradability.

Using the same methods, Kravetz et al. (1991) also compared the primary and ultimate biodegradation of a number of AE differing in hydrophobe branching and chain lengths. Two branched oxo-alcohol ethoxylates contained 2.9 ($C_{12}AE_7$) and 4.0 ($C_{13}AE_7$) internal methyl groups/hydrophobe. In ultimate biodegradation tests and continuous flow-through activated sludge tests, the more highly branched AE biodegraded more slowly and less extensively than did those with less hydrophobe branching. Using 30-day BOD tests they found that the 75% n-pri-$C_{12-15}AE_9$ was 88% biodegraded compared to 44% for br-$C_{13}AE_7$. Results were similar for the CO_2 evolution test.

Hughes et al. (1989) compared biodegradation behavior of linear and branched AE and linear alkylbenzene sulfonates (LAS) by a number of standard laboratory methods. The C_{12} and C_{13} oxo-alcohol portions of the branched AE were derived from mixed olefin feedstock containing primarily propylene, but also some ethylene and butylene. The resultant methyl branching was random, with an average of approximately three methyl groups per molecule; the oxo process leads to 30 to 50% 2-methyl (or ethyl) branching. They found that branched alkyl chains could match the rate and extent of biodegradation of linear alkyl chains. Relative biodegradation rates for seven-mole ethoxylates (EO_7) in the closed-bottle test were: linear AE > branched AE = LAS. In the Sturm test, all materials were equivalent; in the acclimated Gledhill test, linear AE = branched AE = LAS, with linear slightly greater than LAS; in the sludge-inoculated Gledhill test, linear AE > branched AE > LAS.

Secondary alcohols. AE derived from linear secondary alcohols degrade slightly more slowly than the corresponding linear primary AE and the position of secondary attachment may influence the rate of biodegradation. In river water the linear secondary dodecyl AE ($C_{12}AE_8$), branched at the 4-C, required 10 days for 99% disappearance as indicated by surface tension measurements whereas the linear primary isomer required 5-6 days. The 6-C-branched isomer was 70% biodegraded after 13 days (Blankenship and Piccolini, 1963).

Using the 24-hr SCAS test, Han (1967) found removals of 100 and 86% for the n-pri-$C_{14-16}AE_8$ and $C_{14-16}AE_{16}$ as measured by sulfation/methylene blue and removals of 92 and 57% for the n-sec-$C_{11-15}AE_8$ and $C_{11-15}AE_{15}$. Using the same test, the n-pri-$C_{12-14}AE_{10}$ was degraded 99.7% as indicated by foam loss and the n-sec-$C_{11-15}AE_9$ was 96.2% degraded in 17 hours (Mausner et al., 1969). The biodegradation of a linear isomer of $C_{12}AE_9$ derived from 1-dodecanol was significantly faster than linear secondary $C_{13}AE_{10}$ derived from 7-tridecanol, 97% compared to 57% after 20 days as measured by total organic carbon (Kuwamura and Takahaski, 1976). Two cyclic and two tertiary alcohol ethoxylates were even more resistant to biodegradation. Myerly et al. (1964) found the same rate, 97% biodegradation in six days as measured by CTAS, for the linear secondary $C_{11-15}AE_9$ and the corresponding linear primary $C_{11-15}AE_9$ in river water. In river-water dieaway tests lasting 28 days, Vath (1964) found similar removals for primary and secondary $C_{16}AE_n$, 93-98% as measured by CTAS, foaming, and surface tension. Kravetz et al. (1978) found that biodegradation of n-sec-$C_{11-15}AE_9$ was only slightly slower than primary AE. Biodegradability was measured by CO_2 formation, DOC, BOD, and CTAS.

Using the shake-flask test and fresh unacclimated sludge from a municipal sewage treatment plant as the inoculum, Booman (unpublished data) found that n-pri-$C_{12}AE_8$ was completely degraded as indicated by foam loss in seven days, but n-sec-$C_{12-15}AE_9$ was degraded only 12-50% in the same time period. The large difference in biodegradability between primary and secondary AE reported in this study may be due to using a bacterial inoculum that was not acclimated to either of the substrates prior to the test.

Using a die-away screening test and an activated sludge confirmatory test, Birch (1982, 1984) compared the degradation of linear and secondary AE and found that the effect of the secondary structure was small up to an EO length of 20 units. At that point differences due to increased branching became apparent. The work of Birch (1984; 1991b) illustrates the effect of ethoxylate chain length, alkyl branching, and secondary alcohol structure on primary biodegradation (Table 3-2 and Table 3-3). Analysis was by the Wickbold BIAS method. Additional studies on biodegradation of linear primary, linear secondary, and branched AE can be found in the Appendix.

Table 3-2. Primary Biodegradation (Percent) vs Chemical Structure in the OECD Screening Test

Alcohol	EO_{10}	EO_{20}	EO_{30}	EO_{40}	EO_{50}
Linear primary	99	99	98	98	98
75% n-pri-linear oxo	100	98	93	88	92
50% n-pri-linear oxo	84	83	79	—	69
Linear secondary	96	64	59	65	—

Table 3-3. Primary Biodegradation (Percent) vs Chemical Structure in the OECD Confirmatory Test

Alcohol	EO_{10}	EO_{20}	EO_{30}	EO_{40}	EO_{50}
75% n-pri-linear oxo	93	83	85	89	86
50% n-pri-linear oxo	97	89	75	71	68
Linear secondary	87	85	72	54	—

3. Propylene Oxide Addition

Substitution of propylene oxide or higher analogs for ethylene oxide units retards biodegradation, probably as a consequence of methyl branching (Swisher, 1987). Single and double propylene oxide groups are degradable, but multiple groups, i.e., greater than four, do not appear to be practically biodegradable (Naylor et al., 1988).

Using SCAS units, Naylor et al. (1988) studied the primary (as measured by foaming properties) and ultimate (as measured by BOD and DOC) degradation of several commercial and experimental surfactants in which the polyoxypropylene block was located between two polyoxyethylene blocks of equal size and the alcohols varied in extent of branching. Biodegradation was inversely proportional to both the amount of propylene oxide incorporated into the surfactant and the amount of branching in the alcohol. For linear alkyl chains, the maximum PO block size that allowed complete biodegradation was about four units if located in the middle of the ethoxylate chain. These results are summarized in the Appendix.

Two independent investigations into the potential biodegradability of several propylene oxide-containing surfactants have been conducted using SCAS and batch jars (Procter & Gamble, 1985, 1991) and using shake flasks (BASF, 1990). The Pluronic EO/PO surfactants that were tested ranged in molecular weight from 1900 to 8400 and EO weight percents of 10 to 80%. These polyethers have a PO block between two polyoxyethylene blocks, except for Pluronic 25R2 surfactant in which this sequence is reversed. The shake flask data showed appreciable levels of biodegradability (as measured by CO_2 formation) for the Pluronic surfactants up to PO chain lengths of 30. At higher PO chain lengths (54 units), only low levels of mineralization were observed. The batch jar data show the same inverse molecular weight effect, with significantly increased biodegradability observed in 1991 tests as compared to the earlier batch jar tests. The differences between these latter two sets of data could be due to acclimation of bacteria, since the more recent tests were conducted with acclimated activated sludge. Also, tests with Plurafac RA 20, an alkoxylate with a lower molecular weight (550-600), in batch jars showed some removal and biodegradation. Extent of biodegradation of the individual compounds is summarized in the Appendix.

D. EFFECT OF ENVIRONMENTAL VARIABLES

In the field, low temperatures may have a slight inhibiting effect on biodegradation by changing the composition of microbial populations or necessitating acclimation of microorganisms to the new temperature. Results of two field studies, however, show no affect of normal seasonal temperatures on removal of AE surfactants. At a wastewater treatment facility with a trickling filter system, Mann and Reid (1971) found 84 and 89% removals of two linear primary alcohol ethoxylates, as measured by TLC, during winter conditions in the U.K. During all seasons (water temperature range 13.4-20.6°C), Maki et al. (1979) and Sykes et al. (1979) found an overall removal of 90%, as measured by CTAS, of n-pri-$C_{14-15}AE_7$ dosed into an activated sludge treatment facility in Ohio.

In the laboratory the effect of low temperature on biodegradation appears to be minimal. Biodegradation time is related to the source and acclimation of microorganisms and the endpoint of the test, i.e., primary or secondary biodegradation. At temperatures of 8°, 11°, and 15°C, removals of the n-pri-$C_{9-11}AE_8$ and $C_{12-15}AE_9$ were both >94% in porous-pot activated-sludge units (Stiff et al., 1973). Larson and Games (1981) described the rate of ethoxylate chain degradation, as measured by CO_2 evolution, of $C_{16}AE_3$ over a 31° temperature span (3-34°C) with the Arrhenius equation; rate increased by a factor of 1.8 for each 10°C rise. For the longer-chain ethoxylate, $C_{12}AE_9$, this relationship held true only between the temperatures of 14 and 25°C.

Kravetz et al. (1984) examined the biodegradability of $C_{12-15}AE_9$ in closed continuous bench-scale biotreaters under seasonal temperature conditions. From 25°C down to 12°C, primary biodegradation was in excess of 97% for a sludge residence time of ten days. Lowering the temperature in the biotreaters to 8°C retarded the extent of primary biodegradation to 93% after the biotreaters had run for 27 days. Ultimate biodegradation of the radiolabeled hydrophobe chain was only marginally affected by temperature, decreasing from 79% at 25°C to 76% at 8°C. Ultimate biodegradation of the radiolabeled ethoxylate chain decreased to 58%, 50%, and 10% at temperatures of 25°C, 12°C, and 8°C, respectively.

Birch (1984, 1991a) found that temperature had only a marginal effect on biodegradation of AE, even at sludge retention times as short as two days. Using laboratory-scale activated sludge plants, degradation of linear oxo-derived AE with seven EO units at an influent concentration of 5 mg/L exceeded 90% under any conditions.

Schöberl and Mann (1976) studied the effect of temperature on primary degradation of AE in both fresh- and seawater. Percent biodegradation (as measured by BIAS) in 23 days for a linear oxo-alcohol ethoxylate (C_{11-14}) with 7, 9, or 11 EO units decreased with a drop in temperature from 20-28°C to 3-4°C and was slightly faster in seawater than in pond water. Increased length of the ethoxylate chain resulted in decreased rates of degradation, particularly at the low temperature in freshwater.

In a similar study and using the CTAS measurement method, the primary degradation of n-pri-$C_{14-15}AE_7$ in estuarine, ocean, and river water were compared. In both estuarine and river waters, greater than 85% degradation occurred in 11 days or less. Degradation in ocean water was slower, requiring 21-35 days to achieve 85% primary biodegradation (Procter & Gamble Co., 1981).

The extent of biodegradation of mono-, di-, and triethylene glycols, which may have arisen via the central cleavage mechanism, was influenced by the source of river water, presence of sediment, aeration, and temperature, with slower rates at 8° than at 20°C (Evans and David, 1974).

E. METABOLIC PATHWAYS OF BIODEGRADATION

Conflicting results among the many studies that focus on the pathway(s) of degradation suggest that AE may be biodegraded differently by different bacterial systems and chemical structure may affect the rate and pathway. Tagging of compounds with radiotracers, usually ^{14}C and/or ^{3}H, has been particularly useful for identifying degradation pathways.

Three initial points of bacterial attack have been documented for linear AE surfactants: central fission, separating the hydrophilic and hydrophobic groups; attack at the terminal end of the alcohol chain; and attack at the terminal end of the ethylene oxide chain (Swisher, 1987). Central fission by hydrolysis of the ether linkage results in liberation of the hydrophilic ethoxylate chain as polyethylene glycols which are broken down into lower molecular weight polyethylene glycols and ultimately CO_2 and water. The hydrophobic alkyl chain is oxidized via (beta) β-oxidation to CO_2 and water. Alkyl groups attached to the α-carbon of the alcohol may hinder attack by the central cleavage mechanism. Biodegradation may then proceed by oxidation of the terminal methyl group of the fatty alcohol (omega [ω] oxidation followed by β-oxidation) and/or oxidation at the terminal end of the polyoxyethylene chain.

1. Aerobic Biodegradation

Central fission. The following studies document initial central fission which takes place either by direct hydrolysis of the ether linkage or following oxidation of the α carbon of the alkyl or ethylene oxide chain by an enzymic activation mechanism. Using TLC to identify degradation products, Patterson et al. (1970) proposed that linear primary AE biodegrade by initial cleavage of the ether linkage simultaneous with rapid β-oxidation of the alkyl moiety to CO_2 and water and slower oxidation of the polyoxyethylene group to ethylene glycol units and then to CO_2 and water. Beta-oxidation liberates two carbons at a time as acetyl groups and oxidation of the polyoxyethylene moiety liberates ethylene glycol. Information on the biodegradation pathways for highly branched AE is conflicting.

$$CH_3-(CH_2)_x-CH_2-O-(CH_2-CH_2-O)_y-H$$
$$\downarrow$$
$$CH_3-(CH_2)_x-CH_2-OH \ + \ HO-(CH_2-CH_2-O)_y-H$$

$\downarrow$	$\downarrow$
β-oxidation	glycol oxidation
$\downarrow$	$\downarrow$
$CH_3-(CH_2)_{x-2}-CH_2-OH$	$HO-(CH_2-CH_2-O)_{y-1}-H$
$\downarrow$	$\downarrow$
$\downarrow$	$\downarrow$
$CO_2 + H_2O$	$CO_2 + H_2O$

In a series of papers, Scharer et al. (1979) and Kravetz (1981; Kravetz et al., 1982) presented data for biodegradation of essentially linear AE that strongly indicated biodegradation proceeded via initial ester formation in the area of the hydrophobe-hydrophile junction, followed by hydrolysis and then oxidation of the alkyl and polyoxyethylene moieties. They used doubly radio-labeled $C_{12-15}AE_9$ (as synthesized by Shebs and Smith, 1981) containing tritium in the hydrophobic moiety, primarily on the 1- and 3-carbons and uniformly labeled ^{14}C in the hydrophilic moiety. Based on the rapid appearance of ^3H_2O accompanied by very little CO_2 evolution, they proposed that the tritiated portion of the alkyl chain degraded faster than the alkyl chain as a whole. $^{14}CO_2$ from the ethylene oxide chain appeared later.

Additional studies document that the polyoxyethylene chains are more resistant to microbial degradation than the alkyl moiety. Using shake-flask cultures and GC analysis of HBr reaction products, Tobin et al. (1976a; 1976b) showed that, after 97 hours of incubation, alkyl moieties were largely removed while the polyethoxylate moiety had degraded only 25%. The appearance of polyethylene glycol and the low initial CO_2 evolution indicated hydrolysis of the ether linkage. Vashon and Schwab (1982) monitored ultimate biodegradation of trace concentrations of linear AE in water from Escambia Bay, Florida. ^{14}Carbon dioxide production from $^{14}C_{16}AE_3$ labeled at the α-alkyl carbon was rapid and first order with respect to concentration, indicating hydrolysis of the ether linkage. The biodegradation of $C_{12}{}^{14}AE_9$ labeled uniformly in the ethoxylate chain was less rapid and kinetics were dependent on concentration.

If polyoxyethylene chains are more resistant to degradation than the alkyl chain, then one would expect to find an accumulation of PEG. After 5 days of incubation of 19-20 mg/L of n-$C_{14}AE_{8.3}$ in river water, Frazee et al. (1964) found only PEG of undetermined length. Patterson et al. (1967), Pitter (1968a; 1968b), Cook (1979), and Birch (1982) among others also found a temporary buildup of neutral and/or acid PEG-type residues.

Terminal end of ethoxylate chain. Although limited, published literature on mechanistic pathways for the biodegradation of highly branched AE suggest they biodegrade by a different pathway than linear AE. Patterson et al. (1967) found a highly branched $C_{13}AE$ degraded much more slowly than a corresponding linear AE. In addition, TLC analysis revealed virtually no PEG accumulated during the early stages of biodegradation as was found for linear AE. Schöberl et al. (1981) found a "...suite of shorter ethoxylates..." during biodegradation of a highly branched $C_{13}AE_{12}$. Both Patterson's and Schöberl's studies suggest biodegradation of highly branched AE is initiated at the ω end of the polyoxyethylene chain, followed by stepwise shortening of this chain rather than by attack at the hydrophobe-hydrophile junction as appears to be the case for linear AE.

Terminal end of alcohol. Several researchers documented the initial point of attack as the terminal methyl of the alkyl group. Using ^{14}C-radiolabeled linear primary alcohol ethoxylates, labeled (1) uniformly in the alkyl chain, (2) uniformly in the ethoxylate chain or (3) at the α-carbon of the alkyl chain, Nooi et al. (1970) proposed that the terminal methyl group of the alkyl chain is the initial point of attack, followed by β-oxidation. Liberation of $^{14}CO_2$ from the α-carbon was slower than the average for the entire alkyl chain, indicating that oxidation proceeded from the methyl group inward. They found acidic products which contained both carboxyl and ethoxy groups.

In a series of articles, Steber and Wierich (1983, 1985) discussed the fate of radiolabeled stearyl alcohol ethoxylate in a model continuous-flow activated sludge plant. Their results indicated two mechanisms: intramolecular scission as well as ω- and β-oxidation of the alkyl chain, with faster degradation of the alkyl than the polyethylene glycol moiety. Regardless of the position of the label, uncharged and carboxylated, mainly dicarboxylated polyethylene glycols, constituted the major degradation products. When 1-^{14}C-stearyl alcohol ethoxylate ($^{14}C_{18}AE_7$) was added to the system, predominantly carboxylated and dicarboxylated ^{14}C-labeled polyethylene glycols were isolated, indicating a rapid oxidation of the fatty alcohol, beginning with the methyl end, and leaving the label attached to the polyethylene glycol chain. Dicarboxylic polyethylene glycols indicated slow oxidation beginning at the terminal end of the chain. Primary degradation products of the uniformly ^{14}C-labeled ethoxylate chain were a series of PEG.

Schöberl et al. (1981) followed the metabolism of the 40% linear $C_{13-14}AE_{12}$ (60% of the compound had a single methyl branch in the β-position to the ether linkage) in CAS units inoculated with either sludge or air-borne microorganisms. Although they isolated a small amount of $C_{13-14}AE_4$, they found primarily carboxylated [$HOOC-(CH_2)_2-O-(CH_2CH_2O)_4H$] compounds and polydiols with an EO content of 13 moles. Anionic intermediates with <5 carbon atoms could not be isolated with the method they used. For linear AE they proposed that the primary mechanism of biodegradation proceeds by carboxylation of the terminal methyl group of the alkyl chain followed by β-oxidation (cleavage of acetyl-CoA units). Branching of the alkyl chain presented a hinderance to this pathway.

2. Anaerobic Biodegradation

Biodegradation of AE surfactants is much slower under anaerobic conditions than when there is an abundance of oxygen (Swisher, 1987). Vath (1964) reported that n-sec-$C_{11-15}AE_{9.2}$ was rapidly degraded as indicated by foam loss, but the extent of degradation was not specified. Using a laboratory-scale fixed-bed reactor containing anoxic sediment or sludge samples, $C_{12}AE_{23}$ (Brij 35) and $C_{10-12}AE_{7.5}$ (E-LM 75), at concentrations up to 1.0 g/L, were degraded (90%) to methane and CO_2, with small amounts of acetate and propionate present (Wagener and Schink, 1987). Similar results were obtained by Steber and Wierich (1987) using an anaerobic sludge digester. After four weeks of incubation of ^{14}C-labeled stearyl alcohol ethoxylate ($C_{18}AE_7$) uniformly labeled in both the alkyl and EO chains, more than 80% of the initial radioactivity was found as methane and CO_2; 10% was attributable to assimilation into biomass in the sludge. The authors isolated and identified radiolabeled compounds remaining after anaerobic degradation. The metabolites indicated scission into alkyl and polyethylene glycol moieties followed by oxidative or hydrolytic depolymerization steps. Other studies have confirmed the biodegradation of polyethylene glycols by anaerobic bacteria (Schink and Stieb, 1983; Dwyer and Tiedje, 1983).

REFERENCES

Abram, F.S.H., V.M. Brown, H.A. Painter and A.H. Turner. 1977. The biodegradability of two primary alcohol ethoxylate nonionic surfactants under practical conditions, and the toxicity of the biodegradation products to rainbow trout. Paper presented at the IVth Yugoslav Symposium on Surface Active Substances held in Dubrovnik, October 17-21, 1977.

Albanese, P. and R. Capuci. 1974. Biodegradation of nonionic surfactants. Note 2: Measuring biodegradation. Riv. Ital. Sost. Grasse 51:70-81.

Allred, R.C. and R.L. Huddleston. 1967. Microbial oxidation on surface active agents. Southwest Water Works J. 49:26-28.

Ankley, G.T. and L.P. Burkhard. 1991. Identification of surfactants as toxicants in a primary effluent. Environ. Toxicol. Chem. 11:1235-1248.

APHA (American Public Health Association). 1989. *Standard Methods for the Examination of Water and Wastewater*, 17th ed., APHA, New York.

Arpino, A., E. Fedeli, G.B. Bosari, F. Buosi and E.P. Fuochi. 1974. Evaluation of synthetic primary fatty alcohol derivatives in relation to chemical structure. II. Applicability of Wickbold method to the evaluation of the biodegradability of ethoxylate nonionic surfactants. Riv. Ital. Sost. Grasse 51:253-265.

Arthur D. Little, Inc. 1977. Human Safety and Environmental Aspects of Major Surfactants. A Report to the Soap and Detergent Association. Arthur D. Little, Cambridge, MA.

ASTM (American Society for Testing Materials). 1990. *Annual Book of Test Standards*, Volume 11.04, ASTM, Philadelphia, PA.

BASF Corporation. 1990. CO_2 production tests performed for BASF Corporation by Roy F. Weston, Inc. Unpublished data.

Birch, R.R. 1982. The biodegradability of alcohol ethoxylates. XIII Jornadas Com. Español Deterg. 33-48. (Cited in Swisher, 1987).

Birch, R.R. 1984. Biodegradation of nonionic surfactants. J. Am. Oil Chem. Soc. 61:340-343.

Birch, R.R. 1991a. Prediction of the fate of detergent chemicals during sewage treatment. J. Chem. Tech. Biotechnol. 50:411-422.

Birch, R.R. 1991b. Recent developments in the biodegradability testing of nonionic surfactants. Riv. Ital. Sost. Grasse 68:433-437.

Blankenship, F.A. and V.M. Piccolini. 1963. Biodegradation of nonionics. Soap Chem. Special. 39:75-78, 181.

Bock, K.J. 1973. Anionische und nichtionishe tenside in kläranlagen und flüssen. Tenside 10:178.

Booman, K.A. Unpublished data. (Cited in Arthur D. Little, 1977).

Booman, K.A., J. Dupré, and E.S. Lashen. 1967. Biodegradable surfactants for the textile industry. Am. Dyestuff Reptr. 56:P82-P88.

Borstlap, C. and C. Kortland. 1967. Biodegradability of nonionic surfactants under aerobic conditions. Fette Seifen Anstrichmittel 69:736-738. (Cited in Swisher, 1987).

Bosari, G.B., F. Buosi and E.P. Fuochi. 1974. Evaluations of synthetic primary fatty alcohol derivatives in relation to their chemical structure. I. Biodegradability and fish toxicity of surfactants. Riv. Ital. Sost. Grasse 51:193-207.

Boyer, S.L., K.F. Guin, R.M. Kelley, M.L. Mausner, et al. 1977. Analytical method for nonionic surfactants in laboratory biodegradation and environmental studies. Env. Sci. Technol. 11:1167-1171.

Brown, D. 1976. The assessment of biodegradability: a consideration of possible criteria for surface-active substances. In: Proceedings of the Seventh International Congress on Surface-active Substances, USSR National Committee on Surface Active Substances, Moscow, USSR, 4:44-57.

Brown, D., H. de Henau, J.T. Garrigan, P. Gerike, M. Holt, E. Keck, E. Kunkel, E. Matthijs, J. Waters and R. J. Watkinson. 1986. Removal of nonionics in a sewage treatment plant. Removal of domestic detergent nonionic surfactants in an activated sludge sewage treatment plant. Tenside 23:190-195.

Brown, D., H. de Henau, J.T. Garrigan, P. Gerike, M. Holt, E. Kunkel, E. Matthijs, J. Waters and R.J. Watkinson. 1987. Removal of nonionics in sewage treatment plants. II. Removal of domestic detergent nonionic surfactants in a trickling filter sewage treatment plant. Tenside 24:14-19.

Conway, R.A., C.A. Vath and C.E. Renn. 1965. New detergent nonionics . . . . biodegradable. Water Works Waste Eng. 2:28-31.

Cook, K.A. 1979. Degradation of the nonionic surfactant Dobanol 45-7 by activated sludge. Water Res. 13:259-266.

Dai Ichi Kogyo Pharmaceutical Co. 1974. Soft type nonionic surface-active agents. Yukagaku 23:685. (Cited in Arthur D. Little, 1977).

Dwyer, D.F. and J.M. Tiedje. 1983. Degradation of ethylene glycol and polyethylene glycols by methanogenic consortia. Appl. Environ. Microbiol. 46:185-190.

Evans, W.H. and E.J. David. 1974. Biodegradation of mono-, di-, and triethylene glycols in river waters under controlled laboratory conditions. Water Res. 8:97-100.

Federal Register 50:39277-39280, September 27, 1985.

Federle, T.W. and G.M. Pastwa. 1988. Biodegradation of surfactants in saturated subsurface sediments: a field study. Groundwater 26:761-770.

Fischer, W.K. 1971. Testing and evaluation of biodegradability of nonionic surfactants. II. Testing nonionic surfactants for biodegradability in the closed bottle test. Tenside 8:182-188.

Frazee, C.D., Q.W. Osburn and R.O. Crisler. 1964. Application of infrared spectroscopy to surfactant degradation studies. J. Am. Oil Chem. Soc. 41:808-812.

Gerbil, B.A. and H.M. Naim. 1969. Biodegradable nonionic surfactants from chlorinated Egyptian solar. Indian J. Technol. 7:365-369.

Gerike, P. and R. Schmid. 1973. Determination of nonionic surfactants with the Wickbold method in biodegradation research and in river waters. Tenside 10:186-189.

Gledhill, W.E. 1975. Screening test for assessment of ultimate biodegradability: linear alkyl benzene sulfonate. Appl. Microbiol. 30:922-929.

Gledhill, W.E., R.L. Huddleston, L. Kravetz, A.M. Nielsen, R.I. Sedlak and R.D. Vashon. 1989. Treatability of surfactants at a wastewater treatment plant. Tenside 26:276-281.

Goyer, M.M., J.H. Perwak, A. Sivak and P.S. Thayer. 1981. Human Safety and Environmental Aspects of Major Surfactants (Supplement). Report by Arthur D. Little, Inc. to the Soap and Detergent Association.

Han, K.W. 1967. Determination of biodegradability of nonionic surfactants by sulfation and methylene blue extraction. Tenside 4:43-45.

Heinz, H.J. and W.K. Fischer. 1967. International status of methods for determining biodegradability of surfactants, and the possibilities for standardization. II. Current test methods and suggestions for a combined international standard method. Fette Seifen Anstrichmittel 69:188-196.

Howells, W.G., L. Kravetz, D. Loring, C.D. Piper, B.W. Poovaiah, E.C. Seim, V.P. Rasmussen, N. Terry, and L.J. Waldron. 1984. The use of nonionic surfactants for promoting the penetration of water into agricultural soils. In: Proceedings of the Eighth World Surfactants Congress, Munich, West Germany.

Huddleston, R.L. and R.C. Allred. 1964. Evaluation of detergents by using activated sludge. J. Am. Oil Chem. Soc. 41:732-735.

Hughes, A.I., D.R. Peterson and R.K. Markarian. 1989. Comparative biodegradability of linear and branched alcohol ethoxylates. Presented at the American Oil Chemists' Society annual meeting, Cincinnati, OH, May 3-7, 1989.

Huyser, H.W. 1960. Relation between the structure of detergents and their biodegradation. In: Proceedings of the 3rd International Congress on Surface-Active Substances, Cologne, Germany. University of Mainz Press, Germany, 3:295-301. (Cited in Swisher, 1987).

Jones, P. and G. Nickless. 1978. Characterization of non-ionic detergents of the polyethoxylated type from water systems. II. Isolation and examination of polyethoxylated material before and after passage through a sewage plant. J. Chromatog. 156:99-110.

Knaebel, D.B., T.W. Federle and J.R. Vestal. 1990. Mineralization of linear alkylbenzene sulfonate (LAS) and linear alcohol ethoxylate (LAE) in 11 contrasting soils. Environ. Toxicol. Chem. 9:981-988.

Kravetz, L., H. Chung, J.C. Rapean, K.F. Guin and W.T. Shebs. 1978. Ultimate biodegradability of detergent range alcohol ethoxylates. Presented at Amer. Oil Chem. Soc., 69th Ann Meet., St. Louis, May 1978. Shell Chemical Co. SC:321-81.

Kravetz, L., H. Chung, K.F. Guin and W.T. Shebs. 1979. Ultimate biodegradation of alcohol ethoxylates. Surfactant concentration and polyoxyethylene chain effects. Presented at the 70th Annual Meeting of the American Oil Chemist's Society, San Francisco, CA. Shell Chemical Company Technical Bulletin SC:442-80.

Kravetz, L. 1981. Biodegradation of nonionic ethoxylates. J. Am. Oil Chem. Soc. 58:58A-65A.

Kravetz, L., H. K.F. Guin, W.T. Shebs, L.S. Smith and H. Stupel. 1982. Ultimate biodegradation of an alcohol ethoxylate and a nonylphenol ethoxylate under realistic conditions. Soap Cosmet. Chem. Special. April:34-42, 102B.

Kravetz, L., H. Chung, K.F. Guin, W.T. Shebs and L.S. Smith. 1984. Primary and ultimate biodegradation of an alcohol ethoxylate and a nonylphenol ethoxylate under average winter conditions in the United States. Tenside 21:1-6

Kravetz, L., J.P. Salanitro, P.B. Dorn and K.F. Guin. 1991. Influence of hydrophobe type and extent of branching on environmental response factors of nonionic surfactants. J. Am. Oil Chem. Soc. 68:610-618.

Kuwamura, T. and H. Takahaski. 1976. Structural effects of hydrophobe on the surfactant properties of polyglycol monoalkyl ethers. In: Proceedings of the Seventh International Congress on Surface-active Substances. USSR National Committee on Surface Active Substances, Moscow, USSR, 1:33-43. (Cited in Swisher, 1987).

Laboureur, P., M. Bechet and J. Emeraud. 1976. Ultimate biodegradability of ethoxylated and propoxylated alcohols. In: Proceedings of the Seventh International Congress on Surface-active Substances. USSR National Committee on Surface Active Substances, Moscow, USSR, 4:139-148.

Larson, R.J. and L.M. Games. 1981. Biodegradation of linear alcohol ethoxylates in natural waters. Environ. Sci. Technol. 15:1488-1493.

Maki, A.W., A.J. Rubin, R.M. Sykes, R.L. Shanks. 1979. Reduction of nonionic surfactant toxicity following secondary treatment. J. Water Pollut. Control Fed. 51:2301-2313.

Mann, A.H. and V.W. Reid. 1971. Biodegradation of synthetic detergents evaluation by community trials. Part 2. Alcohol and alkylphenol ethoxylates. J. Am. Oil Chem. Soc. 48:794-797.

Mann, A.H. and P. Schöberl. 1976. Der biologische abbau eines nichtionischen Tensides in brackwasser. Arch. Fisch. Wiss. 26:177-180.

Mausner, M., J.H. Benedict, K.A. Booman, et al. 1969. The status of biodegradability testing of nonionic surfactants. J. Am. Oil Chem. Soc. 46:432-444.

Möller, U.J. 1972. Estimation of biological effects of nonionic emulsifiers in split-water phase from metal working emulsions. Erdöl Kohle 25:451-456. (Cited in Swisher, 1987).

Myerly, R.C., J.M. Rector, E.C. Steinle, C.A. Vath and H.T. Zika. 1964. Secondary alcohol ethoxylates as degradable detergent materials. Soap Chem. Spec. 40:78-82.

Naylor, C.G., F.J. Castaldi and B.J. Hayes. 1988. Biodegradation of nonionic surfactants containing propylene oxide. J. Am. Oil Chem. Soc. 65:1669-1676.

Nooi, J.R., M.C. Testa and S. Willemse. 1970. Biodegradation mechanisms of fatty alcohol nonionics. Experiments with some ^{14}C-labelled stearyl alcohol/EO condensates. Tenside 7:61-65.

OECD (Organization for Economic Cooperation and Development). 1981. OECD Guidelines for Testing of Chemicals. The OECD Expert Group on Degradation and Accumulation, Paris.

Patterson, S.J., C.C. Scott and K.B.E. Tucker. 1967. Nonionic detergent degradation. I. Thin-layer chromatography and foaming properties of alcohol polyethoxylates. J. Am. Oil Chem. Soc. 44:407-412.

Patterson, S.J., C.C. Scott and K.B.E. Tucker. 1970. Nonionic detergent degradation: III. Initial mechanism of the degradation. J. Am. Oil Chem. Soc. 47:37-41.

Pitter, P. 1968a. Relation between degradability and chemical structure on nonionic polyethylene oxide compounds. In: Chimie, Physique et Applications Pratiques des Agents de Surface, Ediciones Unidas, Barcelona Spain, 1:115-123. (Cited in Swisher, 1987).

Pitter, P. 1968b. Biodegradability of surface-active alkyl polyethylene glycol ethers. Coll. Czech. Chem. Comm. 33:4083-4088. (Cited in Swisher, 1987).

Pitter, P. 1972. Biodegradation of polyethylene glycols. In: Proceedings of the Sixth International Surfactant Congress held in Zurich, September 11-15, 1972. Carl Hanser Verlag, Munich.

Procter & Gamble. 1981. Unpublished data (Cited in Arthur D. Little, 1981).

Procter & Gamble. 1985. Unpublished data.

Procter & Gamble. 1991. Unpublished data.

Ruschenberg, E. 1963. Structure elements of detergents and their influence on biodegradation. Vom Wasser 30:232-248.

Salanitro, , J.P., G.C. Langston, P.B. Dorn, and L. Kravetz. 1988. Activated sludge treatment of ethoxylate surfactants at high industrial use concentrations. Water Sci. Technol. 20:126-130.

Scharer, D.H., L. Kravetz and J.B. Carr. 1979. Biodegradation of nonionic surfactants. Tappi 62:3-10.

Schick, M.J. (Ed.). 1967. *Nonionic Surfactants*. Volume 1, Surfactant Science Series. Marcel Dekker, New York.

Schink, B. and M. Stieb. 1983. Fermentative degradation of polyethylene glycol by a strictly anaerobic, gram-negative, nonsporeforming bacterium, *Pelobacter venetianus* sp. nov. Appl. Environ. Microbiol. 45:1905-1913.

Schöberl, P., and H. Mann. 1976. Temperature influence on the biodegradation of nonionic surfactants in sea- and freshwater. Arch. Fischereiwiss. 27:149-158. (Cited in Swisher, 1987).

Schöberl, P., E. Kunkel and K. Espeter. 1981. Comparative studies of the microbial metabolism of a nonylphenol ethoxylate and an oxo alcohol ethoxylate. Tenside 18:64-72.

SDA (Soap and Detergent Association). 1965. A procedure and standards for the determination of the biodegradability of alkyl benzene sulfonate and linear alkylate sulfonate. J. Am. Oil Chem. Soc. 42:986-933.

Sedlak, R.I. and K.A. Booman. 1986. LAS and alcohol ethoxylate: a study of their removal at a municipal wastewater treatment plant. Soap Cosmetics Chem. Special. April:44-46, 107.

Shebs, W.T. and L.S. Smith. 1981. Synthesis of two nonionic surfactants labeled with carbon-14 and tritium. Presented at the American Oil Chemists' Society 72nd Annual Meeting, May 1981, New Orleans, LA.

Steber, J. and P. Wierich. 1983. The environmental fate of detergent range fatty alcohol ethoxylates - biodegradation studies with a ^{14}C-labelled model surfactant. Tenside 20:183-187.

Steber, J. and P. Wierich. 1985. Metabolites and biodegradation pathways of fatty alcohol ethoxylates in microbial biocenoses of sewage treatment plants. Appl. Environ. Microbiol. 49:530-537.

Steber, J. and P. Wierich. 1987. The anaerobic degradation of detergent range fatty alcohol ethoxylates. Studies with ^{14}C-labelled model surfactants. Water Res. 21:661-667.

Stiff, M.J., R.C. Rootham and G.E. Gully. 1973. The effect of temperature on the removal of nonionic surfactants during small-scale activated-sludge sewage treatment. I. Comparison of alcohol ethoxylates with a branched-chain alkyl phenol ethoxylate. Water Res. 7:1003-1010.

Stühler, H. and H. Wellens. 1976. Biodegradation of nonionic surfactants (polyglycol ethers). In: Proceedings of the Seventh International Congress on Surface-Active Substances, Moscow, USSR, 4:106-115. (Cited in Swisher, 1987).

Sturm, R.N. 1973. Biodegradability of nonionic surfactants: screening test for predicting rate and ultimate biodegradation. J. Am. Oil Chem. Soc. 50: 159-167.

Swisher, R.D. 1987. *Surfactant Biodegradation*. Marcel Dekker, New York. p. 7, 476-497,

Sykes, R,M., A.J. Rubin, S.A. Rath and M.C. Chang. 1979. Treatability of a nonionic surfactant by activated sludge. J. Water Pollut. Control Fed. 51:71-77.

Tabak, H.H. and R.L. Bunch. 1981. Measurement of nonionic surfactants in aqueous environments. Proceedings of the Purdue Industrial Waste Conference, 36:888-907.

Tobin, R.S., F.I. Onuska, D.H.J. Anthony and M.E. Comba. 1976a. Nonionic surfactants: conventional biodegradation test methods do not detect persistent polyglycol products. Ambio 5:30-31.

Tobin, R.S., F.I. Onuska, B.G. Brownlee, D.H.J. Anthony and M.E. Comba. 1976b. The application of an ether cleavage technique to a study of the biodegradation of a linear alcohol ethoxylate nonionic surfactant. Water Res. 10:529-535.

Vashon, R.D. and B.S. Schwab. 1982. Mineralization of linear alcohol ethoxylates and linear alcohol ethoxy sulfates at trace concentrations in estuarine water. Env. Sci. Technol. 16:433-436.

Vath, C.A. 1964. A sanitary engineer's approach to biodegradation of nonionics. Soap Chem. Special. 40(2):56-58, 182; 40(3):55-58, 108.

Wagener, S. and B. Schink. 1987. Anaerobic degradation of nonionic and anionic surfactants in enrichment cultures and fixed-bed reactors. Water Res. 21:615-622.

Wagner, R. 1978. Behavior of MBAS and BIAS in a municipal sewage treatment plant. Gas-Wasserfach Wasser Abwasser 119:235-242. (Cited in Goyer et al., 1981).

Wee, V.T. 1981. Determination of linear alcohol ethoxylates in waste- and surface water. In: L.H. Keith, Ed., Advances in the Identification & Analysis of Organic Pollutants in Water, Vol. 1, Ann Arbor Science, pp. 467-479.

Wickbold, R. 1974. Analytical comments on surfactant biodegradation. Tenside 11: 137-144.

IV. ENVIRONMENTAL TOXICOLOGY

A. AQUATIC TOXICITY

Acute and chronic toxicity tests have been performed in the laboratory with linear primary, linear secondary, and branched AE using fish, daphnids, oysters, shrimp, crabs, freshwater algae, and aquatic macrophytes. The bioaccumulation of AE in fish has also been assessed.

Linear primary AE surfactants are acutely toxic to freshwater fish and crustaceans at median lethal concentrations (LC_{50} concentrations) ranging from 0.4 to 10 mg/L and 0.29-0.4 to 20 mg/L, respectively. Most LC_{50} values were <5 mg/L. The lowest reported values were for the brown trout (*Salmo trutta*), 0.4 mg/L of $C_{16,18}AE_{14}$ and the cladoceran (*Daphnia magna*), 0.29-0.4 mg/L of $C_{14-15}AE_7$. In chronic studies, survival and reproduction of fish (*Pimephales promelas*) and invertebrates (*D. magna*) were not affected at concentrations of 0.18 mg/L and 0.24 mg/L, respectively. Both the acute toxicity and chronic no-effect values, based on single chemical testing, are greater than detected concentrations of the same or related chemicals in environmental media, but are lower than environmental concentrations for total nonionics measured by chemical non-specific methods. Because measurements of total nonionics overestimate the concentrations of AE present in the environment, no conclusions regarding risk to aquatic organisms can be drawn.

The toxicity of intact surfactants to aquatic organisms is related to chemical structure. Linear, primary AE are more acutely and chronically toxic than AE with branched alkyl structure or secondary attachment of the alkyl group. In addition, toxicity decreases with increasing ethylene oxide chain length. While the linear parent compounds are inherently more toxic than branched compounds, the slower degradation of branched compounds results in less treatability in pilot plants and more effluent toxicity. Bioaccumulation of intact surfactants is high, with factors of several hundred times that of water, but elimination is rapid, with half-lives of ≤3 days.

Standard testing methods were used in most of the following studies. For acute tests, exposure periods last 48 hours for invertebrates and 96 hours for fish, the latter in either static, static-renewal or flow-through systems. Results are comparable among these short-term tests according to Wildish (1974) and Macek and Krzeminski (1975). Chronic toxicity tests for invertebrates and fish typically last 21 days and several months, respectively. Relatively short-term (7 day) tests to estimate long-term effects have also been developed (APHA, 1989; Peltier and Weber, 1985; Horning and Weber, 1985; Weber et al., 1988). Endpoints include lethality, reproductive potential, and growth.

The water flea (*D. magna*), the fathead minnow (*P. promelas*), and the green algae (*Selenastrum capricornutum*), used in many of the following studies, have proven to be sensitive test species in toxicity assays. Unless otherwise stated, test results are based on nominal concentrations of the active ingredients tested as ~99% active.

1. Acute Toxicity to Fish

Data on the acute toxicity of AE to fish are summarized in Table 4-1. Toxicity data were available both for several commonly used test species exposed to a variety of AE and on a commonly used AE in several species.

Overall, LC_{50} values for linear AE ranged from 0.4 mg/L of $C_{16,18}AE_{14}$ for the brown trout (*Salmo trutta*) in a 96-hr assay to >100 mg/L $C_{12}AE_{20}$ for the golden orfe or ide (*Idus melanotus*) in a 1-hr assay. Ninety-six hour LC_{50} values of linear primary AE for the commonly used test species, fathead minnow (*P. promelas*), ranged from 0.48 mg/L of $C_{12-15}AE_7$ to 3.6 mg/L of $C_{12-15}AE_{12}$. Alcohol ethoxylates with branched chains and secondary attachment of the alkyl group were less acutely toxic to this species than linear primary AE, with LC_{50} values of 4.4-6.1 mg/L and 2.8-2.9 mg/L, respectively. The commonly used AE, n-pri-$C_{14-15}AE_7$, was acutely toxic at low concentrations when tested in four species: bluegill sunfish (*Lepomis macrochirus*), 0.7 mg/L; rainbow trout fingerlings (*S. gairdneri*), 0.9 mg/L; fathead minnows (*P. promelas*), 1.2-2.5 mg/L; and the mummichog (*Fundulus heteroclitus*), 1.45 mg/L. Acute no-observed-effects concentrations (NOEC) of $C_{14-15}AE_7$ for fathead minnows ranged from 0.8 to 2.0 mg/L, depending on water source and hardness (Maki et al., 1979). The NOEC was twice as high in secondary effluent as in carbon-filtered tap water and stream water.

Fish react to acutely toxic concentrations of nonionic surfactants with a sequential pattern of increased activity, inactivation, immobilization and, if not removed from the exposure, death. The cause of death is suffocation, probably as a result of both physical and chemical disruption of the gill epithelium (Shell Chemical Company, 1983).

Several investigators monitored the acute toxicity of effluents from laboratory-scale biodegradation units to fish. The bioassays were carried out by exposing the organisms to the biodegrading medium or effluent and comparing survival times to those in known concentrations of the starting materials. Concentrations in the biodegrading medium or effluent were not measured.

Reiff (1976), in laboratory dieaway studies, showed that after 14 days, intermediate biodegradation products of $C_{14-15}AE_7$ and $C_{14-15}AE_{11}$ (Dobanols 45-7 and 45-11) were not acutely toxic to rainbow trout. In a field trial using the same surfactants, effluents from biological filters containing domestic sewage and dosed with toxic levels (30-35 mg/L) were not acutely toxic to juvenile rainbow trout (Turner et al., 1985). Maki et al. (1979) found the same result following dosing of $C_{14-15}AE_7$ into the influent of an activated sludge sewage treatment plant and monitoring toxicity with fathead minnows.

Bench scale biotreaters fed 100 mg/L influent dosages of linear AE ($C_{12-15}AE_7$) produced non-acutely toxic effluent whereas a highly branched APE showed acute effects (Salanitro et al., 1988). Effective concentrations for death of *P. promelas* were >100% effluent. When linear ($C_{12-15}AE_9$) or branched ($C_{13}AE_7$) AE were dosed into continuous flow-through activated sludge units, effluents from these units were not acutely toxic to fathead minnows (Dorn et al., 1990; 1993; Kravetz et al., 1991). Effluents from SCAS units treated with linear

Table 4-1. Acute Toxicity to Fish

Species/ Common Name	Surfactant/ Trade Name	LC_{50} (mg/L)	Test Duration	Reference
Brachydanio rerio zebra fish	$C_{14-15}AE_7$ (Dobanol 45-7)	1.3	96-hr	Procter & Gamble, 1990
Carassius auratus goldfish	n-pri-$C_{12-14}AE_{6.3}$ n-pri-$C_{12-14.5}AE_{7.4}$ n-sec-$C_{11-15}AE_7$ n-sec-$C_{11-15}AE_9$	1.4 1.4 2.1-2.5 2.1	24-hr 24-hr 24-hr 24-hr	Monsanto Company, 1977
Carassius auratus goldfish	$C_{16-18}AE_{14}$ $C_{12-14}AE_{11}$ $C_{14}AE_{14}$	10.0-12.5[a] 0.0-5.0[a] 0.0-5.0[a]	— — —	Unilever Research Laboratories, 1977
Carassius auratus goldfish	$C_{12}AE_4$	5.2	6-hr	Marchetti, 1964
Carassius auratus goldfish, juvenile	C_nAE_{10}	7.1	48-hr	Fisher and Gode, 1978
Carassius auratus goldfish, juvenile	sec-$C_{12-14}AE_7$ (SEC-7) sec-$C_{12-14}AE_9$ (SEC-9) sec-$C_{12-14}AE_{12}$ (SEC-12) $C_{12-15}AE_9$ (LA-9) $C_{12-15}AE_9$ (OXO-9)	3.3 5.1 12.0 1.9 1.4	48-hr 48-hr 48-hr 48-hr 48-hr	Kurata et al., 1977
Carassius auratus goldfish	$C_{12-14}AE_8$ $C_{12-14}AE_{10-11}$ $C_{16-18}AE_{14}$	1.8 4.3 7.9	6-hr 6-hr 6-hr	Reiff et al., 1979

Table 4-1. (cont.)

Species/Common Name	Surfactant/Trade Name	LC$_{50}$ (mg/L)	Test Duration	Reference
Carassius auratus goldfish	C$_{13}$AE$_5$	8.5	6-hr	Shell Chemical Company, 1983
Fundulus heteroclitus mummichog	C$_{14-15}$AE$_7$	1.45	96-hr	Maki, 1979b
Gadus morhua cod	C$_{16,18}$AE$_{10}$	0.5-1.0	96-hr	Swedmark et al., 1971
Ictalurus punctatus channel catfish	C$_{12-15}$AE$_9$ (Neodol 25-9) C$_{12-15}$AE$_{12}$ (Neodol 25-12)	1.2 1.8	96-hr	Shell Chemical Company, 1983
Lebistes reticulatus	Pluronic surfactant (EO/PO=0.3-4.9)	7,000-41,800	96-hr	Karpinska-Smulikowska, 1984
Idus melanotus[b] golden orfe	C$_{12}$AE$_2$ C$_{12}$AE$_4$ C$_{12}$AE$_6$ C$_{12}$AE$_8$ C$_{12}$AE$_{10}$ C$_{12}$AE$_{12}$ C$_{12}$AE$_{14}$ C$_{12}$AE$_{16}$ C$_{12}$AE$_{18}$ C$_{12}$AE$_{20}$	1.9 4 5 7 10 20 30 40 100 150	1-hr 1-hr 1-hr 1-hr 1-hr 1-hr 1-hr 1-hr 1-hr 1-hr	Gloxhuber and Fischer, 1968; Shell Chemical Company, 1983
Idus melanotus golden orfe	C$_{12-14}$AE$_8$ C$_{12-14}$AE$_{10.5}$ C$_{12-14}$AE$_{11}$ C$_{16,18}$AE$_{14}$	1.8-2.7 4.1-4.5 2.7 2.3-2.5	96-hr 96-hr 48-hr 96-hr	Shell Chemical Company, 1983

Table 4-1. (cont.)

Species/Common Name	Surfactant/Trade Name	LC$_{50}$ (mg/L)	Test Duration	Reference
Lepomis macrochirus bluegill sunfish	pri-C$_{12-15}$AE$_3$ (Neodol 25-3)	1.5	96-hr	Macek and Krzeminski, 1975
	pri-C$_{12-15}$AE$_9$ (Neodol 25-9)	2.1	96-hr	
	pri-C$_{10-12}$AE$_6$ (Alfonic 1016-60)	6.4	24-hr	
	pri-C$_{13}$AE$_9$ (Surfonic TD-90)	7.8	24-hr	
	sec-C$_{11-15}$AE$_9$ (Tergitol 15-S-9)	4.7	24-hr	
Lepomis macrochirus bluegill sunfish	pri-C$_{12-15}$AE (60% EO by weight)	2.8	24-hr	Hendricks et al, 1974
Lepomis macrochirus bluegill sunfish	C$_{14-15}$AE$_7$ (Neodol 45-7)	0.7	96-hr	Bishop and Perry, 1981; Lewis and Perry, 1981
Lepomis macrochirus bluegill sunfish	C$_{12-13}$AE$_{6.5}$ (Neodol 23-6.5)	2.0-2.4	96-hr	Shell Chemical Company, 1983
	C$_{12-15}$AE$_9$ (Neodol 25-9) (75% linear primary)	7.8	96-hr	
	C$_{12-15}$AE$_9$ (Neodol 25-9) (98% linear primary)	11.0	96-hr	
	C$_{14-18}$AE$_9$	10.0	96-hr	
	C$_{12,14,16,18}$AE$_{7.5}$ (coconut alcohol EO$_{7.5}$)	12.3	96-hr	
Lepomis macrochirus bluegill sunfish	Surfonic JL-70X[c]	2.3	96-hr	Naylor et al, 1988

Table 4-1. (cont.)

Species/Common Name	Surfactant/Trade Name	LC$_{50}$ (mg/L)	Test Duration	Reference
Lepomis macrochirus bluegill sunfish [juveniles]	$C_{12,14,16}AE_{12-14}$ $C_{16-18}AE_{18}$ $C_{16-18}AE_{30}$ Tergitol 80-L-97B	2.2 2.8 3.6 6.4	96-hr 96-hr 96-hr 96-hr	Procter & Gamble, 1985
Macrones vittatus minnow	Swanic 6L	0.5	96-hr	Verma et al., 1978
Phoxinus phoxinus minnow	$C_{16-18}AE_{14}$	3.4	24-hr	Unilever Research Laboratories, 1977
Pimephales promelas fathead minnow	n-pri-$C_{12-14}AE_{6.3}$ n-pri-$C_{12-14}AE_{7.4}$ n-sec-$C_{11-15}AE_7$ n-sec-$C_{11-15}AE_9$	1.8 1.8 2.9 2.8	24-hr 24-hr 24-hr 24-hr	Monsanto Company, 1977
Pimephales promelas fathead minnow	$C_{14-15}AE_7$ (Neodol 45-7)	1.2-2.5^d	96-hr	Maki et al., 1979
Pimephales promelas fathead minnow	$C_{12-15}AE_9$ (Neodol 25-9) $C_{12-15}AE_{12}$ (Neodol 25-12) $C_{12-14}AE_{6.3}$ $C_{12-14}AE_{7.4}$	1.6 3.6 1.8 1.8	96-hr 96-hr 24-hr 24-hr	Shell Chemical Company, 1983; Mayer and Ellersieck, 1986
Pimephales promelas fathead minnow	$C_{11}AE_5$ (Neodol 1-5)	>1.6	96-hr	Shell Chemical Company, 1988
Pimephales promelas fathead minnow	$C_{12-15}AE_7$ (Neodol 25-7)	0.48	96-hr	Salanitro et al., 1988
Pimephales promelas fathead minnow	n-$C_{13-15}AE_7$ br-$C_{12}AE_7$ br-$C_{13}AE_7$	1.5 6.3 4.4	96-hr 96-hr 96-hr	Markarian et al., 1989

Table 4-1. (cont.)

Species/Common Name	Surfactant/Trade Name	LC_{50} (mg/L)	Test Duration	Reference
Pimephales promelas fathead minnow	n-pri-$C_{12,15}AE_9$ br-$C_{13}AE_7$ (2 methyl branches/hydrophobe) br-$C_{13}AE_7$ (4 methyl branches/hydrophobe)	1.6[e]; 1.2[f] 4.5[e]; 3.3[f] 6.1[e]; 5.0[f]	96-hr 96-hr 96-hr	Kravetz et al., 1991; Dorn et al, 1993
Pimephales promelas fathead minnow	Polytergent SLF-18[c]	13	96-hr	Procter & Gamble, 1985
Pleuronectes flesus flounder	$C_{16,18}AE_{10}$ (tallow alcohol EO_{10})	0.5-1.0	96-hr	Swedmark et al., 1971
Poecilis reticulata guppy	$C_{16,18}AE_{14}$	0.7	24-hr	Unilever Research Laboratories, 1977
Rasbora heteromorpha harlequin fish	$C_{16,18}AE_{14}$ (tallow alcohol EO_{14}) $C_{12-14}AE_{10-11}$ $C_{12-14}AE_8$	0.7 1.6-2.8 1.2	96-hr 96-hr 48-hr	Reiff et al., 1979
Rasbora heteromorpha harlequin fish	$C_{16,18}AE_{14}$	1.5	24-hr	Unilever Research Laboratories, 1977
Rasbora heteromorpha harlequin fish	$C_{12,14,16,18}AE_3$ (coconut alcohol EO_3)	10-100	48-hr	Shell Chemical Company, 1983
Salmo gairdneri[g] rainbow trout	$C_{12,15}AE_7$ (Dobanol 25-7) $C_{16-18}AE_{14}$ $C_{12-14}AE_{11}$ $C_{14}AE_8$	2.0-2.7 0.7 6.2 2.4	96-hr 48-hr 48-hr 48-hr	Unilever Research Laboratories, 1977

Table 4-1. (cont.)

Species/ Common Name	Surfactant/ Trade Name	LC_{50} (mg/L)	Test Duration	Reference
Salmo gairdneri rainbow trout, [fingerlings]	$C_{9-11}AE_{2.5}$ (Dobanol 91-2.5)	5-7	96-hr	Shell Chemical Company, 1983; Mayer and Ellersieck, 1986
	$C_{9-11}AE_5$ (Dobanol 91-5)	8-9	96-hr	
	$C_{12-13}AE_2$ (Dobanol 23-2)	1.0-2.0	96-hr	
	$C_{12-15}AE_3$ (Dobanol 25-3)	1.3-1.7	96-hr	
	$C_{12-15}AE_3$ (Dobanol 25-3)	1.0	96-hr	
	$C_{12-15}AE_9$ (Neodol 25-9)	1.2	96-hr	
	$C_{12-14}AE_{10.5}$	0.8-1.8	96-hr	
	$C_{14-15}AE_7$ (Neodol 45-7)	0.9	96-hr	
	$C_{14-15}AE_{11}$ (Dobanol 45-11)	1.8-2.5	96-hr	
	$C_{14-15}AE_{18}$ (Dobanol 45-18)	5.0-6.3	96-hr	
	$C_{16-18}AE_{14}$	0.8	24-hr	
Salmo gairdneri rainbow trout	$C_{14-15}AE_7$ (Dobanol 45-7)	0.91 0.71	96-hr 7-day	Turner et al, 1985
	$C_{14-15}AE_{11}$ (Dobanol 45-11)	1.12 0.98	96-hr 7-day	
Salmo salar Atlantic salmon [parr]	$C_{12}AE_4$	1.5	96-hr	Wildish, 1974
Salmo trutta brown trout	$C_{16-18}AE_{14}$	1.0	24-hr	Unilever Research Laboratories, 1977

Table 4-1. (cont.)

Species/ Common Name	Surfactant/ Trade Name	LC_{50} (mg/L)	Test Duration	Reference
Salmo trutta brown trout	$C_{16,18}AE_{14}$ $C_{12-14}AE_{10-11}$ $C_{12-14}AE_8$	0.4 0.2-1.8 0.8	96-hr 96-hr 96-hr	Reiff et al., 1979

[a] $LC_0 - LC_{100}$.
[b] Renamed ide *(Leuciscus idus)*.
[c] Contains propylene oxide.
[d] Various dilution media.
[e] Measured concentration.
[f] Nominal concentration.
[g] Renamed *Oncorhynchus mykiss*.
A dash (—) indicates no data.

AE containing propylene oxide (PO) blocks were used in bioassays with several aquatic species (Naylor et al., 1988). The effluents showed little or no toxicity to fathead minnows.

2. Acute Toxicity to Aquatic Invertebrates

Invertebrates differ in their sensitivity to surfactants depending on size, age, and thickness of body covering. Some larger, marine species were less sensitive than smaller, freshwater species and early life stages were more sensitive than adults (Swedmark et al., 1971). Crabs and adult molluscs appear to be the least sensitive species.

Toxicity data on aquatic invertebrates are summarized in Table 4-2. For linear AE, the LC_{50} values for *D. magna* in 48-hr static tests ranged from 0.29-0.4 mg/L for $C_{14-15}AE_7$ to 20 mg/L for a tallow alcohol ethoxylate ($C_{16,18}AE_{18}$). For linear primary AE, toxicity appears to be chemical specific. In *D. magna*, branched alkyl chain AE were less acutely toxic (LC_{50} values of 5.9-11.6 mg/L) than the corresponding linear AE (LC_{50} values of 0.6-1.3 mg/L); secondary alcohol ethoxylates were less toxic (LC_{50} values of >5-13 mg/L) than similar primary alcohol ethoxylates (LC_{50} values of 2.3-2.4 mg/L); and the alcohol EO/PO derivative, Polytergent SLF-18, at 72 mg/L, was the least toxic.

Bode et al. (1978) tested a series of fatty alcohol ethoxylates (C_{10} to C_{18} and EO_5 to EO_{18}) on the freshwater coelenterate *Hydra attenuata*. A concentration of 0.002 mM (0.76-2.12 mg/L) had no effect on budding rate. At a concentration of 0.02 mM (7.6-21.2 mg/L), 10 of the 12 compounds were lethal within 24 hours and at a concentration of 0.2 mM (76-212 mg/L) all were lethal. The two surfactants $C_{10}AE_{6.7}$ and $C_{18}AE_7$ were less toxic than their homologs. Lethal concentrations coincided with a surface tension of 49 dynes/cm.

The effluents from laboratory-scale biotreater units were tested for acute toxicity to aquatic organisms. Bench scale biotreaters fed 100 mg/L influent dosages of linear AE ($C_{12-15}AE_7$) produced non-acutely toxic effluent whereas a highly branched APE showed acute effects (Salanitro et al., 1988). Effective concentrations (EC_{50}) for immobilization of *D. magna* were >100% effluent. When linear ($C_{12-15}AE_9$) or branched ($C_{13}AE_7$) AE were dosed into continuous flow-through activated sludge units, effluents from these units were not acutely toxic to water fleas (Dorn et al., 1990; 1993; Kravetz et al., 1991). Effluents from SCAS units treated with linear AE containing propylene oxide (PO) blocks showed little or no toxicity to water fleas (Naylor et al., 1988). Effluents from tests using $C_{12}AE_7$ and $C_{17}AE_{25}$ were not toxic to *Daphnia* sp. (Janicke, 1988).

3. Toxicity to Algae

Depending on the concentration of the surfactant, acute exposures of algae may stimulate or inhibit cell division and growth. Although the concept of toxicity is difficult to apply to algal cultures because of the rapid, often logarithmic, growth following removal to fresh media, several effect parameters are commonly used. The EC_{50} and EC_{100} values (the latter also referred to as the algistatic concentration) are those concentrations that reduce an

Table 4-2. Acute Toxicity to Aquatic Invertebrates

Species/Common Name	Surfactant/Trade Name	LC$_{50}$ (mg/L)	Test Duration	Reference
Platyhelminthes				
Dugesia sp.	C$_{14-15}$AE$_7$ (Neodol 45-7)	1.0	48-hr	Lewis and Suprenant, 1983
Nematoda				
Rhabditis sp.	C$_{14-15}$AE$_7$ (Neodol 45-7)	6.8	48-hr	Lewis and Suprenant, 1983
Annelida				
Dero sp.	C$_{14-15}$AE$_7$ (Neodol 45-7)	2.6	48-hr	Lewis and Suprenant, 1983
Arthropoda				
Daphnia magna water flea	C$_{12-18}$AE$_{7.4}$ (Alfonic 1218-60)	3.3	24-hr	Vista Chemical Company, 1977
Daphnia magna water flea	C$_{14-15}$AE$_7$	0.7	48-hr	Bishop and Perry, 1981
Daphnia magna water flea	C$_{14-15}$AE$_7$ (Neodol 45-7)	0.29-0.4	48-hr	Lewis and Perry, 1981
Daphnia magna water flea	C$_{14}$AE$_1$ C$_{14}$AE$_2$ C$_{14}$AE$_3$ C$_{14}$AE$_4$ C$_{14}$AE$_6$ C$_{14}$AE$_9$	0.83 1.53 0.73 1.76 4.19 10.07	48-hr 48-hr 48-hr 48-hr 48-hr 48-hr	Maki and Bishop, 1979

Table 4-2. (cont.)

Species/Common Name	Surfactant/Trade Name	LC_{50} (mg/L)	Test Duration	Reference
Daphnia magna water flea	$C_{12-13}AE_{6.5}$ $C_{14-15}AE_7$	1.14 0.43	96-hr 96-hr	Maki, 1979b; 1979c
Daphnia magna water flea	Pluronic surfactant (EO/PO=0.3-4.9)	490-10,000	48-hr	Karpinska-Smulikowska, 1984
Daphnia magna water flea	$C_{11}AE_5$	3.3	48-hr	Shell Chemical Company, 1988
Daphnia magna water flea	Surfonic JL-80-X[a]	2.6	48-hr	Naylor et al., 1988
Daphnia magna water flea	n-$C_{13-15}AE_7$ (Exxal-L-1315-7) br-$C_{12}AE_7$ (Exxal-12-based) br-$C_{13}AE_7$ (Exxal-13-based)	0.6 6.8 5.9	48-hr 48-hr 48-hr	Markarian et al., 1989
Daphnia magna water flea	$C_{12-18}AE_{14}$ $C_{12-18}AE_{11}$ $C_{14}AE_8$	1.1 5.1 2.0	24-hr 24-hr 24-hr	Unilever Research Laboratories, 1977
Daphnia magna water flea	n-pri-$C_{12-15}AE_9$ br-$C_{13}AE_7$ (2 methyl branches/hydrophobe) br-$C_{13}AE_7$ (4 methyl branches/hydrophobe)	1.3[b]; 1.6[c] 9.8[b]; 8.6[c] 11.6[b]; 9.4[c]	48-hr 48-hr 48-hr	Kravetz et al., 1991; Dorn et al., 1993

Table 4-2. (cont.)

Species/Common Name	Surfactant/Trade Name	LC_{50} (mg/L)	Test Duration	Reference
Daphnia magna water flea	$C_{18}AE_n$ (Brij 93)	0.19	48-hr	Procter & Gamble, 1985
	$C_{16-18}AE_{18}$	20	48-hr	
	$C_{16-18}AE_{30}$	18	48-hr	
	Polytergent SLF-18	72	48-hr	
Daphnia magna/ *Daphnia pulex* water flea	$C_{12-15}AE_7$ (Neodol 25-7)	0.76	48-hr	Salanitro et al., 1988
	$C_{12-13}AE_{6.5}$ (Neodol 23-6.5)	0.57, 0.90	96-hr	Shell Chemical Company, 1983
	$C_{12-15}AE_9$ (Neodol 25-9)	1.23	96-hr	
	$C_{12-14}AE_{6.3}$	1.5	96-hr	
Daphnia pulex water flea	$C_{9-11}AE_6$	5.4	48-hr	Burlington Research Inc., 1985; Moore et al., 1987
Daphnia sp. water flea	n-pri-$C_{12-14}AE_{6.3}$	2.4	48-hr	Monsanto Company, 1977
	n-pri-$C_{12-14}AE_{7.4}$	2.3	48-hr	
	n-sec-$C_{11-15}AE_7$	>5	48-hr	
	n-sec-$C_{11-15}AE_9$	13	48-hr	
Asellus sp. isopod	$C_{14-15}AE_7$	6.2	48-hr	Lewis and Suprenant, 1983
Gammarus sp. amphipod	$C_{14-15}AE_7$	1.4	48-hr	Lewis and Suprenant, 1983
Balanus balanoides barnacle [larvae]	$C_{16-18}AE_{10}$	1.2	96-hr	Swedmark et al., 1971

Table 4-2. (cont.)

Species/Common Name	Surfactant/Trade Name	LC_{50} (mg/L)	Test Duration	Reference
Callinectes sapidus blue crab	$C_{14-15}AE_7$	30.9	96-hr	Maki, 1979a
Carcinus maenas shore crab	$C_{16,18}AE_{10}$	>100	96-hr	Swedmark et al., 1971
Carcinus maenas shore crab	$C_{12-15}AE_9$ (Dobanol 25-9)	>640	6-hr	Shell Chemical Company, 1983
Crangon crangon brown shrimp	$C_{12-15}AE_9$ (Dobanol 25-9)	>3300	48-hr	Shell Chemical Company, 1983
	$C_{14-15}AE_1$ (Dobanol 45-1)	500	48-hr	
	$C_{14-15}AE_3$ (Dobanol 45-3)	200	48-hr	
	$C_{14-15}AE_{11}$ (Dobanol 45-11)	3300	48-hr	
Eupagurus bernhardus hermit crab [adults, larvae]	$C_{16,18}AE_{10}$ (tallow alcohol EO_{10})	>100	96-hr	Swedmark et al., 1971
		800	96-hr	
Hyas araneus spider crab	$C_{16,18}AE_{10}$ (tallow alcohol EO_{10})	>100	96-hr	Swedmark et al., 1971
Leander adspersus decapod	$C_{16,18}AE_{10}$ (tallow alcohol EO_{10})	>100	96-hr	Swedmark et al., 1971
Leander squilla decapod	$C_{16,18}AE_{10}$ (tallow alcohol EO_{10})	>100	96-hr	Swedmark et al., 1971

Table 4-2. (cont.)

Species/Common Name	Surfactant/Trade Name	LC_{50} (mg/L)	Test Duration	Reference
Mysidopsis bahia mysid shrimp	$C_{10}AE_4$ (decyl alcohol based) $C_{13}AE_{10}$ (tridecyl alcohol based)	5.57 2.24	48-hr 48-hr	Hall et al., 1989
Mysidopsis bahia mysid shrimp	$C_{14-15}AE_7$ (Neodol 45-7)	0.2; 0.3	96-hr	Procter & Gamble, 1985
Penaeus duorarum pink shrimp	$C_{14-15}AE_7$ (Neodol 45-7)	0.8	96-hr	Maki, 1979a
Penaeus duorarum pink shrimp [juveniles; adults]	$C_{14-15}AE_7$	1.2; 1.1	96-hr	Procter & Gamble, 1985
Aedes aegypti mosquito [larvae]	pri-$C_{12-15}AE_9$	200 (LC_{100})	24-hr	Van Emden et al., 1974
Culex pipens mosquito [pupae]	$C_{8-10}AE_{15}$ $C_{13}AE_3$ (tridecyl alcohol-based) $C_{13}AE_6$ $C_{13}AE_9$ $C_{13}AE_{12}$	>180 13 29 44 64	— — — — —	Maxwell and Piper, 1968
Isonychia sp. mayfly	$C_{12-15}AE_n$	6.0	24-hr	Dolan et al., 1974
Paratanytarsus parthenogenica midge	$C_{14-15}AE_7$	5.0	48-hr	Lewis and Suprenant, 1983

Table 4-2. (cont.)

Species/ Common Name	Surfactant/ Trade Name	LC_{50} (mg/L)	Test Duration	Reference
Mollusca				
Biomphalaria glabrata snail	pri-$C_{12-15}AE_9$	11	24-hr	Van Emden et al., 1974
Cardium edule cockle	$C_{16,18}AE_{10}$ (tallow alcohol EO_{10})	<5	96-hr	Swedmark et al., 1971
Goniobasis sp. snail	pri-$C_{12-15}AE$ (50% EO)	15.9	24-hr	Hendricks et al., 1974
Mya arenaria clam	$C_{16,18}AE_{10}$ (tallow alcohol EO_{10})	100	96-hr	Swedmark et al., 1971
Mytilus edulis mussel	$C_{16,18}AE_{10}$ (tallow alcohol EO_{10})	50	96-hr	Swedmark et al., 1971
Ostrea edulis/ Crassostrea virginica oyster [larvae] [adults]	$C_{14-15}AE_7$ $C_{12-15}AE_9$ (Dobanol 25-9)	0.11 20,000	48-hr 6-hr	Shell Chemical Company, 1983

[a]Contains propylene oxide.
[b]Measured concentration.
[c]Nominal concentration.
A dash (—) indicates no data.

effect, usually growth, by 50% or 100%, respectively, compared to the control population. Growth resumes upon removal to fresh medium. The algicidal concentration is the concentration that causes cell death. Although the tests in the studies summarized here are referred to as acute, in some instances the 96-hour length of the toxicity test period spans several generations and, therefore, can also be considered chronic (Horning and Weber, 1985). Thus all studies involving algae, regardless of time limit are summarized in Table 4-3.

Studies on the effects of AE surfactants on several freshwater and one marine species of algae were located. For the green algae, *S. capricornutum*, exposed to linear AE, EC_{50} values for growth range from 0.09 mg/L for $C_{14-15}AE_6$ to 10 mg/L for $C_{12-14}AE_{13}$. The secondary and branched AE were less toxic to this species than the linear primary AE, with LC_{50} values of 52-56 mg/L and 7.5-39 mg/L, respectively.

Ernst et al. (1983) also studied the effect of chemical structure on toxicity. They tested a series of linear AE surfactants with alkyl chain lengths ranging from 10 to 18 carbons and EO units ranging from 5 to 40 on the green alga, *Chlamydomonas reinhardii*. Concentrations ranged from 0.02-2 mM (7.6 mg/L for $C_{10}AE_5$ to 4000 mg/L for $C_{18}AE_{40}$). Growth decreased with increasing ethoxylate chain length and increased with increasing alkyl chain length. Greatest growth stimulation compared to controls was seen with $C_{18}AE_7$ at 0.02 mM (11.6 mg/L).

The response of freshwater algae to the same surfactant is species-specific. In the laboratory, blue-green algae (*Microcystis aeruginosa*) appear to be more tolerant of several surfactants than the green algae (*S. capricornutum*) and diatoms (*Nitzschia fonticola* and *Navicula* sp.) (Payne and Hall, 1979; Yamane et al., 1984; Lewis and Hamm, 1986).

Batch additions of 50 mg/L of C_nAE_7 or $C_nAE_{7.4}$ to pond water stimulated the growth of species of blue-green algae, but depressed the numbers of green algae (Davis and Gloyna, 1967). The authors also found these effects on growth when testing individual populations of green and blue-green algae in the laboratory.

Monthly *in situ* monitoring studies were conducted by Lewis and Hamm (1986) by suspending bottles containing native phytoplankton and surfactants in a freshwater lake for three hours. The 3-hr EC_{50} for growth depression for $C_{14-15}AE_6$ ranged from 0.8-6.5 mg/L (mean 2.1 mg/L) and varied with water temperature and the relative abundance of the major phytoplankton groups present during the seasons. Lower EC_{50} values were recorded during summer (mean water temperature 27°C). In contrast to laboratory studies, diatoms, which predominated in May and October, were less sensitive than green and blue-green algae. In the laboratory, the 96-hr EC_{50} values for *S. capricornutum*, *M. aeruginosa*, and *Navicula pelliculosa* were slightly lower than in the lake community: 0.09, 0.60, and 0.28 mg/L, respectively. It was concluded that the toxicity of most surfactants to natural assemblages of algae under natural conditions is less than that predicted from laboratory tests (Lewis, 1990).

Effluents from SCAS units treated with linear AE containing propylene oxide (PO) blocks had a stimulatory effect on the growth rate of a green alga, *S. capricornutum* (Naylor et al., 1988).

Table 4-3. Effects on Algae

Surfactant	Test Species	Effect	Concentration (mg/L)	Reference
n-pri-$C_{12-16}AE_{6.3}$	*Selenastrum* sp.	growth stimulation	0.7	Monsanto Company, 1977
n-pri-$C_{12-14}AE_{7.4}$		EC_{50} (growth)	5.0	
n-sec-$C_{11-15}AE_{7}$		EC_{50} (growth)	3.8	
n-sec-$C_{11-15}AE_{9}$		EC_{50} (growth)	56	
		EC_{50} (growth)	52	
n-$C_{12-14}AE_{6}$	*Selenastrum capricornutum*	5-day algistatic algicidal	50 1000	Hall, 1973; Payne and Hall, 1979
	Navicula seminulum	5-day algistatic algicidal	5-10 100	
	Microcystis aeruginosa	no effect	1000	
$C_{12,14,16,18}AE_{n}$	*Gymnodinium breve*	2-day growth stimulation 33% mortality	0.0125 0.05	Kutt and Martin, 1974
$C_{10}AE_{5}$-$C_{18}AE_{40}$	*Chamydomonas reinhardi*	lipophile, hydrophile - dependent changes in growth	7.6-4000	Ernst et al., 1983
$C_{18}AE_{7}$		growth stimulation	11.6	
EO/PO block copolymer Pluronic (EO/PO=0.3-4.9)	*Scenedesmus quadricauda*	change in photosynthesis, growth	<0.1-10,000	Karpinska-Smulikowska, 1984
$C_{12-14}AE_{4}$	*Selenastrum capricornutum*	EC_{50} (growth)	2-4	Yamane et al., 1984
$C_{12-14}AE_{9}$			4-8	
$C_{12-14}AE_{13}$			10	
$C_{12-14}AE_{9}$	*Microcystis aeruginosa*	EC_{50} (growth)	10-50	
	Nitzschia fonticola	EC_{50} (growth)	5-10	
$C_{14-15}AE_{6}$	*Selenastrum capricornutum*	96-hr EC_{50} (growth)	0.09	Lewis and Hamm, 1986
	Navicula pelliculosa	96-hr EC_{50} (growth)	0.28	
	Microcystis aeruginosa	96-hr EC_{50} (growth)	0.60	

Table 4-3. (cont.)

Surfactant	Test Species	Effect	Concentration (mg/L)	Reference
$C_{11}AE_5$	*Selenastrum capricornutum*	96-hr EC_{50} (growth) no-observed effect lowest-observed effect	2.9; 3.5[a] 1.2; 2.1 3.0; 3.4	Shell Chemical Company, 1988
$C_{12}AE_4$ $C_{12}AE_{23}$ $C_{16}AE_{10}$	*Selenastrum capricornutum*	100% mortality 100% mortality 97.4% mortality	10 300 10	Nyberg, 1988
n-$C_{12-15}AE_7$ n-$C_{13-15}AE_7$ br-$C_{12}AE_7$ br-$C_{13}AE_7$	*Selenastrum capricornutum*	96-hr EC_{50} (growth) 96-hr EC_{50} (growth) 96-hr EC_{50} (growth) 96-hr EC_{50} (growth)	1.3 0.9 38.7 37.2	Markarian et al., 1989
$C_{14-15}AE_7$	*Navicula pelliculosa* *Navicula seminulum* *Microcystis aeruginosa*	EC_{50} (cell count) EC_{50} (cell count) EC_{50} (cell count)	0.28 1.34 0.59	Procter & Gamble, 1985
n-pri-$C_{12-15}AE_9$ br-$C_{13}AE_7$ (2 methyl branches/ hydrophobe) br-$C_{13}AE_7$ (4 methyl branches/ hydrophobe)	*Selenastrum capricornutum*	96-hr EC_{50} (growth) 96-hr NOEC 96-hr LOEC MATC[b] 96-hr EC_{50} (growth) 96-hr NOEC 96-hr LOEC MATC 96-hr EC_{50} (growth) 96-hr NOEC 96-hr LOEC MATC	0.70 1.0 0.6 0.8 7.5 10 20 14.2 10.0 4.0 10.0 6.3	Dorn et al., 1993

[a]Numbers refer to results from 5% and 10% stock solutions used to prepare dilutions.
[b]MATC or chronic value is the geometric mean of the NOEC and LOEC.

4. Sublethal Toxicity and Behavioral Responses

Sublethal exposures of aquatic organisms to AE may result in reduced energy for growth and behavioral effects, the latter leading to increased susceptibility to predation. In several of the following studies, the effects of sublethal exposures are difficult to weight because known toxicity endpoints such as growth, reproduction, and increased mortality were not measured.

Observed sublethal effects of AE on fish have included increased and erratic swimming activity, loss of equilibrium, and immobilization with slow respiration. Recovery occurred with removal to clean water. These symptoms occurred in rainbow trout at <2 mg/L of $C_{12-15}AE_7$ (Unilever Research Laboratories, 1977) and in cod at 0.5 mg/L of $C_{16-18}AE_{10}$ (Swedmark et al., 1971).

In 48-hr tests, the respiratory rate of bluegill sunfish was significantly suppressed ($p<0.01$) as measured concentrations of $C_{12-13}AE_{6.5}$ were increased from 0 mg/L to 1.2 mg/L. The effect occurred over the normal diurnal activity pattern of bluegills. A no-effect level on respiratory rate was determined to be between 0.26 and 0.54 mg/L. Respiration was not significantly suppressed by $C_{14-15}AE_7$ over the concentration range of 0.56 to 1.2 mg/L (Maki, 1979a).

Swedmark et al. (1971) also observed adverse effects in invertebrates following exposures to $C_{16-18}AE_{10}$ at concentrations less than the LC_{50}. Sublethal effects included impaired locomotory activity and breathing rate in crustaceans, impaired byssus activity and shell closure in *Mytilus edulis*, and impaired burrowing activity in burrowing species of bivalves. Larval stages were affected at concentrations as low as 1 mg/L, while adults were affected at 10 mg/L.

The 30-min EC_{50} for immobilization of barnacle nauplii (*Elinius modestus*) was 580 mg/L of $C_{10}AE_{20}$ (Wright, 1976). The 48-hr EC_{50} for inhibition of larval development of Eastern oysters (*Crassostrea virginica*) was 0.11 mg/L of $C_{14-15}AE_7$ (Maki, 1979b). At concentrations of 1.75 and 1.60 mg/L of alkyl polyether alcohol, Hidu (1965) observed a 50% reduction in the development of fertilized clam (*Mercenaria mercenaria*) and oyster (*C. virginica*) eggs to normal 48-hr straight-hinge veliger.

5. Chronic Toxicity

Daphnia sp. and fathead minnows (*P. promelas*) are the species of choice in chronic studies because of their respective short life cycles (eight days for *Daphnia* sp. and 10-12 months for fathead minnows) and ease in handling and culture and because they are representative of the most sensitive invertebrate and fish species.

Chronic toxicity of linear primary AE to aquatic species occurs at concentrations ≥ 0.5 mg/L. Endpoints of chronic tests are survival, growth, and reproduction and are usually expressed as no- and lowest-observed-effects concentrations (NOEC and LOEC). Results from chronic studies on fish and invertebrates are presented in Table 4-4. Data on algae were presented in Table 4-3.

Table 4-4. Chronic Tests with Aquatic Species

Species	Chemical	Effect Studied	NOEC (mg/L)	LOEC (mg/L)	Reference
Lepomis macrochirus bluegill sunfish	$C_{12-13}AE_{6.5}$	respiratory rate	0.32-1.0		Maki, 1979a; Shell Chemical Co., 1983
	$C_{14-15}AE_7$	respiratory rate	0.21-0.28		
Fundulus heteroclitus mummichog	$C_{14-15}AE_7$	survival	1.0		Maki, 1979b; Shell Chemical Co., 1983
Pimephales promelas fathead minnow	$C_{12-13}AE_{6.5}$	1-year survival; reproduction	0.32		Maki, 1979c; Shell Chemical Co., 1983
	$C_{14-15}AE_7$	30-day larval survival	0.18		
Pimephales promelas fathead minnow	n-pri-$C_{12-15}AE_9$	7-day survival	1.0	2.0	Dorn et al., 1993
		7-day growth	0.4	1.0	
	br-$C_{13}AE_7$ (2 methyl branches/hydrophobe)	7-day survival	1.0	4.0	
		7-day growth	1.0	>1.0	
	br-$C_{13}AE_7$ (4 methyl branches/hydrophobe)	7-day survival	1.0	2.0	
		7-day growth	1.0	2.0	
Daphnia magna water flea	$C_{12-13}AE_{6.5}$	21-day reproduction	0.24		Maki, 1979c; Shell Chemical Co., 1983
	$C_{14-15}AE_7$	21-day reproduction	0.24		
Daphnia magna water flea	n-pri-$C_{12-15}AE_9$	7-day survival	2.0	4.0	Dorn et al., 1993
		7-day growth	1.0	>1	
	br-$C_{13}AE_7$ (2 methyl branches/hydrophobe)	7-day survival	2.0	4.0	
		7-day growth	2.0	>2.0	
	br-$C_{13}AE_7$ (4 methyl branches/hydrophobe)	7-day survival	4.0	6.0	
		7-day growth	4.0	>4.0	
Ceriodaphnia dubia cladoceran	$C_{14-15}AE_7$	7-day survival	0.70-1.4[a]		Masters et al., 1991
		7-day reproduction	0.17-0.70[a]		

Table 4-4. (cont.)

Species	Chemical	Effect Studied	NOEC (mg/L)	LOEC (mg/L)	Reference
Ceriodaphnia sp. cladoceran	$C_{9-11}AE_6$	7-day reproduction		>2.0	Burlington Research, Inc., 1985
Penaeus duorarum pink shrimp	$C_{14-15}AE_7$	survival	0.56		Maki, 1979b; Shell Chemical Co., 1983
Callinectes sapidus blue crab	$C_{14-15}AE_7$	survival	10.0		Maki, 1979b; Shell Chemical Co., 1983
Crassostrea virginica oyster	C_nAE_n	larval growth	0.5	1.6	Hidu, 1965
Oyster larvae	$C_{14-15}AE_7$	48-hr larval development	0.06		Maki, 1979b
Mercenaria mercenaria clam	C_nAE_n	larval growth	1.0	1.75	Hidu, 1965
Mytilus edulis mussel	$C_{16,18}AE_{10}$	5-month spawning, fertilization, and development	1.5		Granmo and Jorgensen, 1975

[a]Chronic value (NOEC/FOEC).

Data in the table show that most species of invertebrates and fish were able to survive and reproduce at linear AE concentrations of less than 1 mg/L. *Pimephales promelas* larvae and *Daphnia* sp. were the most sensitive species, with NOEC for survival and reproduction of 0.18 and 0.24 mg/L, respectively. The NOEC for the respiratory rate of bluegill sunfish was also low. Studies on the chronic and sublethal toxicities of surfactants to aquatic species were recently reviewed by Lewis (1991). The author points out that most data on chronic effects are based on nominal concentrations. The data of Maki (1979c), Shell Chemical Company (1988), and Dorn et al. (1993) in Table 4-4 are based on measured test concentrations.

In chronic tests, effluents from flow-through activated sludge units, dosed with n-C_{12-15} AE_9, br-$C_{13}AE_7$ (3.6 internal methyl groups/hydrophobe), or $C_{12-15}AE_7$ (4.6 internal methyl groups/hydrophobe) were tested for toxicity to fathead minnows and water fleas (Dorn et al., 1990; 1993; Kravetz et al., 1991). The units were treated with surfactant at a rate of 50 mL/day and sludge residence time was 45-60 days. Effluents from the units treated with the highly branched AE tended to be more surface active and more toxic than those originating from units treated with the linear AE. The toxicity of the branched AE in effluents was greater than predicted by CTAS analysis of effluents.

The influence of sediment characteristics on toxicity was assessed during partial life-cycle studies with the midge, *Chironomus riparius* (Marshall and McInnes, 1990). Organic matter in the tested sediments ranged from 1-14%. In addition to biodegradation, which occurred during the test period, adsorption of the surfactant (C_nAE_7) to clay particles reduced toxicity. Survival was affected at concentrations below 1 g/kg in the sediment, but toxicity was several orders of magnitude less than that in water without sediment (acute EC_{50} of 1-8 mg/L). When using an artificial sediment spiked at a concentration of 10 g/kg, there was no effect on survival or development through four instars.

6. Structure-Activity Relationship

Although toxicity of AE to aquatic organisms is compound and species-specific, several generalizations concerning chemical structure and toxicity can be made based on the data in Tables 4-1 through 4-4. Toxicity generally decreases with increasing EO chain length (decreasing liposolubility), branched alkyl chains are less toxic than linear alkyl chains, secondary attachment of the alcohols reduces toxicity compared to primary alcohols, and surfactants containing EO/PO block copolymers are less toxic than those containing only EO.

Toxicity decreased with increasing EO chain length for *Daphnia* sp., shrimp, algae, and several species of fish (Gloxhuber and Fischer, 1968; Maki and Bishop, 1979; Shell Chemical Company, 1983; Yamane et al., 1984; Markarian et al., 1989). The study by Gloxhuber and Fischer using golden orfe (*I. melanotus*) most clearly illustrates this point although exposure periods were extremely short (1 hour) (Figure 4-1). The authors used products which were pure with respect to alcohol content (C_{12}) and containing 2 to 20 ethylene oxide units. Median lethal concentrations increased from 1.9 mg/L for $C_{12}AE_2$ to 150 mg/L for $C_{12}AE_{20}$. The LC_{50} values coincided with a surface tension of 45 dynes/cm. Wildish (1974) also related toxicity to

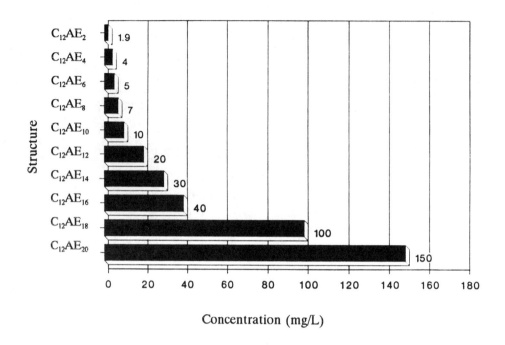

Figure 4-1. Relationship between Toxicity to Fish and Ethoxylate Chain Length. Data are 1-Hour LC_{50} Values for Golden orfe (*Idus melanotus*). Source Gloxhuber and Fischer, 1968.

fish with liposolubility; with increasing EO number, the lipid solubility and toxicity of polyoxyethylene esters, ethers and amines decreased. The incipient lethal level of $C_{12}AE_4$ and $C_{12}AE_{23}$ to Atlantic salmon (*Salmo salar*) parr were 1.5 and 22.5 mg/L, respectively.

Different effects were observed for several species of algae. Using a series of AE consisting of ten to eighteen carbons in the alkyl chain and five to forty EO units, Ernst et al. (1983) found decreased growth of the green alga, *C. reinhardi* with increasing EO length. The most lipophilic compound ($C_{18}AE_7$) showed growth-promoting effects.

Data reviewed by Markarian et al. (1989; 1990) indicate that increases in alkyl-chain length increase toxicity to the aquatic organisms *S. capricornutum*, *D. magna*, and *P. promelas*. Branching of the alkyl chain decreased toxicity relative to the linear alkyl chain. The branched AE, $C_{12}AE_7$ and $C_{13}AE_7$, were less toxic (up to five times for *D. magna* and up to 40 times for *S. capricornutum*) than a linear AE, $C_{13-15}AE_7$. *Ceriodaphnia* sp. had similar acute sensitivities to linear and branched AE, but reproductive NOEC and LOEC values were 5-6 times higher for the branched AE compared to the linear AE.

Dorn et al. (1990, 1993) and Kravetz et al. (1991) determined that the intact branched $C_{13}AE_7$ was less acutely and chronically toxic to *D. magna* and *P. promelas* than the linear C_{12-15} AE_9. Differences (high/low LC_{50} values) between linear and branched toxicities were: 5 times for fish, 11 times for *D. magna*, and 17 times for algae. Likewise, linear secondary AE may be less toxic than linear primary AE to *Daphnia* sp., fathead minnows, and goldfish (Monsanto Company, 1977; Kurata et al., 1977).

As discussed earlier, toxicity tests of effluents from activated sludge units on aquatic biota have been used as an index of surfactant removal via a combination of biodegradation, chemical hydrolysis, and adsorption. In these tests the removal of AE reduces their toxicity to fish and aquatic invertebrates. The data of Reiff (1976), Abram et al. (1977), Kurata et al. (1977), Maki et al. (1979), Turner et al. (1985), Dorn et al. (1990; 1993) and Kravetz et al. (1991) all indicate that the effluents from activated sludge units are less toxic than the parent compounds. For a series of AE with hydrophobe branching ranging from 0.1 to 4 internal methyl groups, effluent toxicity increased with increased branching, reflecting their slower biodegradation (Dorn et al., 1990; 1993; Kravetz et al., 1991).

7. Effect of Environmental Variables on Toxicity

Under natural conditions in the environment, the toxicity of surfactants based on laboratory results, may be reduced. Differences in temperature and water hardness within normal ambient ranges, however, exert little effect on toxicity (Lewis, 1992). Several studies addressed the effect of water hardness on toxicity. Rainbow trout (*S. gairdneri*) and goldfish (*Carassius auratus*) were slightly less susceptible to $C_{12}AE_{3.25}$ in hard water than in soft water, even when acclimated in soft water. Survival times for rainbow trout acclimated in soft water and exposed to 5 mg/L of $C_{12}AE_{3.25}$ were 152 minutes in hard water and 107 minutes in soft water (Tovell et al., 1975). Maki and Bishop (1979) and Maki et al. (1979) suggested a slight decrease in $C_{14-15}AE_7$ toxicity for *Daphnia* sp. as water hardness increased. The 48-hour LC_{50} increased from 0.36 mg/L at 50 mg/L $CaCO_3$ to 0.90 mg/L at 350 mg/L $CaCO_3$. However, no such trends were found in other tests (Maki and Bishop, 1979; Lewis and Perry, 1981; Procter & Gamble Company, 1981).

In their comparison of the acute toxicity of $C_{14-15}AE_7$ (Neodol 45-7) to fathead minnows in different dilution media, Maki et al. (1979) found that the surfactant was least toxic in secondary effluent from a sewage plant (suspended solids of 50 mg/L and COD of 33-124 mg/L), intermediately toxic in stream water (suspended solids of 3-20 mg/L and COD of 5 mg/L), and most toxic in filtered tap water (no suspended solids or COD). The 96-hour LC_{50} values were 2.5, 1.4, and 1.2 mg/L, respectively.

Results of an *in situ* monitoring study using the photosynthetic response of lake plankton to several surfactants including $C_{14-15}AE_6$ showed that surfactants were less toxic during periods of lower water temperature (range of 17-28°C), diatom dominance, and low algal density (Lewis and Hamm, 1986). No studies were located on the response of fish to AE surfactants at different temperatures.

8. Bioaccumulation

AE are rapidly taken up and distributed throughout the body by fish. Bioaccumulation factors (concentration in body/concentration in water) may be several hundred, but elimination is rapid upon removal to clean water. Chemical-specific half-lives of 27-75 hours were measured.

In the two following studies, bioaccumulation of AE surfactants in fish was assessed on the basis of ^{14}C activity measurements; no distinction was made between the administered compounds and metabolites. Thus when bioaccumulation factors were calculated, they represent maximum values. In 28-day tests, Bishop and Maki (1980) exposed juvenile bluegill sunfish (*L. macrochirus*) to ^{14}C-labeled $C_{14}AE_7$ at concentrations of 0.02 or 0.20 mg/L. Uptake of ^{14}C was rapid, with the whole-body concentration 700-800 times that of the ambient solution. In an unpublished report (Procter & Gamble Company, 1981), the whole body concentration of $C_{14}AE_7$ from a solution containing 0.016 mg/L was 613 times that in the ambient solution in small bluegill (0.6 g) and 445 times that in the ambient solution in larger fish (4 g). During enterohepatic circulation, the concentration in the gall bladder reached a factor of 35,056 times that in the ambient solution.

Wakabayashi et al. (1987), in their study of the uptake, distribution, and clearance of several ^{14}C-labeled compounds ($C_{12}AE_4$, $C_{12}AE_8$, and $C_{12}AE_{16}$) used thin-layer chromatography to examine metabolites in several tissues of carp (*Cyprinus carpio*) exposed to concentrations of 0.2-0.6 mg/L. Autoradiograms and liquid scintillation counting showed that radioactivity was rapidly absorbed into the body and, within two hours, was distributed in skin, nasal and oral cavities, gills, brain, hepatopancreas, kidney, gall bladder, and intestinal tract. After two hours of exposure, between 50 and 70% of the radioactivity of the parent compounds was present in these organs; after 24 hr, <20% was unmetabolized. In the gall bladder, almost all of the radioactivity was in the form of metabolites. Bioconcentration was dependant on the degree of ethoxylation. Uptake and distribution of the label was much less for $C_{12}AE_{16}$ than for the other two compounds. Whole-body wet weight bioconcentration factors for $C_{12}AE_4$, $C_{12}AE_8$, and $C_{12}AE_{16}$ at 72 hours were 310, 220, and 4.3, respectively. Half-lives of the respective compounds were 27, 70, and 75 hr.

B. EFFECTS ON MICROORGANISMS

Few data on the effects of surfactants on soil microorganisms were located. Surfactants may interact with soil microorganisms, causing inhibition of growth or damage to cellular components. In this respect, nonionic surfactants are generally less toxic than cationic and anionic surfactants. Inhibitory activity depends on specific environmental conditions: the type(s) and concentrations of surfactants present, the types and acclimation history of bacterial species, and the presence of foods and other materials. In general, toxicity to bacteria increases with increasing hydrophobe chain length and decreases with increasing ethoxylate chain length (Swisher, 1987).

Arthur D. Little (1977) discussed the following studies. Vandoni and Goldberg-Federico (1973) reported that the nitrification process in two soil types was slightly inhibited by a 0.1% concentration of $C_{12}AE_n$ (EO = 0.1% by weight). In soil watered daily with 50 mL of a solution containing 1000 mg/L of n-sec-$C_{11-15}AE_9$ (Tergitol 15-S-9), there was a 16% reduction in number of microfungi, but no reduction in number of species (Lee, 1970). The growth of some species was stimulated at a concentration of 10 mg/L, indicating the utilization of the surfactant as a carbon source.

In a more recent study, the effect of high concentrations of $C_{12-15}AE_7$ (Neodol 25-7) on nitrification was studied. In bench scale biotreater units containing activated sludge, concentrations of 80-100 mg/L did not inhibit the growth or activity of nitrifying microbes present (Salanitro et al., 1988). Conversion of ammonia to nitrate was similar to that in the control unit.

At low concentrations, $C_{8-14}AE_{4.5}$ (Alfonic 810-60) and $C_{12-14}AE_{5.6}$ (Novel II 1412-56), stimulated biodegradation of sorbed hydrocarbons in aquifer sands and soil slurries (Aronstein and Alexander, 1992). At concentrations of 10 or 100 μg/g of aquifer solids or soil, the mineralization of phenanthrene and biphenyl was enhanced. It was suggested that surfactants may be useful for the bioremediation of sites contaminated with sorbed aromatic hydrocarbons.

The 5-min EC_{50} values of n-pri-$C_{12-15}AE_9$, br-$C_{13}AE_7$ (2 methyl branches/hydrophobe) and br-$C_{13}AE_7$ (4 methyl branches/hydrophobe) for the marine luminescent bacterium, *Photobacterium phosphoreum*, were 1.5, 11.4, and 8.1 mg/L, respectively (Dorn et al., 1993).

C. EFFECTS ON HIGHER PLANTS

At high concentrations (>10 mg/L), AE have little effect on the growth of higher plants. Based on frond count and root length, the 7-day EC_{50} for growth inhibition of duckweed (*Lemna minor*) in flow-through exposures was 21 mg/L (Bishop and Perry, 1981).

Ernst et al. (1971) studied the effect of surfactants (concentrations of 4-1000 mg/L) on orchid seedlings (*Phalaenopsis* sp.) and seeds (*Epidendrum* sp.). A solution of 10 mg/L did not affect the survival of *Phalaenopsis* sp. At 1000 mg/L, survival after five months was 40% in $C_{12-14}AE_6$ (Alfonic 1214-6) and n-pri-$C_{12-15}AE_9$ (Neodol 25-9) and 70% in n-sec-$C_{11-15}AE_9$ (Tergitol 15-S-9) and $C_{13}AE_9$ (Surfonic TD-90). Treated seedlings weighed less than control seedlings. There was little effect on germination and growth of *Epidendrum* sp. at 10 or 100 mg/L. The authors hypothesized that the site of surfactant action is the cytomembrane.

Valoras and Letey (1978) reported on the effects of ten linear primary surfactants (n-pri-C_{12-15} with 3, 7, 9, 12, 20, or 30 EO units, n-pri-C_{9-11} with 2.5, 6, or 9 EO units and n-pri-$C_{12-13}AE_3$) on the growth of two species of grass. Barley and rye grasses were exposed to 50 mg/L or 100 mg/L solutions of each surfactant. At a concentration of 50 mg/L, growth for both species was equal to or greater than controls. Growth for both species was inhibited at a concentration of 100 mg/L. At this concentration, $C_{12-13}AE_3$ and $C_{12-15}AE_3$ were the most toxic, reducing growth in barley by 25 and 29% respectively and in rye by 50 and 80%, respectively. The least phytotoxic compounds for both species were $C_{12-15}AE_{20}$, $C_{9-11}AE_6$, and $C_{9-11}AE_9$. At AE (formula not given) concentrations of 4 to 8 mg/L, oat and barley growth was stimulated (Hartmann, 1966). However, growth of root hairs was inhibited.

Shell Chemical Co. (1981) investigated the ability of $C_{9-11}AE_8$ (Neodol 91-8) to reduce soil hydrophobicity in burned areas following reseeding. Application rates of 16 or 32 lb/acre following seeding with ryegrass produced no phytotoxicity. The area was rapidly revegetated with perennials and annuals. Dilute solutions of $C_{12-15}AE_7$ (Neodol 25-7) increased infiltration of water into hydrophobic soils resulting in an increase in crop growth (Howells et al., 1984). In field trials, application of the surfactant at a rate of 9.5 liters/hectare (1 gallon/acre) increased the yield of silage corn by 29% compared to a control plot. The same treatment did not increase the tuber yield of potato plants.

As part of a study on wetting agents for agricultural sprays, leaves of apple, plum, and peach trees were immersed in 0.03-1.0% (300-10,000 mg/L) aqueous solutions of $C_{16}AE_{18}$ (cetylpolyethylene glycol) for 10-15 seconds. Damage was assessed after two hours and up to two months later. At all concentrations, cetylpolyethylene glycol caused virtually no damage (Furmidge, 1959a). Using a series of 0.05 and 0.5% solutions of $C_{12}AE$ (7, 9, and 16 EO units) and $C_{16}AE$ (10, 16, 24, and 60 EO units), Furmidge (1959b) again found little phytotoxicity to apple and plum leaves. The $C_{16}AE_n$ were less toxic than $C_{12}AE_n$, and wetting ability and toxicity decreased with increasing ethoxylate chain length.

Effects of applications of 0.1% solutions (1000 mg/L) of $C_{12-15}AE_3$, $C_{12-15}AE_7$, $C_{12-15}AE_9$, $C_{12-15}AE_{12}$, and $C_{12-15}AE_{30}$ (Neodol 25 ethoxylates) to the adaxial surface of 10-day-old cowpea (*Vigna unguiculata*) leaves were studied by Lownds and Bukovac (1988). Phytotoxicity, as evidenced by discoloration and slight necrosis, resulted from exposure to ethoxylate chain lengths of 3-12; $C_{12-15}AE_{30}$ produced no tissue damage.

Concentrations giving 50% growth inhibition for cell suspensions of soybean (*Glycine max*) were as follows: $C_{12}AE_0$, >400 uM (>75 mg/L); $C_{12}AE_2$, ~100 uM (27 mg/L); $C_{12}AE_6$, ~41 uM (19 mg/L); $C_{12}AE_8$, ~13 uM (7 mg/L); $C_{12}AE_{17}$, ~28 uM (26 mg/L); and $C_{10}AE_{6.7}$, ~6 uM (3 mg/L). Cell metabolism of the surfactants occurred, lowering the effective concentration (Davis et al., 1982; 1984). For $C_{12}AE_8$, growth was inhibited at 15 mg/L, intermediate at 9 mg/L, and only slightly inhibited at 2 mg/L (Davis and Stolzenberg, 1986).

D. MODE OF ACTION

The mode of action of AE on aquatic organisms is not well understood. Many nonionic surfactants interact with proteins, changing the shape and activity of enzymes and solubilizing structural proteins, which may result in changes in cell permeability (Helenius and Simons, 1975; Swisher, 1987). A chain length of 12 carbons ($C_{12}AE_n$) appears to exert maximal effects on membrane permeability (Florence et al., 1984).

In fish and invertebrates, surfactants act both physically and chemically on respiratory structures. Surfactants emulsify the natural mucus and oils on gill surfaces. The gill membrane may be chemically penetrated by the hydrophobe portion of the molecule, resulting in alteration of the membrane structure and interference with gas and ion exchange. Respiratory function may be further disrupted by the swelling and secretion of mucous by the gill epithelium, which interrupts the diffusion of oxygen through the gill. The animals die of suffocation (Swedmark et al., 1971; Wildish, 1974; Shell Chemical Co., 1983). According to Swedmark et al. (1971), invertebrates with lower respiratory rates and thicker cuticle around the gill epithelium may be less susceptible to surfactant toxicity.

The observed decrease in toxicity with increasing EO chain length for aquatic species may be attributed to the hydrophilicity of the molecule. A longer EO chain makes the molecule less fat soluble, thus hindering emulsification of or penetration into the gill membrane (Wildish, 1974; Shell Chemical Co., 1983). This effect was not seen for algae where growth decreased with increasing EO chain length and increased with lipophilicity (Davis et al., 1982; Ernst et al., 1983). Algae may use the more lipophilic AE as a carbon source.

Gloxhuber and Fischer (1968) suggested that in addition to reduction in surface tension, symptoms observed with exposure of fish to surfactants are consistent with those of absorbed anesthetics. Wildish (1974) suggested that surfactants may associate with lipoproteins in membranes of nerve cells, thus blocking nerve function in the gill area.

In higher plants, AE surfactants altered the ultrastructure of cells by emulsifying membranes (Healy et al., 1971; Davis and Stolzenberg, 1986). At concentrations below the CMC, some organelles of soybean cells appeared to have reduced or missing internal membranes and myelin-like membrane whorls were present in some cells.

REFERENCES

Abram, F.S.H., V.M. Brown, H.A. Painter and A.H. Turner. 1977. The biodegradability of two primary alcohol ethoxylate nonionic surfactants under practical conditions, and the toxicity of the biodegradation products to rainbow trout. Paper presented at the IVth Yugoslav Symposium on Surface Active Substances held in Dubrovnik, October 17-21, 1977.

APHA (American Public Health Association). 1989. *Standard Methods for the Examination of Water and Wastewater*, 17th ed., APHA, New York.

Aronstein, B.N. and M. Alexander. 1992. Surfactants at low concentrations stimulate biodegradation of sorbed hydrocarbons in samples of aquifer sands and soil slurries. Environ. Toxicol. Chem. 11:1227-1233.

Arthur D. Little. 1977. Human Safety and Environmental Aspects of Major Surfactants. A Report to the Soap and Detergent Association. Arthur D. Little, Inc., Cambridge, MA.

Bishop, W.E. and R.L. Perry. 1981. Development and evaluation of a flow-through growth inhibition test with duckweed (*Lemna minor* L.). In: Aquatic Toxicology and Hazard assessment: Fourth Conference, ASTM STP 737, D.R. Branson and K.L. Dickson, Eds.

Bishop, W.E. and A.W. Maki. 1980. A critical comparison of two bioconcentration test methods. In: Aquatic Toxicology, J.G. Eaton, P.R. Parrish, and A.C. Hendricks, Eds., American Society for Testing and materials, ASTM STP 707:61-77.

Bode, H., R. Ernst and J. Arditti. 1978. Biological effects of surfactants. III. Hydra as a highly sensitive assay animal. Environ. Pollut. 17:175-185.

Burlington Research Inc. 1985. Acute and chronic bioassays of industrial surfactants, PPP Grant Final Report.

Davis, D.G. and R.L. Stolzenberg. 1986. Effects of a homogeneous linear alcohol ethylene oxide surfactant on the ultrastructure of soybean cell suspension cultures. Can. J. Bot. 64:618-625.

Davis, D.G., R.L. Stolzenberg and J.A. Dusky. 1984. A comparison of various growth parameters of cell suspension cultures to determine phytotoxicity of xenobiotics. Weed Sci. 32:235-242.

Davis, D.G., R.L. Stolzenberg and G.E. Stolzenberg. 1982. Phototoxicity of selected non-ionic surfactants to soybean *Glycine max* cell suspensions. Environ. Pollut. Ser A 27:197-206.

Davis, E.M. and E.F. Gloyna. 1967. Biodegradability of nonionic and anionic surfactant by blue-green and green algae. Report to the Center for Research in Water Resources. (Cited in Arthur D. Little, 1977).

Dolan, I.M. III, B.C. Gregg, J. Cairns, Jr., K.L. Dickson and A.C. Hendricks. 1974. The acute toxicity of three new surfactant mixtures to a mayfly larvae. Arch. Hydrobiol. 1:123-132.

Dorn, P.B., P.J. Downey, J.P. Salanitro, B.J. Venables, L.K. Kravetz and J.H. Rodgers, Jr. 1990. Assessing the aquatic hazard of some nonionic surfactants by biodegradation and toxicity. Proceedings of the Society of Environmental Toxicology and Chemistry, November 11-15, Arlington, VA. (Abstract 386).

Dorn, P.B., J.P Salanitro, S.H. Evans and L. Kravetz. 1993. Assessing the aquatic hazard of some linear and branched nonionic surfactants by biodegradation and toxicity. Environ. Toxicol. Chem. (in press).

Ernst, R., J. Arditti and P.L. Healey. 1971. Biological effects of surfactants. I. Influence on the growth of orchid seedlings. New Phytol. 70:457-475.

Ernst, R., C.J. Gonzales and J. Arditti. 1983. Biological effects of Surfactants: Part 6 - Effects of anionic, non-ionic and amphoteric surfactants on a green alga (*Chlamydomonas*). Environ. Pollut. Ser. A 31:159-175.

Fischer, W.K. and P. Gode. 1978. Comparative investigation of various methods of examining fish toxicity, with special consideration of the German golden orfe test and the iso zebra fish test. Z.f. Wasser Abwasser Forsch 11:99-105.

Florence, A.T., I.G. Tucker and K.A. Walters. 1984. Interactions of nonionic polyoxyethylene alkyl and aryl esters with membranes and other biological systems. In: M.J. Rosen (Ed.), Structure/Performance Relationships in Surfactants. American Chemical Society, Washington, DC.

Furmidge, C.G.L. 1959a. Physico-chemical studies on agricultural sprays. II. The phytotoxicity of surface-active agents on leaves of apple and plum trees. J. Sci. Food Agric. 10:274-282.

Furmidge, C.G.L. 1959b. Physico-chemical studies on agricultural sprays. III. Variation of phytotoxicity with the chemical structure of surface-active agents. J. Sci. Food Agric. 10: 419-425.

Gloxhuber, C. and W.K. Fischer. 1968. Studies on the action of high concentrations of alkylpolyglycol ethers to fish (German translation). Food Cosmet. Toxicol. 6:469-477.

Granmo, A. and G. Jorgensen. 1975. Effects on fertilization and development of the common mussel *Mytilus edulis* after long-term exposure to a nonionic surfactant. Marine Biol. 33:17-20.

Hall, R.H. 1973. An algal toxicity test used in the safety assessment of detergent components. Presented at the thirty-sixth annual meeting of the American Society of Limnology nd Oceanography, Inc., Salt Lake City, Utah, June 12, 1973. (Cited in Arthur D. Little, 1977).

Hall, W. S., J.B. Patoczka, R.J. Mirenda, B.A. Porter and E. Miller. 1989. Acute toxicity of industrial surfactants to *Mysidopsis bahia*. Arch. Environ. Contam. Toxicol. 18:765-722.

Hartmann, L. 1966. Die toxicität neuer tenside gegen autotrophe organismen. Gas-Wasserfach 107:251-255. (Cited in Arthur D. Little, 1977).

Helenius, A. and K. Simons. 1975. Solubilization of membranes by detergents. Biochim. Biophy. Acta 415:29-79.

Healy, P.L., R. Ernst and J. Arditti. 1971. Biological effects of surfactants. II. Influence on the ultrastructure of orchid seedlings. New Phytol. 70:477-482.

Hendricks, A.C., M. Dolan, F. Camp, J. Cairns, Jr., and K.L. Dickson. 1974. Comparative toxicities of intact and biodegraded surfactants to fish, snail, and algae. Center for Environmental Studies, Virginia Polytechnic Institute and State University, Blacksburg, VA.

Hidu, H. 1965. Effects of synthetic surfactants on the larvae of clams (*M. mercenaria*) and oysters (*C. virginica*). J. Water Poll. Contr. Fed. 37:262-270.

Horning, W.B. and C.I. Weber. 1985. Short-Term Methods for Estimating the Chronic Toxicity of Effluents and Receiving Waters to Freshwater Organisms. EPA-600/4-85-014. Environmental Monitoring and Support Laboratory, U.S. Environmental Protection Agency, Cincinnati, OH.

Howells, W.G., L. Kravetz, D. Loring, C.D. Piper, B.W. Poovaiah, E.C. Seim, V.P. Rasmussen, N. Terry, and L.J. Waldron. 1984. The use of nonionic surfactants for promoting the penetration of water into agricultural soils. In: Proceedings of the Eighth World Surfactants Congress, Munich, West Germany.

Janicke, W. 1988. Biologic degradation of nonionic surfactants: Biologic degradation of surfactants of ethoxylates with the hydrophobic portion of the molecule free of aromatics in the confirmatory tests. Communication 10. On the behavior of synthetic organic compounds for wastewater treatment.

Karpinska-Smulikowska, J. 1984. Studies on the relationship between composition and molecular mass of nonionic surfactants of the pluronic type and their biotoxic activity. Tenside 21:243-246.

Kravetz, L., J.P. Salanitro, P.B. Dorn and K.F. Guin. 1991. Influence of hydrophobe type and extent of branching on environmental response factors of nonionic surfactants. J. Am. Oil Chem. Soc. 68:610-618.

Kurata, N., K. Koshida, and T. Fujii. 1977. Biodegradation of surfactants in river water and their toxicity to fish. Yukagaku 26:115-118.

Kutt, E. and D. Martin. 1974. Effect of selected surfactants on the growth characteristics of *Gymnodinium breve*. Marine Biol. 28:253-259.

Lee, B.K.H. 1970. The effect of anionic and nonionic detergents on soil microfungi. Can. J. Bot. 48:583-589. (Cited in Arthur D. little, 1977).

Lewis, M.A. 1990. Chronic toxicities of surfactants and detergent builders to algae: A review and risk assessment. Ecotoxicol. Environ. Safety 20:123-140.

Lewis, M.A. 1991. Chronic and sublethal toxicities of surfactants to aquatic animals: A review and risk assessment. Water Res. 25:101-113.

Lewis, M.A. 1992. The effects of mixtures and other environmental modifying factors on the toxicities of surfactants to freshwater and marine life. Water Res. 26:1013-1023.

Lewis, M.A. and B.G. Hamm. 1986. Environmental modification of the photosynthetic response of lake plankton to surfactants and significance to a laboratory-field comparison. Water Res. 20:1575-1582.

Lewis, M.A. and R.L. Perry. 1981. Acute lethalities of equimolar and equitoxic surfactant mixtures to *Daphnia magna* and *Lepomis macrochirus*. In: Aquatic Toxicology and Hazard Assessment: Fourth Conference, D.R. Branson and K.L. Dickson, Eds., ASTM STP 737.

Lewis, M.A. and D. Suprenant. 1983. Comparative acute toxicities of surfactants to aquatic invertebrates. Ecotoxicol. Environ. Safety 7:313-322.

Lownds, N.K. and M.J. Bukovac. 1988. Studies on octylphenoxy surfactants: V. Toxicity to cowpea leaves and effects of spray application parameters. J. Amer. Soc. Hort. Sci. 113(2): 205-210.

Macek, K.J. and S.F. Krzeminski. 1975. Susceptibility of bluegill sunfish (*Lepomis macrochirus*) to nonionic surfactants. Bull. Environ. Contam. Toxicol. 13:377-384.

Maki, A.W. 1979a. Respiratory activity of fish as a predictor of chronic fish toxicity values for surfactants. In: Aquatic Toxicology, L.L. Marking and R.A. Kimerle, Eds., American Society for Testing and Materials, ASTM ATP 667:77-95.

Maki, A.W. 1979b. An environmental safety evaluation of detergent chemicals in estuarine ecosystems. Proc. 14th Intl. Mar. Biol. Symp., Protection of Life in the Sea, held in Helgoland, Germany, Sept. 24028, 1979. Biologische Anstalt Helgoland, Hamburg, Germany. (Cited in Shell Chemical Company, 1983).

Maki, A.W. 1979c. Correlations between *Daphnia magna* and fathead minnow (*Pimephales promelas*) chronic toxicity values for several classes of test substances. J. Fish Res. Bd. Can. 36:411-421.

Maki, A.W. and W.E. Bishop. 1979. Acute toxicity studies of surfactants to *Daphnia magna* and *Daphnia pulex*. Arch. Environ. Contam. Toxicol. 8:599-612.

Maki, A.W., A.J. Rubin, R.M. Sykes and R.L. Shank. 1979. Reduction of nonionic surfactant toxicity following secondary treatment. J. Water Pollut. Control Fed. 51:2301-2313.

Marchetti, R. 1964. Toxicity of some surfactants to fish. La Riv. Ital. Delle Sos. Grasse 41:533-542. (Cited in Shell Chemical Co., 1983).

Markarian, R.K., K.W. Pontasch, D.R. Peterson and A.I. Hughes. 1989. Comparative toxicities of selected surfactants to aquatic organisms. In: Review and Analysis of Environmental Data on Exxon Surfactants and Related Compounds. Technical Report, Exxon Biomedical Sciences, Inc., East Millstone, NJ.

Markarian, R.K., M.L. Hinman and M.E. Targia. 1990. Acute and chronic aquatic toxicity of selected nonionic surfactants. Proceedings of the Society of Environmental Toxicology and Chemistry, November 11-15, Arlington, VA. (Abstract P326).

Marshall, S.J. and A.D. McInnes. 1990. The influence of natural and artificial sediment characteristics on the toxicity of a nonionic surfactant to *Chironomus riparius*. Proceedings of the Society of Environmental Toxicology and Chemistry, November 11-15, Arlington, VA. (Abstract P333).

Masters, J.A., M.A. Lewis, D.H. Davidson, and R.D. Bruce. 1991. Validation of a four-day *Ceriodaphnia* toxicity test and statistical considerations in data analysis. Environ. Toxicol. Chem. 10:47-55.

Maxwell, K.E. and W.D.Piper. 1968. Molecular structure of nonionic surfactants in relation to laboratory insecticidal activity. J. Econ. Entomol. 61:1633-1636.

Mayer, F.L., Jr. and M.R. Ellersieck. 1986. Manual of acute toxicity: interpretation and data base for 410 chemicals and 66 species of freshwater animals. U.S. Department of the Interior, Fish and Wildlife Service, Columbia, MO. Resource Publication 160.

Monsanto Company. 1977. (Unpublished data cited in Arthur D. Little, 1977).

Moore, S.B., R.A. Diehl, J.M. Barnhardt and G.B. Avery. 1987. Aquatic toxicities of textile surfactants. Text. Chem. Col. 19(5):29-32.

Naylor, C.G., F.J. Castaldi and B.J. Hayes. 1988. Biodegradation of nonionic surfactants containing propylene oxide. J. Am. Oil Chem. Soc. 65:1669-1676.

Nyberg, H. 1988. Growth of *Selenastrum capricornutum* in the presence of synthetic surfactants. Wat. Res. 22(2):217-223.

Payne, A.G. and R.H. Hall. 1979. A method for measuring algal toxicity and its application to the safety assessment of new chemicals. In: Aquatic Toxicology, L.L. Marking and R.A. Kimerle, Eds, ASTM STP 667:171-180.

Peltier, W.H. and C.I. Weber. 1985. Methods for Measuring the Acute Toxicity of Effluents to Freshwater and Marine Organisms. EPA-600/4-85-013. Environmental Monitoring and Support Laboratory, U.S. Environmental Protection Agency, Cincinnati, OH.

Procter & Gamble, Co. 1981. Unpublished data (Cited in Goyer et al., 1981).

Procter & Gamble, Co. 1985. Unpublished data.

Procter & Gamble, Co. 1990. Unpublished data.

Reiff, B. 1976. The effect of biodegradation of three nonionic surfactants on their toxicity to rainbow trout. In: Proceedings of the Seventh International Congress on Surface-Active Substances, Moscow, USSR, September 12-18.

Reiff, B., R. Lloyd, M.J. How, D. Brown and J.S. Alabaster. 1979. The acute toxicity of eleven detergents to fish: results of an interlaboratory exercise. Water Res. 13:207-210.

Salanitro, J.P., G.C. Langston, P.B. Dorn, and L. Kravetz. 1988. Activated sludge treatment of ethoxylate surfactants at high industrial use concentrations. Water Sci. Technol. 20:126-130.

Shell Chemical Co. 1981. (Unpublished data cited in Goyer et al., 1981).

Shell Chemical Company. 1983. The Aquatic Safety of Neodol* Products. Brochure SC:612-83.

Shell Chemical Company. 1988. Aquatic Toxicity of Neodol* 1-5 to Algae, Daphnids and Fish. Technical Progress Report WRC 225-87. Shell Development Company, Houston, TX. Unpublished data.

Swedmark, M., B. Braaten, E. Emanuelsson and A. Granmo. 1971. Biological effects of surface active agents on marine animals. Marine Biol. 9:183-201.

Swisher, R.D. 1987. *Surfactant Biodegradation*, 2nd ed. Surfactant Science Series, Vol. 18. Marcel Dekker, Inc., New York.

Tovell, P.W.A., C. Newsome and D. Howes. 1975. Effect of water hardness on the toxicity of a nonionic detergent to fish. Water Res. 9:31-36.

Turner, A.H., F.S. Abram, V.M. Brown and H.A. Painter. 1985. The biodegradability of two primary alcohol ethoxylate nonionic surfactants under practical conditions and the toxicity of the biodegradation products to rainbow trout. Water Res. 19:45-51.

Unilever Research Laboratories. 1977. (Unpublished data cited in Arthur D. Little, 1977).

Vandoni, M.V. and L. Goldberg-Federico. 1973. Sul comportamento dei detergenti di sintest nel terreno agrario--Nota VIII. Riv. Ital. Sost. Grasse. 50:185-192. (Cited in Arthur D. Little, 1977).

Van Emden, H.M., C.C.M. Kroon, E.N. Schoeman and H.A. Van Seventer. 1974. The toxicity of some detergents tested on *Aedes aegypti* L., *Lebistes reticulatus* Peters, and *Biomphalaria glabrata* (Say). Environ. Pollut. 6:297-308.

Valoras, N. and J. Letey. 1978. Screening of Neodol chemicals for potential use in erosion control. Report of the University of California, Riverside, to Shell Chemical Co., Houston, Texas, unpublished. (Cited in Goyer et al., 1981).

Verma, S.R., D. Mohan, and R.C. Dalela. 1978. Studies on the relative toxicity of a few synthetic detergents to a fish *Macrones vittatus*. Acta Hydrochem. Hydrobiol. 6:121-128.

Vista Chemical Company. 1977. (Unpublished data cited in Arthur D. Little, 1977, as Continental Oil Co., unpublished data).

Wakabayashi, M., M. Kikuchi, A. Sato and T. Yoshida. 1987. Bioconcentration of alcohol ethoxylates in carp (*Cyprinus carpio*). Ecotoxicol. Environ. Safety 13:148-161.

Weber, C.I., W.B. Horning, D,J, Klemm, T.W. Neiheisel, P.A. Lewis, E.L. Robinson, J. Menkedick and F. Kessler. 1988. Short-Term Methods for Estimating the Chronic Toxicity of Effluents and Receiving Waters to Marine and Estuarine Organisms. EPA-600/4-87-028. Environmental Monitoring and Support Laboratory, U.S. Environmental Protection Agency, Cincinnati, OH.

Wildish, D.J. 1974. Lethal response by Atlantic salmon parr to some polyethoxylated cationic and nonionic surfactants. Water Res. 8:433-437.

Wright, A. 1976. The use of recovery as a criterion for toxicity. Bull. Environ. Contam. Toxicol. 15:747-749.

Yamane, A.N., M. Okada and R. Sudo. 1984. The growth inhibition of planktonic algae due to surfactants used in washing agents. Water Res. 18:1101-1105.

V. HUMAN SAFETY

An evaluation of the data up to 1981 dealing with various measures of mammalian toxicity as indicators of potential human safety as well as actual human exposures indicate that AE do not represent a hazard to human health (Arthur D. Little, 1977; Goyer et al., 1981). The data reviewed below indicate that for both acute and chronic exposures, AE exhibit a low order of toxicity by the oral, dermal, or inhalation routes of intake. Alcohol ethoxylates are not skin sensitizers and at dilutions of $\leq 0.1\%$ are usually not skin or eye irritants. AE surfactants test negative in *in vitro* genotoxicity assays. In mammalian species, AE have not been found to cause reproductive, genotoxic, or carcinogenic effects.

A. ANIMAL STUDIES

1. Acute Toxicity

In laboratory animals, AE exhibit a low order of acute toxicity by the oral, dermal, and inhalation routes of exposure. Oral toxicity to rats appears to be related to chemical structure, with toxicity increasing with increasing EO length up to 11.6 units. In contrast, dermal toxicity to rabbits appeared to be independent of chemical structure.

Oral. In the fasting rat, the LD_{50} values of linear primary alcohol ethoxylates with varying ethylene oxide chain lengths range from 544 mg/kg for $C_{14-15}AE_{11}$ in female rats to >25,000 mg/kg for $C_{18}AE_2$ (oleyl and stearyl AE) in both species combined (Table 5-1). The linear secondary AE, $C_{11-15}AE_3$ and $C_{11-15}AE_9$ (Tergitols 15-S-3 and 15-S-9) and the PO-EO containing AE, with LD_{50} values of 410 µg/kg (~410 mg/kg[1]) to 11,700 mg/kg and 810 to >16,000 mg/kg, respectively, also fall close to this range. In some of the studies that treated the data for males and females separately, females appeared to be more sensitive to AE surfactants than males. Based on the toxicity rating scale of Gosselin et al. (1984), AE surfactants can be considered slightly to moderately toxic by the oral route of exposure.

Some relationships between structure and toxicity are apparent. The data show a trend of increasing oral toxicity with increments in the length of the ethoxylate chain up to 11.6 EO units (Figure 5-1). The alkyl chain length does not appear to affect toxicity. The PEG, environmental biodegradation products of AE, are practically nontoxic to the rat, with LD_{50} values increasing with increasing chain length from 8540 mg/kg for ethylene glycol to >50,000 mg/kg for some high molecular weight (≥ 1250) PEG (Schick, 1967).

Treated animals were observed for several days for signs of toxicity. Clinical signs of rats dying from acute doses included diarrhea, diuresis, weight loss, piloerection, lethargy, ataxia, and abnormal posture. Necropsies of animals that succumbed revealed one or more of

---

[1] The specific gravity of most AE = ~1

Table 5-1. Acute Oral Toxicity to Mammals

Species	Surfactant	How Dosed	LD$_{50}$ (mg/kg) Male	LD$_{50}$ (mg/kg) Female	LD$_{50}$ (mg/kg) Combined	Reference
Rat	$C_{12}AE_7$	not specified			4,150	Grubb et al., 1960
	$C_{12}AE_4$ (Brij 30)	undiluted	8,600	9,070		Schick, 1967
	$C_{12}AE_{23}$ (Brij 35)	20% in water	8,600	9,350		
	$C_{16}AE_2$	50% in water	25,100	22,100		
	$C_{16}AE_{10}$ (Brij 52)	25% in water	3,490	2,460		
	$C_{16}AE_{20}$ (Brij 56)	25% in water	3,510	3,950		
	$C_{18}AE_2$ (Brij 58)	40% in water	>25,000^c	>25,000^c		
	$C_{18}AE_{10}$ (Brij 72)	25% in water	2,910	2,000		
	$C_{18}AE_{20}$ (Brij 76)	25% in water	1,920	2,330		
	$C_{18}AE_2$ (Brij 78)	undiluted	25,900	25,800		
	$C_{18}AE_{10}$ (Brij 92)	25% in water	3,580	2,700		
	$C_{18}AE_{20}$ (Brij 96)	25% in water	3,100	2,770		
	n-sec-$C_{12-15}AE_9$ (Brij 98) (Tergitol 15-S-9)	not specified			2,380	
	n-pri-$C_{12-13}AE_{6.5}$ (Neodol 23-6.5)	33% in water	2,100 1,400		2,600 (weanling)	Benke et al., 1977
	n-pri-$C_{14-15}AE_7$ (Neodol 45-7)	33% in water	3,300			
	$C_{9-11}AE_{2.5}$ (Neodol 91-2.5)	not specified			2,700	Shell Chemical Company, 1984
	$C_{9-11}AE_5$ (Dobanol 91-5)	undiluted			2,900	Shell Research Limited, 1980b

Table 5-1. (cont.)

Species	Surfactant	How Dosed	LD$_{50}$ (mg/kg) Male	LD$_{50}$ (mg/kg) Female	LD$_{50}$ (mg/kg) Combined	Reference
Rat (cont.)	C$_{9-11}$AE$_6$ (Dobanol 91-6) (Neodol 91-6)	undiluted 3.2–33% in water not specified			3,100 1,378 1,200	Shell Research Limited, 1980b; Shell Development Company, 1981b; Gingell and Lu, 1991; Shell Chemical Company, 1984
	C$_{9-11}$AE$_8$ (Neodol 91-8) (Dobanol 91-8)	not specified undiluted			1,000 2,700	Shell Chemical Company, 1984; Shell Research Limited, 1980b
	C$_{12-13}$AE$_{6.5}$ (Neodol 23-6.5) (Dobanol 23-6.5)	25% in water 50% m/v in corn oil undiluted	2,360 ~2,500 1,738	>1,250<2,500 1,637 1,206	2120 1,439	Shell Oil Company, 1990; Shell Research Limited, 1986; Shell Research Limited, 1990
	C$_{12-15}$AE$_3$ (Neodol 25-3)	not specified			2,500	Shell Chemical Company, 1984
	C$_{12-15}$AE$_7$ (Dobanol 25-7)	undiluted	2,000–3,145	1,321	1,642	Shell Research Limited, 1984a
	C$_{12-15}$AE$_9$ (Neodol 25-9)	not specified			1,600 3,200 5,600	Shell Chemical Company, 1984
	C$_{12-15}$AE$_{12}$ (Neodol 25-12)	not specified			1,800	Shell Chemical Company, 1984
	C$_{14-15}$AE$_{11}$ (Dobanol 45-11)	undiluted 50% m/v in corn oil	1,077 1,963	544 1,684	722 1,772	Shell Research Limited, 1984b; Shell Research Limited, 1986

Table 5-1. (cont.)

Species	Surfactant	How Dosed	LD$_{50}$ (mg/kg) Male	LD$_{50}$ (mg/kg) Female	LD$_{50}$ (mg/kg) Combined	Reference
Rat (cont.)	C$_{12-13}$AE$_7$ (Neodol 23-7P7)	undiluted			1,600	Shell Oil Company, 1989b
	n-pri-C$_{12-14}$AE$_{6.5}$ (Tergitol 24-L-45)	undiluted	2,710^b	1,870^b		Union Carbide Corporation, 1987c
	n-pri-C$_{12-14}$AE$_{7.2}$ (Tergitol 24-L-60)	undiluted	2,240^b	1,230^b		Union Carbide Corporation, 1987d
	n-pri-C$_{12-14}$AE$_{3.5}$ (Tergitol 24-L-3N)	undiluted	6,170^b	5,660^b		Union Carbide Corporation, 1988a
	n-pri-C$_{12-14}$AE$_{5.2}$ (Tergitol 24-L-25N)	undiluted	6,500^b	2,140^b		Union Carbide Corporation, 1987f
	n-pri-C$_{12-14}$AE$_{6.4}$ (Tergitol 24-L-50N)	undiluted	2,830^b	1,190^b		Union Carbide Corporation, 1987h
	n-pri-C$_{12-14}$AE$_7$ (Tergitol 24-L-60N)	undiluted	2,140^b	1,070^b		Union Carbide Corporation, 1986d
	n-pri-C$_{12-14}$AE$_9$ (Tergitol 24-L-75N)	undiluted, melted	1,230^b	930^b		Union Carbide Corporation, 1987i
	n-pri-C$_{12-14}$AE$_{11.6}$ (Tergitol 24-L-98N)	undiluted	930^b	620^b		Union Carbide Corporation, 1987g
	n-pri-C$_{12-16}$AE$_{1.6}$ (Tergitol 26-L-1.6)	undiluted	>16,000^{a,b}	>8,000^b		Union Carbide Corporation, 1987e
	n-pri-C$_{12-16}$AE$_3$ (Tergitol 26-L-3)	undiluted	6,500^b	4,920^b		Union Carbide Corporation, 1986e
	n-pri-C$_{12-16}$AE$_5$ (Tergitol 26-L-5)	undiluted	4,290^b	2,530^b		Union Carbide Corporation, 1987a
	n-sec-C$_{11-15}$AE$_3$ (Tergitol 15-S-3)	undiluted	11,700^b	5,600^b		Union Carbide Corporation, 1986b

Table 5-1. (cont.)

Species	Surfactant	How Dosed	LD$_{50}$ (mg/kg) Male	LD$_{50}$ (mg/kg) Female	LD$_{50}$ (mg/kg) Combined	Reference
Rat (cont.)	n-sec-C$_{11-15}$AE$_9$ (Tergitol 15-S-9)	10% v/v in distilled water	1,620[b]	410[b]		Union Carbide Corporation, 1986c
	C$_4$H$_9$O(CH$_2$CHCH$_3$)$_x$(CH$_2$CH$_2$O)$_y$H[c] (Tergitol XH)	undiluted, melted	>16,000[a,b]	>16,000[a,b]		Union Carbide Corporation, 1986f
	C$_4$H$_9$O(CH$_2$CHCH$_3$O)$_x$(CH$_2$CH$_2$O)$_y$H[c] (Tergitol XD)	undiluted, melted	>16,000[a,b]	>16,000[a,b]		Union Carbide Corporation, 1986g
	C$_4$H$_9$O(CH$_2$CHCH$_3$O)$_x$(CH$_2$CH$_2$O)$_y$H[c] (Tergitol XJ)	25% w/v in distilled water	2,830	810		Union Carbide Corporation, 1986h
	C$_{8-10}$H$_{17-21}$O(CH$_2$CH$_2$O)$_x$(CH$_2$CHCH$_3$O)$_y$H[c] (Tergitol 80-L-50N)	undiluted	1,660[b]	810[b]		Union Carbide Corporation, 1987b
	C$_{12-14}$O(CH$_2$CH$_2$O)$_x$(CH$_2$CHCH$_3$O)$_y$H[c] (Tergitol Mechanical Dish Surfactant)	undiluted	5,410[b]	2,140[b]		Union Carbide Corporation, 1986a
	C$_{8,10,12}$AE$_{3.1}$ (Alfonic 610-50)	undiluted			5,050	Vista Chemical Company, 1977
	C$_{8,10}$AE$_{2.2}$ (Alfonic 810-40)	not specified			5,150	Vista Chemical Company, 1985
	C$_{10,12}$AE$_{5.4}$ (Alfonic 1012-60)	undiluted			3,300[b]	Vista Chemical Company, 1964b
	C$_{10,12,14}$AE$_{2.7}$ (Alfonic 1014-40)	undiluted			9,800[b]	Vista Chemical Company, 1964a
	C$_{10,12,14}$AE$_{6.3}$ (Alfonic 1014-60)	undiluted			2,180	Vista Chemical Company, 1967
	C$_{12,14}$AE$_2$ (Alfonic 1214-HA-30)	not specified			>5,000	Vista Chemical Company, 1985
	C$_{12,14}$AE$_{6.5}$ (Alfonic 1214-HB-58)	undiluted	3,130	2,190	2,570	Vista Chemical Company, 1979d

Table 5-1. (cont.)

Species	Surfactant	How Dosed	LD$_{50}$ (mg/kg)			Reference
			Male	Female	Combined	
Rat (cont.)	C$_{12,14}$AE$_{10.6}$ (Alfonic 1214-70)	not specified			2,900	Vista Chemical Company, 1985
	C$_{12,14,16}$AE$_{1.3}$ (Alfonic 1216-22)	undiluted			>5,250[a]	Vista Chemical Company, 1980b
	C$_{12,14,16}$AE$_{3.2}$ (Alfonic 1218-40)	undiluted			~10,500[b]	Vista Chemical Company, 1964c
	C$_{12,14,16,18}$AE$_{11.5}$ (Alfonic 1218-70)	not specified			3,300	Vista Chemical Company, 1985
	C$_{12,14}$AE$_{3}$ (Alfonic 1412-40)	not specified			~9,800	Vista Chemical Company, 1985
	C$_{12,14}$AE$_{7}$ (Alfonic 1412-60)	not specified			2,800	Vista Chemical Company, 1985
	C$_{12}$AE$_{4}$ (Laureth-4)	not specified	8,600	9,100		CIR Expert Panel, 1983
	C$_{18}$AE$_{2}$ (Steareth-2)	25% in corn oil unspecified 40% in water	>25,100		21,000 >16,000 >25,000	CIR Expert Panel, 1988
	C$_{18}$AE$_{10}$ (Steareth-10)	unspecified	2,910		>16,000	CIR Expert Panel, 1988
	C$_{18}$AE$_{20}$ (Steareth-20)	25% in water 25% in corn oil unspecified	1,920		2,070 2,100	CIR Expert Panel, 1988
	C$_{12,14}$AE$_{3}$ (Surfonic L24-3)	undiluted			>5000	Texaco Chemical Company, 1992
	C$_{12,14}$AE$_{9}$ (Surfonic L24-9)	undiluted	2627	1789	2140	Texaco Chemical Company, 1991a

Table 5-1. (cont.)

Species	Surfactant	How Dosed	LD$_{50}$ (mg/kg) Male	LD$_{50}$ (mg/kg) Female	LD$_{50}$ (mg/kg) Combined	Reference
Rabbit	$C_{12-14}AE_{6.5}$ (Alfonic 1214-HB-58)	undiluted			1,180	Vista Chemical Company, 1979c
	n-pri-$C_{12-14}AE_n$ (Tergitol 24-L-60N)	undiluted	710[b]	930[b]		Union Carbide Corporation, 1988e
	n-sec-$C_{11-15}AE_9$ (Tergitol 15-S-9)	undiluted	1,190[b]	1,230[b]		Union Carbide Corporation, 1988e
Mouse	n-pri-$C_{12}AE_{11.9}$	not specified			1,170	Zipf et al., 1957
	n-pri-$C_{12}AE_9$ (Laureth-9)	distilled water	3,300			Duprey and Hoppe, 1960
	$C_{12}AE_7$	not specified			1,170	Grubb et al., 1960
	$C_{12}AE_4$ (Laureth-4)	not specified	4,900	7,600		CIR Expert Panel, 1983
Dog	n-pri-$C_{12-13}AE_{6.5}$ (Neodol 23-6.5)	25% in water			>1,650	Benke et al., 1977
Monkey	n-pri-$C_{12-13}AE_{6.5}$ (Neodol 23-6.5)	25% in water			>1,500 (young) >3,300 (young)[d] ≥10,000 (young)[d]	Benke et al., 1977
	n-pri-$C_{14-15}AE_7$ (Neodol 45-7)	33% in water				

[a] No deaths.
[b] μL/kg.
[c] Contains propylene oxide.
[d] Different species.

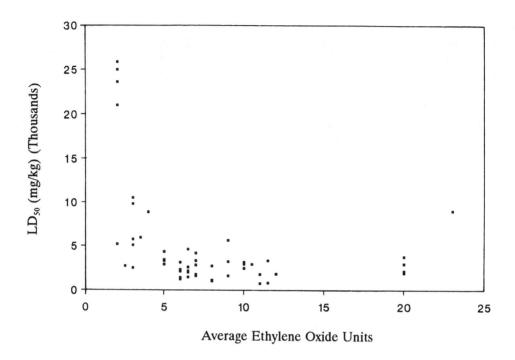

Figure 5-1. Acute Oral LD_{50} Values in the Rat

the following gross abnormalities: congestion of the kidneys, adrenals, liver, lungs and gastrointestinal tract; hemorrhages and proliferation of the gastric mucosa; adhesions of the abdominal viscera; reddish colored urine in the urinary bladder; and paleness of the liver. No emesis occurred in the rat (Benke et al., 1977; Shell Research Limited, 1984a; 1984b; Union Carbide Corporation, 1989; Vista Chemical Company, 1964a; 1967; 1977; 1979d; Texaco Chemical Company, 1991a; 1992). In the case of $C_{9-11}AE_6$ (Neodol 91-6), toxic effects occurred within 1 hour of treatment and death generally occurred between 1 and 6 hours after dosing. Surviving rats recovered within 48 hours (Gingell and Lu, 1991).

Beagle dogs dosed by gavage with 1650 mg/kg of $C_{12-13}AE_{6.5}$ and monkeys dosed with 1500 mg/kg of $C_{12-13}AE_{6.5}$ or up to 6700 mg/kg of $C_{14-15}AE_7$ showed no effects other than emesis and diarrhea (Benke et al., 1977). One of two monkeys administered 10,000 mg/kg of $C_{14-15}AE_7$ died.

Zerkle et al. (1987) studied the oral toxicity of AE in animals and assessed its relevance to humans. Oral administration of $C_{12-13}AE_{6.5}$ to rats and mice resulted in effects suggestive of general anesthesia, but only at doses exceeding those anticipated from accidental ingestion of household products containing AE. Furthermore, in contrast to mice and rats, AE are likely to induce emesis in humans, limiting absorption and the potential for AE-induced effects. Thus, AE-induced anesthetic effects are not predicted in humans following accidental ingestion of household cleaning products containing AE.

Dermal. For acute dermal toxicity studies, the undiluted test material was applied to the clipped, intact skin of the trunk and covered with plastic or an impervious sheeting for 24 hours. After 24 hours, the occlusive material was removed and the remaining material washed away. Observations were made up to 14 days following treatment.

The acute dermal toxicity of undiluted AE when applied to the intact, shaved backs of rabbits, rats, or guinea pigs is low (Table 5-2). For a series of linear primary AE, the 24-hr percutaneous LD_{50} values in rabbits ranged from ~930 mg/kg to 11,800 mg/kg, with most values ranging between 1,000 and 5,000 mg/kg. Two secondary AE were of a similar order of toxicity. Three C_4 propylene oxide-containing AE were nontoxic, with no deaths occurring at doses >16,000 μL/kg. Abrasion of the skin at the site of application did not greatly affect toxicity values. Application of some of the same primary linear AE to the shaved backs of rats resulted in dermal LD_{50} values of >2000 to >5000 mg/kg. Based on these LD_{50} data, the AE surfactants can be classified as slightly to practically nontoxic by the dermal route of application (Weiss, 1980).

For the linear primary and secondary AE, there appeared to be no relationship between dermal toxicity and chemical structure with regard to alkyl chain structure and ethylene oxide content. As noted, the C_4 ethoxylate/propoxylate copolymer surfactants did not exhibit toxicity under similar experimental conditions.

Necropsy findings for rabbits at 14 days post-treatment included damaged skin at the site of application with varying degrees of dryness and sloughing. Some of the animals that succumbed showed pulmonary congestion and blanching, accumulation of fluid in the peritoneal cavity, erosion of the mucosa of the gastrointestinal tract, and occasional liver and kidney damage. Gross necropsies of animals which survived were unremarkable (Vista Chemical Company, 1964c; 1967; 1977; 1978a; 1979a; Gingell and Lu, 1991).

Union Carbide Corporation (1981) has reported a pattern of delayed deaths and visual evidence of lung injury following 24-hr occluded cutaneous application of Tergitol AE to rabbits. Median time till death was 3 to 4.5 days. Microscopic findings of lung injury included bronchopneumonia, pneumonitis, alveolar histiocytosis, edema, congestion, and necrosis. The injuries appeared to be associated with the presence of aspirated feed particles in the lower respiratory tract and lung. It was suspected that stomach material was regurgitated followed by aspiration into the lungs. The relevance of these findings to human health is unknown.

Because of concern over this pattern of lung injury and delayed death in rabbits, Union Carbide Corporation conducted an extensive series of acute cutaneous tests. Tests with two surfactants of moderate acute cutaneous toxicity, $C_{12-14}AE_7$ (Tergitol 24-L-60N) and sec-$C_{11-15}AE_9$ (Tergitol 15-S-9), were repeated, comparing routes of administration and using lower doses, aqueous dilutions, restricted food intakes, longer-term applications, and a second species, the rat. Temporal toxicity was also followed in the rabbit over the 14-day post-exposure period. The same test protocol involving 24-hr occluded exposures was used in all of the studies. All studies were followed by histopathological examinations of the lungs. With the exception of the studies using the rat as the test animal, time to death was delayed for many of the treated

Table 5-2. Acute Dermal Toxicity to Mammals[a]

Surfactant	LD$_{50}$ (mg/kg)			Reference
	Male	Female	Combined	
Rabbit				
n-pri-C$_{12\text{-}13}$AE$_{6.5}$ (Neodol 23-6.5)			>2,000	Benke et al., 1977
n-pri-C$_{14\text{-}15}$AE$_7$ (Neodol 45-7)			~2,000	
C$_{9\text{-}11}$AE$_{2.5}$ (Neodol 91-2.5)			>5,000	Shell Chemical Company, 1984
C$_{9\text{-}11}$AE$_6$ (Neodol 91-6)			5,000 >2,000	Shell Chemical Company, 1984 Shell Development Company, 1981a; Gingell and Lu, 1991
C$_{12\text{-}13}$AE$_3$ (Neodol 23-3)			3,300	Shell Chemical Company, 1984
C$_{12\text{-}13}$AE$_{6.5}$ (Neodol 23-6.5)			2,000	Shell Chemical Company, 1984
C$_{12\text{-}13}$AE$_7$ (Neodol 23-7P7)			>2,000	Shell Oil Company, 1989a
C$_{12\text{-}15}$AE$_3$ (Neodol 25-3)			3,000	Shell Chemical Company, 1984
C$_{12\text{-}15}$AE$_7$ (Neodol 25-7)			2,300 2,500-5,000	Shell Chemical Company, 1984
C$_{12\text{-}15}$AE$_9$ (Neodol 25-9)			2,500 3,400	Shell Chemical Company, 1984
C$_{12\text{-}15}$AE$_{12}$ (Neodol 25-12)			2,500	Shell Chemical Company, 1984

HUMAN SAFETY

Table 5-2. (cont.)

Surfactant	LD$_{50}$ (mg/kg)			Reference
	Male	Female	Combined	
C$_{14,15}$AE$_7$ (Neodol 45-7)			<5,000	Shell Chemical Company, 1984
C$_{14,15}$AE$_{11}$ (Neodol 45-11)			5,000	Shell Chemical Company, 1984
C$_{14,15}$AE$_{13}$ (Neodol 45-13)			5,000	Shell Chemical Company, 1984
C$_{6,8,10}$AE$_{3.1}$ (Alfonic 610-50)			1,500	Vista Chemical Company, 1977
C$_{8,10,12}$AE$_{5.4}$ (Alfonic 1012-60)			>4,000	Vista Chemical Company, 1978a
C$_{10,12,14}$AE$_{2.8}$ (Alfonic 1014-40)			2,000[b]	Vista Chemical Company, 1964a
C$_{10,12,14}$AE$_{6.3}$ (Alfonic 1014-60)			1,000-2,000	Vista Chemical Company, 1967
C$_{12,14,18}$AE$_{3.2}$ (Alfonic 1218-40)			4,000[b]	Vista Chemical Company, 1964c
C$_{12,14}$AE$_{6.5}$ (Alfonic 1214-HB-58)			2,270	Vista Chemical Company, 1979a
C$_{12,14}$AE$_a$ (Alfonic 1214-60-C)			>4,000	Colgate-Palmolive Company, 1979
C$_{12,14}$AE$_{6.3}$ (Tergitol 24-L-45)	2,000[b]	2,240[b]		Union Carbide Corporation, 1987c
C$_{12,14}$AE$_{7.2}$ (Tergitol 24-L-60)	2,380[b]	1,830[b]		Union Carbide Corporation, 1987d

Table 5-2. (cont.)

Surfactant	LD$_{50}$ (mg/kg)			Reference
	Male	Female	Combined	
C$_{12-14}$AE$_{3.5}$ (Tergitol 24-L-3N)	1,190[b]	1,870[b]		Union Carbide Corporation, 1988a
C$_{12-14}$AE$_{5.2}$ (Tergitol 24-L-25N)	1,780[b]	1,620[b]		Union Carbide Corporation, 1987f
C$_{12-14}$AE$_{6.4}$ (Tergitol 24-L-50N)	1,410[b]	1,870[b]		Union Carbide Corporation, 1987h
C$_{12-14}$AE$_{7}$ (Tergitol 24-L-60N)	930[b]	1,780[b]		Union Carbide Corporation, 1986d
C$_{12-14}$AE$_{8}$ (Tergitol 24-L-75N)	1,230[b]	3,250[b]		Union Carbide Corporation, 1987i
C$_{12-14}$AE$_{11.6}$ (Tergitol 24-L-98N)	1,120[b]	1,190[b]		Union Carbide Corporation, 1987g
C$_{12-16}$AE$_{1.6}$ (Tergitol 26-L-1.6)	6,500[b]	9,500[b]		Union Carbide Corporation, 1987e
C$_{12-16}$AE$_{3}$ (Tergitol 26-L-3)	2,380[b]	2,140[b]		Union Carbide Corporation, 1986e
C$_{12-16}$AE$_{5}$ (Tergitol 26-L-5)	1,780[b]	3,250[b]		Union Carbide Corporation, 1987a
n-sec-C$_{12-15}$AE$_{3}$ (Tergitol 15-S-3)	4,800[b]	4,900[b]		Union Carbide Corporation, 1986b
n-sec-C$_{12-15}$AE$_{9}$ (Tergitol 15-S-9)	1,120[b]	2,380[b]		Union Carbide Corporation, 1986c
C$_4$H$_9$O(CH$_2$CHCH$_3$)$_x$(CH$_2$CH$_2$O)$_y$H[c] (Tergitol XH)	>16,000[b,d]	>16,000[b,d]		Union Carbide Corporation, 1986f

Table 5-2. (cont.)

Surfactant	LD$_{50}$ (mg/kg)			Reference
	Male	Female	Combined	
$C_4H_9(CH_2CH_2CH_3)_x(CH_2CH_2O)_yH^c$ (Tergitol XD)	>16,000[b,d]	>16,000[b,d]		Union Carbide Corporation, 1986g
$C_4H_9O(CH_2CHCH_3O)_x(CH_2CH_2O)_yH^c$ (Tergitol XJ)	>16,000[b,d]	>16,000[b,d]		Union Carbide Corporation, 1986h
$C_{8-10}H_{17-21}O(CH_2CH_2O)_x(CH_2CHCH_3)_yH^c$ (Tergitol 80-L-50N)	1,680[b]	3,250[b]		Union Carbide Corporation, 1987b
$C_{12-14}O(CH_2CH_2O)_x(CH_2CHCH_3O)_yH^c$ (Tergitol Mechanical Dish Surfactant)	11,300[b]	5,700[b]		Union Carbide Corporation, 1986a
$C_{12-14}AE_3$ (Surfonic L24-3)			>3000	Texaco Chemical Company, 1990a
$C_{12-14}AE_9$ (Surfonic L24-9)			>3000	Texaco Chemical Company, 1991b
Rat				
$C_{9-11}AE_{2.5}$ (Neodol 91-2.5)			>2,000	Shell Chemical Company, 1984
$C_{9-11}AE_8$ (Neodol 91-8)			>4,000	Shell Chemical Company, 1984
$C_{12-15}AE_7$ (Neodol 25-7)			>2,000[e]	Shell Chemical Company, 1984
$C_{14-15}AE_7$ (Neodol 45-7)			>5,000	Shell Research Limited, 1984a
$C_{14-15}AE_{11}$ (Neodol 45-11)			>2,000[e]	Shell Research Limited, 1984b

Table 5-2. (cont.)

Surfactant	LD$_{50}$ (mg/kg)			Reference
	Male	Female	Combined	
$C_{12,14}AE_{6.5}$ (Alfonic 1214-HB-58)	11,800[f]	10,150[f]	10,980[f]	Vista Chemical Company, 1979b
$C_{12,14}AE_7$ (Tergitol 24-L-60N)	11,300[b]	11,300[b]		Union Carbide Corporation, 1988e
n-sec-$C_{11-15}AE_9$ (Tergitol 15-S-9)	>16,000[b]	14,200[b]		Union Carbide Corporation, 1988f
Guinea pig				
n-pri-$C_{12,13}AE_{6.5}$ (Neodol 23-6.5)			>2,000	Benke et al., 1977
n-pri-$C_{14,15}AE_7$ (Neodol 45-7)			>2,000	

[a] All AE were applied undiluted.
[b] μL/kg.
[c] Contains propylene oxide.
[d] No deaths.
[e] Maximum dose that could be applied.
[f] Applied to abraded skin.

rabbits. A significant finding in all of the rabbit studies was the presence of lung lesions including pneumonia and feed particles in some of the treated animals.

When undiluted $C_{12\text{-}14}AE_7$ (Tergitol 24-L-60N) was applied percutaneously to rabbits at doses <LD_{50}, 0.075 to 0.25 mL/kg (74.3 to 248 mg/kg), and food was not restricted, histopathological examinations of the lungs revealed pneumonia and feed particle infiltrates at both the highest and lowest doses. (The dosage of 0.075 mL/kg was the lowest dose tested in order to determine a NOEL.) No deaths were observed. One of six rabbits injected subcutaneously had bronchoalveolar necrosis and feed particle infiltration of the lung. These effects were not present following oral or subcutaneous administration to the rat (Union Carbide Corporation, 1989). On the basis of actual dosed amounts, 10% and 25% aqueous dilutions of $C_{12\text{-}14}AE_7$ (Tergitol 24-L-60N) were similar in toxicity to the undiluted material (see Table 5-2), with LD_{50} values of 950 mg/kg (males) and >1600 mg/kg (females) for the 10% dilution and 1300 mg/kg (males) and 2600 mg/kg (females) for the 25% dilution. Histological examinations revealed lung lesions associated with feed particles (Union Carbide Corporation, 1988d).

Female rabbits received a single cutaneous dose of either of the undiluted surfactants at doses of 0, 250, 500, or 1000 mg/kg body weight (15 animals/group) and toxicity was evaluated following interim sacrifices on days 1, 3, 6, 9, and 13 (Union Carbide Corporation, 1988j). Seven animals from the intermediate and high-dose groups died between study days 0 and 6. Treatment resulted in gross and microscopic skin lesions, clinical signs of toxicity, decreased food consumption and body weight, increased lung weight, and tracheal and lung lesions. Tracheal and lung lesions were present after one day of treatment and became more severe by day 3. The lesions were associated with the presence of feed particles.

In two separate studies, $C_{12\text{-}14}AE_7$ (Tergitol 24-L-60N) and sec-$C_{11\text{-}15}AE_9$ (Tergitol 15-S-9), were administered percutaneously to rabbits maintained on a restricted diet (Union Carbide Corporation 1988g; 1988h). The test rabbits were given 50 g of feed the first day followed by 75 g the second day and 100 g on the third day. On the following days 200-250 g of feed were provided. As in the earlier studies, there were delayed deaths. Lung lesions which included pneumonia and contained feed particles were also present, indicating no effect of feed restriction on aspiration.

Over a nine-day period, male and female white rabbits were exposed to $C_{12\text{-}14}AE_7$ (Tergitol 24-L-60N) by occluded cutaneous exposures at doses of 0, 100, 250, or 500 mg/kg body weight/day (0, 0.1, 0.25, or 0.5 mL/kg/day) (Union Carbide Corporation, 1988k). A dose-related increase in lung lesions associated with the presence of feed particles was observed. Toxicity appeared to be cumulative when compared to lung lesions following a single dose. Hematologic and serum chemistry changes as well as moderate to severe local skin lesions were present at doses of 100 mg/kg/day and above.

Because of the delayed deaths and lung lesions in rabbits, the toxicity of both surfactants was followed over a 14-day period in a second species, the rat (Union Carbide Corporation, 1988i). Single cutaneous applications at doses of 0, 1000, 2000, or 8000 mg/kg body weight produced no treatment-related deaths or gross or microscopic lung lesions. Some alterations in motor and reflex functions in some of the treated animals during the first 48 hours after exposure were observed, but no clear dose-response relationship was apparent.

Percutaneous application of $C_{12-14}AE_7$ (Tergitol 24-L-60N) at volumes of 0.5, 1, and 2 mL/kg body weight (495, 990, and 1980 mg/kg/day) for nine days (over an 11-day dosing period) to rats produced no treatment-related pathological changes. There was no evidence of systemic toxicity, including neurotoxicity (Union Carbide Corporation, 1990).

In conclusion, large doses of AE dermally applied for 24 hours to rabbits resulted in severe skin irritation and ataxia which may be a pharmacologic effect of AE. At high doses, lung lesions also were observed which were apparently caused secondary to food particles in the lungs. Inhalation or aspiration of feed particles may have occurred as a secondary effect of a pharmacologic disturbance of the swallowing mechanism (See Section 10. Anesthetic and Analgesic Effects).

Inhalation. AE surfactants were not acutely toxic to rats at concentrations less than or equal to their saturated vapor concentrations in air. Acute toxic thresholds were reached only when animals were exposed to the undiluted test chemical in the form of a respirable mist or aerosol. Under these conditions, 1- or 4-hr inhalation LC_{50} values ranged from 1.5 to 20.7 mg/L (Table 5-3). In some studies no mortalities (1-hr LC_0) occurred at concentrations as high as 52 mg/L. A 14-day observation period followed the exposures. Concentrations were determined by a gravimetric method as indicated in Table 5-3 or were given as nominal concentrations based on the delivery flow concentrations of the undiluted material (calculated from the weight of the undiluted surfactant used and the volume of air passed through the nebulizer chamber).

Acutely exposed animals exhibited one or more of the following symptoms: labored breathing, rales, inactivity, and bloody nasal discharge. Gross necropsies revealed corneal opacities, congestion and mottling of the lungs, and in some cases, paleness or congestion of the liver, kidneys, and adrenals (Benke et al., 1977; Vista Chemical Company, 1978b; 1979g; 1981; 1982). Necropsy findings in survivors were unremarkable.

Grubb et al. (1960) exposed nine rats to steam generated from a 20% aqueous solution of n-pri-$C_{12}AE_7$, two hours/day, for ten days. Two rats had mild laryngeal irritation and one of the two had diffuse peripheral hemorrhages of the lungs. No further details were given. In an *in vitro* study, concentrations of 0.002% (200 mg%) of $C_{12}AE_3$ did not inhibit the ciliary action of rat tracheal tissue (Grubb et al., 1960).

Other Routes. Studies on the acute toxicity of AE by the subcutaneous, intravenous, intrapleural, and intraperitoneal routes of administration were reviewed by Arthur D. Little (1977). AE administered by these routes are generally more toxic than when administered by the oral route, with LD_{50} values for linear primary AE ranging from ≤ 100 mg/kg (intravenous, several species) to 1050 mg/kg (subcutaneous, mice). Intraperitoneal doses to mice, rats and monkeys in the lethal range produced responses of convulsions and/or depression (Benke et al., 1977). These are not common routes of exposure for humans.

Table 5-3. Acute Inhalation Toxicity to Rats

Chemical	Duration/ Test	Concentration (mg/L)	Reference
$C_{12-13}AE_{6.5}$ (Neodol 23-6.5)	4-hr LC_{50}	1.5-3.0[a]	Benke et al., 1977
$C_{14-15}AE_7$ (Neodol 45-7)	4-hr LC_{50}	1.5-3.0[a]	Benke et al., 1977
$C_{9-11}AE_5$ (Dobanol 91-5)	4-hr LC_0	0.22[a,b]	Shell Research Limited, 1980a
$C_{6,8,10}AE_{3.1}$ (Alfonic 610-50)	4-hr LC_0	8	Vista Chemical Company, 1981
$C_{8,10,12}AE_{5.4}$ (Alfonic 1012-60)	4-hr LC_0	5	Vista Chemical Company, 1978b
$C_{10,12,14}AE_{2.8}$ (Alfonic 1014-40)	1-hr LC_0	50[a]	Vista Chemical Company, 1964a
$C_{10,12,14}AE_{6.3}$ (Alfonic 1014-60)	1-hr LC_0	52	Vista Chemical Company, 1967
$C_{12,14}AE_2$ (Alfonic 1214-HA-30)	4-hr LC_0	>1.2[a]	Vista Chemical Company, 1982
$C_{12,14}AE_{6.5}$ (Alfonic 1214-HB-58)	1-hr LC_{50}	20.7	Vista Chemical Company, 1979g
$C_{12,14}AE_{10.6}$ (Alfonic 1214-70)	1-hr LC_{50}	9.4	Vista Chemical Company, 1984
$C_{12,14,16}AE_{1.3}$ (Alfonic 1216-22)	1-hr LC_0	6.3	Vista Chemical Company, 1980a
$C_{12,14}AE_3$ (Alfonic 1412-40)	1-hr LC_0	>50	Vista Chemical Company, 1984
$C_{12,14}AE_7$ (Alfonic 1412-60)	4-hr LC_{50}	>6.6	Vista Chemical Company, 1985
$C_{12,14,16,18}AE_{3.2}$ (Alfonic 1218-40)	1-hr LC_0	70	Vista Chemical Company, 1964c
$C_{12,14,16,18}AE_n$ (Alfonic 1218-70)	1-hr LC_{50}	3.4-5.4	Vista Chemical Company, 1984
$C_{12-14}EO/PO$[c] (Tergitol Mechanical Dish Surfactant)	6-hr LC_0	"substantially saturated vapor"	Union Carbide Corporation, 1986a

[a] Measured concentration.
[b] Saturated vapor concentration in air.
[c] Contains propylene oxide.

$C_{18}AE_2$ (steareth-2) was administered intraperitoneally as a 10% weight/volume solution in normal saline to groups of five female and five male Wistar SPF rats. Doses ranged from 500 to 1260 mg/kg. Following a 14-day observation period, an LD_{50} value of 760 mg/kg was calculated. In another study, $C_{18}AE_{20}$ (steareth-20) was similarly administered at doses of 126 to 251 mg/kg. An LD_{50} value of 190 mg/kg was calculated (CIR Expert Panel, 1988).

A 1% weight/volume solution of $C_{18}AE_2$ (steareth-2) in propylene glycol was administered intravenously to fasted groups of five male and five female Wistar SPF rats. Doses ranged from 15.9 to 100 mg/kg. Following a 7-day observation period, an LD_{50} value of 41 mg/kg was calculated. Using a similar protocol, $C_{18}AE_{20}$ (steareth-20) was administered in a 10% weight/volume solution of sodium chloride. An LD_{50} value of 164 mg/kg was calculated (CIR Expert Panel, 1988).

In order to evaluate the direct effects of surfactants on the lungs and to assess the potential hazards of aspiration, 10 male and 10 female Sprague-Dawley rats were administered a number of surfactants and PEGs by single endotracheal injection (Tyler et al. 1988; Union Carbide Corporation, 1988b; 1988c). Approximately 0.5 mL of the test material was administered undiluted or as a mixture in saline. In order to evaluate lung lesions, 2 males and 2 females were sacrificed at one, two and three days after dosing; survivors were sacrificed at 14 days. The minimum lethal doses for $C_{12-14}AE_{6.3}$ (Tergitol 24-L-45), $C_{12-14}AE_{11.6}$ (Tergitol 14-L-98N), $C_{12-16}AE_3$ (Tergitol 26-L-3), br-C_9AE_6 (Tergitol TMN-6), sec-$C_{11-15}AE_3$ (Tergitol 15-S-3), sec-$C_{11-15}AE_9$ (Tergitol 15-S-9), and $C_{12,14}$EO-PO (Tergitol Mechanical Dish Surfactant) ranged from 0.02 to 0.32 mL/kg (~20 to 320 mg/kg). The PEGs were less toxic than the Tergitols, with minimum lethal doses of >1000 mg/kg. Lung weights expressed as absolute weights and as a percentage of final body weights were elevated, primarily in females. Gross and histologic changes included color change associated with inflammatory infiltrates, edema, fibrinopurulent pneumonia, fibrosis and atelectasis accompanied with mucus accumulation. Recovery took place, with only a few lesions remaining by day 14.

Zerkle et al. (1987) state that ethoxylates can be naturally emetic in man, making aspiration a possible route of lung exposure. However, intratracheal instillation used by Tyler et al. (1988) above may not be the best method for predicting aspiration potential. Alcohol ethoxylates have not been evaluated by the more appropriate Gerarde (1963) technique for aspiration hazard. Osterberg et al. (1976) appraised existing methodology for aspiration toxicity testing and concluded that the Gerarde technique in which the material is placed in the mouth is superior for predicting aspiration hazard and toxicity. The Gerarde technique allows a larger dose to enter the lungs and is consistent in dose-response. Low viscosity and low surface tension increase the aspiration hazard of chemicals.

2. Subchronic Exposures

During short-term and subchronic exposures lasting up to 91 days, oral doses of up to 500 mg/kg/day and dermal exposures of 50 mg/kg/day produced no significant effects.

Oral. In early studies, repeated oral exposures of AE at low doses produced little or no effects on test animals (Arthur D. Little, 1977). Rats fed a diet containing 0.0471, 0.2355, or 1.1775% of n-pri-$C_{12}AE_7$ for four weeks showed no adverse effects (Grubb et al., 1960). Using a food factor of 0.05 (fraction of body weight that is consumed per day as food) for rats (U.S. EPA, 1985), these intakes convert to doses of 23.6, 117.8, and 588.8 mg/kg/day. Rats tolerated doses of 195 and 290 mg/kg/day of n-pri-$C_{12}AE_9$ when administered 18 times over a 22 day period; however, at a dose of 780 mg/kg/day, two of 10 rats died. At this dose, symptoms of inactivity, dyspnea, and excess salivation were present. The five-day LD_{50} was 1190 mg/kg/day (Berberian, 1965a). Brown and Benke (1977) administered up to 500 ppm (25 mg/kg/day) of $C_{12-13}AE_{6.5}$ in the diet to rats for 91 days without adverse effects. Rats fed 1000, 5000, or 10,000 ppm (50, 250, or 500 mg/kg/day) of a commercial formulation had reduced final body weights, growth values and food consumption. These effects may have been related to the unpalatability of the food. Rats fed diets containing 1000, 5000, or 10,000 ppm (50, 250, or 500 mg/kg/day) of $C_{14-15}AE_7$ for 91 days exhibited no adverse effects.

$C_{14-15}AE_7$ (Dobanol 45-7) was fed to groups of 12 male and 12 female rats at dietary concentrations of 0, 300, 1000, 3000 or 10,000 ppm (0, 15, 50, 150, or 500 mg/kg/day) for 13 weeks (Shell Research Limited, 1982a). No deaths occurred in the treated groups, and no effects on the general health and behavior of treated rats were evident. Mean body weights of male rats fed 10,000 ppm (500 mg/kg/day) and of females fed 10,000 and 3000 ppm (500 and 150 mg/kg/day) were significantly lower than controls throughout the exposure period; reduced food intake and increased food spillage were observed in these groups. After correction for reduced terminal body weights, there were significant increases in liver weights of males in the 3000 and 10,000 ppm (150 and 500 mg/kg/day) groups and in females in the 1000 ppm (50 mg/kg/day) and higher groups. Kidney weight of females in the 1000 ppm (50 mg/kg/day) group, and spleen weight of males in the 10,000 ppm (500 mg/kg/day) group were also significantly increased.

At the high dose there were changes in serum and blood chemistry. In rats fed 10,000 ppm (500 mg/kg/day), plasma concentrations of urea, potassium, chloride, calcium and cholesterol and the plasma activity of alkaline phosphatase were increased in one or both sexes when compared to controls. Also in one or both sexes fed 10,000 ppm, hematological alterations including increases in total leukocyte and absolute lymphocyte numbers and decreases in neutrophils, mean cell volume, and mean cell hemoglobin were observed when compared to controls. Males in the two higher dose groups showed shortened prothrombin times. No compound-related gross or histopathological lesions were identified at any dose level. The changes reported were considered minor and not of toxicological significance.

In 91-day feeding studies with ethoxylated tallow alcohols, rats were administered either $C_{16,18}AE_9$ at dietary levels of 0, 0.25, 0.5, or 1.0% (0, 125, 250, or 500 mg/kg/day) or $C_{16,18}AE_{20}$ at dietary levels of 0, 0.01, 0.1 or 1.0% (0, 50, or 500 mg/kg/day). No treatment-related effects were noted for either chemical at any dose (Procter & Gamble Co., 1985).

Dermal. The subchronic toxicity of dermally applied $C_{9-11}AE_6$ (Neodol 91-6) was studied in Fischer 344 rats at concentrations of 0.0, 1.0, 10.0, or 25.0% in deionized water (w/v) (0.5 mL/kg) (Shell Development Company, 1985a; Gingell and Lu, 1991). These concentrations were equivalent to doses of 0, 5, 50, or 125 mg/kg/treatment. Groups of 30 male and 30 female rats were treated on their shaved backs, three times a week for 13 weeks. No mortalities occurred as a result of the treatment. Clinical signs of toxicity were absent as were skin edema and erythema. At the two higher doses, there was some dryness and flaking of the skin at the site of application; males exhibited a yellow to brown skin discoloration over the treatment area at all doses including the control group, whereas in females, the presence of this discoloration was observed primarily in the two higher dose groups.

There were some statistically significant differences in treated rats compared to controls. For example, relative kidney weights were slightly higher in both male and female rats treated with the 25% concentration. However, no pathologic lesions were noted. Serum phosphorus, potassium, and calcium were higher in female rats, but there was no dose-response relationship.

Rabbits treated percutaneously for 91 days with ethoxylated coconut alcohols ($C_{12,14,16}AE_{12}$ or $C_{12,14,16}AE_{20}$) at a dose level of 50 mg/kg/day showed no treatment-related effects except for mild to moderate irritation at the site of exposure. Likewise no effects other than mild skin irritation were observed with ethoxylated tallow alcohol ($C_{16,18}AE_{30}$) at a dose of 50 mg/kg day when the material was applied as a 2.5% aqueous solution (2 mL/kg/day) (Procter & Gamble Co., 1985).

3. Chronic Exposures

Chronic oral and dermal exposures to AE surfactants produced no significant treatment-related effects.

Oral. In a Procter & Gamble report cited by Goyer et al. (1981), Sprague-Dawley rats were fed $C_{12-13}AE_{6.5}$ in the diet for 104 weeks at levels of 0, 0.1, 0.5, or 1.0% (0, 50, 250, or 500 mg/kg/day). Reduced food consumption at the higher dose levels (0.5 and 1.0% for females and 1.0% for males) resulted in a lower body weight gain compared to the control group. After 104 weeks, elevated organ to body weight ratios were observed for females fed the 0.5 and 1.0% dose (liver, kidney, and brain), females fed the 1% dose (heart), and males fed the 1% dose (liver). In male rats a dose-related focal myocarditis was the only pathology observed. Although this is a common spontaneous type of lesion in aging rats, incidences were higher in the treated group than in the control group. No tumors or other treatment-related lesions were observed.

In a second feeding study, Charles River rats were fed $C_{14-15}AE_7$ at the above dose levels for two years (Procter & Gamble Co., 1981). As in the above study, a reduced body weight gain, attributed to diet palatability, was observed. Behavior, appearance, survival, and hematological and biochemical parameters were similar to controls. In females fed the 1%

concentration, the absolute weights of liver, kidneys, heart, and thyroid/parathyroid glands were decreased and in males fed the 1% concentration, the absolute weights of brain and adrenals were decreased. Focal myocarditis appeared to be dose-related at 12 months, but at 24 months, when considering the severity of the lesions, the effect was not treatment related.

Rats exposed for two years and dogs exposed for one year to a polyethylene glycol (PEG 200) at a level of 2% (1000 mg/kg/day) in the diet showed no adverse effects (Smyth et al., 1950; 1955).

Dermal. In an 18-month study, 0.1 ml of an aqueous solution of 0, 0.2, 1.0 or 5.0% $C_{12-13}AE_{6.5}$ was applied to the backs of ICR Swiss mice three times per week (Procter & Gamble Co., 1981). No treatment-related lesions were observed.

4. Acute Irritation

The skin and eye irritation of AE is concentration dependent; undiluted materials are usually moderate to severe skin and eye irritants, whereas 1% aqueous solutions are mildly irritating and 0.1% aqueous solutions are usually not irritating. Actual use concentrations based on detergent formulations (Galante and Dillan, 1981) and directions for use (Shell Chemical Company, 1991) are ≤0.04%. Thus, the concentrations used in animal studies usually far exceed actual use concentrations of household products.

Primary Skin Irritation. The primary skin irritation potential of AE has been tested based on the patch test method of Draize et al. (1944). In this test 0.5 mL of the undiluted or diluted test material is applied to the shaved backs of six albino rabbits. In a later Draize procedure (Draize, 1959), the solution is placed on the gauze patch. Usually four sites, two intact and two abraded are prepared. Patches are taped over the exposure sites and the trunk of the animal is wrapped with plastic. After 24 hours, the wrappings are removed and the sites are scored for irritation. The backs are then washed and the sites are rescored after 72 hours. The sites are scored for both erythema (reddening) and edema (swelling) on a basis of weighted scores of 0 (non-irritating) to 4 (severe reaction) for a total score, referred to as the Primary Irritation Index, of up to 8. The Primary Irritation Index is the average score of the test group as a whole. Depending on the severity of the skin reaction, the sites may be examined up to 14 days post-treatment.

Rabbit skin irritation categories of undiluted and aqueous dilutions of AE surfactants are listed in Table 5-4. When tested undiluted, most test materials were moderately to severely irritating. An aqueous dilution of 1% was minimally to mildly irritating and an aqueous dilution of 0.1% was generally non-irritating.

There are some variations in the test methods used in Table 5-4 that may modify the reported irritancy. These include applying the test material for less than 24 hours, covering the test material with a semi-occluded wrap (which may allow some evaporation) instead of an impervious wrap, and scoring the skin response at additional times. In a few studies, test

Table 5-4. Rabbit Skin Irritation

Surfactant/ Trade Name	Concentration [a,b]	Irritation Category	Reference
$C_{6,8,10}AE_{3.1}$ (Alfonic 610-50)	undiluted	moderate	Vista Chemical Company, 1977
$C_{8,10,12}AE_{5.4}$ (Alfonic 1012-60)	undiluted	mild	Vista Chemical Company, 1964b
$C_{10,12,14}AE_{2.7}$ (Alfonic 1014-40)	undiluted	moderate	Vista Chemical Company, 1964a
$C_{10,12,14}AE_{6.3}$ (Alfonic 1014-60)	undiluted	moderate	Vista Chemical Company, 1967
$C_{12,14}AE_{6.5}$ (Alfonic 1214-HB-58)	undiluted	moderate	Vista Chemical Company, 1979e
$C_{12,14,16}AE_{1.3}$ (Alfonic 1216-22)	undiluted	moderate	Vista Chemical Company, no date
$C_{12,14,16,18}AE_{3.2}$ (Alfonic 1218-40)	undiluted	slight	Vista Chemical Company, 1964c
$C_{9-11}AE_{2.5}$ (Neodol 91-2.5)	undiluted	severe	Shell Chemical Company, 1984
$C_{9-11}AE_{5}$ (Neodol 91-5)	undiluted 10% 1% 0.1%	severe slight minimal non-irritating	Shell Chemical Company, 1984
$C_{9-11}AE_{6}$ (Neodol 91-6)	undiluted 1% 0.1%	severe slight non-irritating	Shell Development Company, 1981d; Shell Chemical Company, 1984; Gingell and Lu, 1991
$C_{9-11}AE_{8}$ (Neodol 91-8)	undiluted 10% 1% 0.1%	severe moderate mild minimal	Shell Chemical Company, 1984
$C_{12-13}AE_{3}$ (Neodol 23-3)	undiluted	severe	Shell Chemical Company, 1984
$C_{12-13}AE_{6.5}$ (Neodol 23-6.5)	undiluted 10% 1% 0.1%	severe moderate mild non-irritating	Shell Chemical Company, 1984
$C_{12-13}AE_{7}$ (Neodol 23-7P7)	undiluted	moderate	Shell Oil Company, 1989c
$C_{12-15}AE_{3}$ (Neodol 25-3)	undiluted	moderate-extreme	Shell Chemical Company, 1984
$C_{12-15}AE_{7}$ (Neodol 25-7)	undiluted 10% 1% 0.1%	moderate moderate mild mild	Shell Chemical Company, 1984 Shell Research Limited, 1984a

Table 5-4. (cont.)

Surfactant/ Trade Name	Concentration [a,b]	Irritation Category	Reference
$C_{12-15}AE_9$ (Neodol 25-9)	undiluted 1% 0.1%	severe non-irritating non-irritating	Shell Chemical Company, 1984
$C_{12-15}AE_{12}$ (Neodol 25-12)	50%	minimal	Shell Chemical Company, 1984
$C_{14-15}AE_7$ (Neodol 45-7)	undiluted 10% 1% 0.1%	severe moderate mild minimal	Shell Chemical Company, 1984
$C_{14-15}AE_{11}$ (Neodol 45-11)	undiluted 10% 1% 0.1%	mild-severe moderate-severe slight non-irritating	Shell Chemical Company, 1984; Shell Research Limited, 1984b
$C_{14-15}AE_{13}$ (Neodol 45-13)	undiluted	moderate	Shell Chemical Company, 1984
$C_{14-15}AE_{18}$ (Neodol 45-18)	undiluted 10% 1% 0.1%	mild slight minimal non-irritating	Shell Chemical Company, 1984
$C_{12,14,16,18}AE_9$ (Coconut alcohol EO_9)	undiluted	moderate	Shell Chemical Company, 1984
n-pri-$C_{12-14}AE_{6.3}$ (Tergitol 24-L-45)	undiluted	severe	Union Carbide Corporation, 1987c
n-pri-$C_{12-14}AE_{7.2}$ (Tergitol 24-L-60)	undiluted	moderate	Union Carbide Corporation, 1987d
n-pri-$C_{12-14}AE_{3.5}$ (Tergitol 24-L-3N)	undiluted	moderate	Union Carbide Corporation, 1988a
n-pri-$C_{12-14}AE_{5.2}$ (Tergitol 24-L-25N)	undiluted	moderate	Union Carbide Corporation, 1987f
n-pri-$C_{12-14}AE_{6.4}$ (Tergitol 24-L-50N)	undiluted	moderate	Union Carbide Corporation, 1987h
n-pri-$C_{12-14}AE_7$ (Tergitol 24-L-60N)	undiluted	moderate	Union Carbide Corporation, 1986d
n-pri-$C_{12-14}AE_8$ (Tergitol 24-L-75N)	undiluted	severe	Union Carbide Corporation, 1987i
n-pri-$C_{12-14}AE_{11.6}$ (Tergitol 24-L-98N)	undiluted	moderate	Union Carbide Corporation, 1987g
n-pri-$C_{12-16}AE_{1.6}$ (Tergitol 26-L-1.6)	undiluted	moderate	Union Carbide Corporation, 1987e

Table 5-4. (cont.)

Surfactant/ Trade Name	Concentration [a,b]	Irritation Category	Reference
n-pri-$C_{12-16}AE_3$ (Tergitol 26-L-3)	undiluted	severe	Union Carbide Corporation, 1986e
n-pri-$C_{12-16}AE_5$ (Tergitol 26-L-5)	undiluted	moderate	Union Carbide Corporation, 1987a
n-sec-$C_{11-15}AE_3$ (Tergitol 15-S-3)	undiluted	severe	Union Carbide Corporation, 1986b
n-sec-$C_{11-15}AE_9$ (Tergitol 15-S-9)	undiluted	severe	Union Carbide Corporation, 1986c
$C_4H_9(CH_2CHCH_3O)_x(CH_2CH_2O)_yH$ [c] (Tergitol XD)	undiluted	moderate	Union Carbide Corporation, 1986g
$C_4H_9O(CH_2CHCH_3O)_x(CH_2CH_2O)_yH$ [c] (Tergitol XH)	undiluted	moderate	Union Carbide Corporation, 1986f
$C_4H_9O(CH_2CHCH_3O)_x(CH_2CH_2O)_yH$ [c] (Tergitol XJ)	undiluted	moderate	Union Carbide Corporation, 1986h
$C_{8-10}H_{17-21}O(CH_2CH_2O)_x(CH_2CHCH_3O)_yH$ [c] (Tergitol 80-L-50N)	undiluted	moderate	Union Carbide Corporation, 1987b
$C_{12-14}O(CH_2CH_2O)_x(CH_2CHCH_3O)_yH$ [c] (Tergitol Mechanical Dish Surfactant)	undiluted	no reaction	Union Carbide Corporation, 1986a
$C_{12-14}AE_3$ (Surfonic L24-3)	undiluted	slight	Texaco Chemical Company, 1990b
$C_{12-14}AE_9$ (Surfonic L24-9)	undiluted	non-irritating	Texaco Chemical Company, 1991c
$C_{12}AE_4$ (Laureth-4)	undiluted	very slight	CIR Expert Panel, 1983
$C_{12}AE_{23}$ (Laureth-23)	undiluted	non-irritating	CIR Expert Panel, 1983
$C_{18}AE_2$ (Steareth-2)	60% 40% 10%	mild mild non-irritating	CIR Expert Panel, 1988
$C_{18}AE_{10}$ (Steareth-10)	60%	mild	CIR Expert Panel, 1988
$C_{18}AE_{20}$ (Steareth-20)	60%	non-irritating	CIR Expert Panel, 1988

Table 5-4. (cont.)

Surfactant/ Trade Name	Concentration [a,b]	Irritation Category	Reference
$C_{12}AE_9$ (Lauryl alcohol EO_9)	undiluted 20% 15%	slight - moderate slight - non-irritating slight - non-irritating	Berberian et al., 1965a
n-pri-$C_{12}AE_7$	100 - 5120 mg/kg	severe	Grubb et al., 1960

[a]Volume of 0.5 mL.
[b]Dilutions are aqueous solutions.
[c]Contains propylene oxide.

material was applied to the intact skin only. In some of the studies reported by the CIR Expert Panel (1988), the Draize 1959 procedure was used. The test material was removed after 6 hours in the Union Carbide Corporation studies. The Shell Chemical Company (1984) assigned descriptive skin ratings to the Draize Primary Irritation Index values. The classifications and scores were non-irritating, 0; minimally irritating, 0.0-0.5; slightly irritating, 0.6-1.5; mildly irritating, 1.6-3.0; moderately irritating, 3.1-5.0; severely irritating, 5.1-6.5; and extremely irritating, 6.6-8.0.

Additional studies reported irritation under longer-term use conditions. Applications of 50 mg/kg/day, five days/week, of $C_{12-13}AE_{6.5}$ to the abraded (four weeks) or intact (13 weeks) skin of rabbits produced slight to moderate skin irritation. Application of $C_{14-15}AE_7$ under the same conditions produced pronounced irritation with papular eruption (Brown and Benke, 1977).

Guinea pigs immersed in a 10% aqueous solution of $C_{14-15}AE_7$, four hours/day, for five consecutive days, suffered fissured skin and dermatologic changes characterized by hyperkeratosis, acanthosis, and infiltration of the superficial dermis (Brown and Benke, 1977).

Rabbits treated percutaneously for 91 days with $C_{12,14,16}AE_{12}$ or $C_{12,14,16}AE_{20}$ (ethoxylated coconut alcohols) at a dose level of 50 mg/kg/day or with $C_{16,18}AE_{30}$ (ethoxylated tallow alcohol) at the same dose level but as a 2.5% solution showed no treatment-related effects except for mild to moderate irritation at the site of exposure (Procter & Gamble Co., 1985).

Eye Irritation. The potential of AE to irritate the eye has been assessed using the standard Draize test (Draize, 1959; Draize et al., 1944) with the albino rabbit as the animal model. A volume of 0.1 mL of the test material is introduced into the conjunctival sac of one eye while the other eye serves as the control. The unwashed eye is then examined for ocular injury at one hour and at daily intervals up to 14 days after instillation. The cornea, iris, and conjunctiva are graded on the basis of 0 (no visible response) to 4 (severe response). A solution of ophthalmic fluorescein is used for detection of corneal lesions. Descriptive eye irritation ratings may be assigned to the Draize indices as noted below.

In the studies in Table 5-5, the irritation categories range from 0 to 4 (Union Carbide Corporation) or from 0 to 110 (Shell, Texaco, and Vista Chemical Companies), the latter being an expanded descriptive rating which considers incidence and persistence of the injury (Kay and Calandra, 1962; Lehman, 1965). Group mean scores for corneal opacity, iritic effects, and conjunctival redness and chemosis were calculated at 24, 48, and 72 hours and the results averaged. If the mean score for any parameter exceeded the final group score, the higher irritant category was assigned to the chemical.

Although many of the surfactants listed in Table 5-5 were severely irritating to the rabbit eye when tested undiluted, concentrations of 0.1% were generally non-irritating. As noted earlier, actual use concentrations of $\leq 0.04\%$ indicate a low hazard potential.

In additional tests with undiluted surfactants, washing the eyes 20-30 seconds after instillation, a simulation of therapy under accident conditions, reduced the toxicity rating. For the tested compounds, irritation categories generally changed from severe to mild-moderate (Shell Chemical Company, 1984; Vista Chemical Company, 1979f; 1979h). For $C_{18}AE_2$, $C_{18}AE_{10}$, and $C_{18}AE_{20}$, a 20 mL water rinse two seconds after instillation reduced irritation of 60% aqueous solutions from minimally irritating to non-irritating (CIR Expert Panel, 1988).

The eye irritancy of three synthetic and one natural fatty alcohol-based nonionic surfactants were compared (Shell Research Limited, 1984c). One percent aqueous solutions of $C_{9-11}AE_6$ (Dobanol 91-6), $C_{12-13}AE_{6.5}$ (Dobanol 23-6.5), $C_{14-15}AE_7$ (Dobanol 45-7), and $C_{12,14,16}AE_7$ (coconut alcohol ethoxylate) were all mildly irritating to the rabbit eye, with no appreciable differences among the materials. At this concentration, none of the materials caused initial pain or discomfort when instilled into the eye.

Balls et al. (1991) present results for $C_{12}AE_{23}$ tested using two *in vitro* test methods for eye irritancy based on physicochemical and cytotoxicity testing strategies. In the neutral red release test, mouse fibroblast-like cells are first exposed to the vital dye, neutral red. The percentage release of the dye during a one-minute exposure to a surfactant or other test chemical is then measured by optical density. The kenacid blue method, which utilizes the same cell line, measures the reduction in cellular protein by optical density following exposure to a test chemical. Of 19 surfactants tested, $C_{12}AE_{23}$ (Brij 35), the only AE tested, was reported to have a low order of toxicity based on results from the first test and intermediate toxicity based on results from the second test.

Mucosal Irritation. Repeated applications of n-pri-$C_{12}AE_9$ to the vaginal mucosa of dogs resulted in no irritation (Berberian, 1965a).

5. Skin Sensitization

Alcohol ethoxylates have not been shown to be skin sensitizers. Skin sensitization, an allergic contact dermatitis reaction mediated by the immune system, has been tested using the guinea pig as the model system. In these tests, induction, consisting of repeated topical applications (0.5 mL) or intradermal injections (0.1 mL) of the tested substance is followed,

Table 5-5. Rabbit Eye Irritation

Surfactant/ Trade Name	Concentration[a,b]	Irritation Category	Reference
$C_{6,8,10}AE_{3.1}$ (Alfonic 610-50)	undiluted	severe	Vista Chemical Company, 1977
$C_{8,10,12}AE_{5.4}$ (Alfonic 1012-60)	undiluted	moderate	Vista Chemical Company, 1964b
$C_{10,12,14}AE_{2.7}$ (Alfonic 1014-40)	undiluted	moderate	Vista Chemical Company, 1964a
$C_{10,12,14}AE_{6.3}$ (Alfonic 1014-60)	undiluted	moderate	Vista Chemical Company, 1967
$C_{12,14}AE_{6.5}$ (Alfonic 1214-HB-58)	undiluted	severe	Vista Chemical Company, 1979f
$C_{12,14,16}AE_{1.3}$ (Alfonic 1216-22)	undiluted	minimal	Vista Chemical Company, 1979h
$C_{12,14,16,18}AE_{3.2}$ (Alfonic 1218-40)	undiluted	moderate	Vista Chemical Company, 1964c
$C_{9-11}AE_{2.5}$ (Neodol 91-2.5)	undiluted	severe	Shell Chemical Company, 1984
$C_{9-11}AE_{5}$ (Neodol 91-5)	undiluted 10% 1% 0.1%	severe moderate non-irritating non-irritating	Shell Chemical Company, 1984
$C_{9-11}AE_{6}$ (Neodol 91-6)	undiluted 1% 0.1%	severe non-irritating non-irritating	Shell Chemical Company, 1984
$C_{9-11}AE_{8}$ (Neodol 91-8)	undiluted 10% 1% 0.1%	severe severe slight non-irritating	Shell Chemical Company, 1984
$C_{12-13}AE_{3}$ (Neodol 23-3)	undiluted	moderate - extreme	Shell Chemical Company, 1984
$C_{12-13}AE_{6.5}$ (Neodol 23-6.5)	undiluted 10% 1% 0.1%	severe moderate non-irritating non-irritating	Shell Chemical Company, 1984
$C_{12-15}AE_{3}$ (Neodol 25-3)	undiluted	severe	Shell Chemical Company, 1984
$C_{12-15}AE_{7}$ (Neodol 25-7)	undiluted 10% 1% 0.1%	moderate mild minimal non-irritating	Shell Chemical Company, 1984; Shell Research Limited, 1984a

Table 5-5. (cont.)

Surfactant/ Trade Name	Concentration[a,b]	Irritation Category	Reference
$C_{12-15}AE_9$ (Neodol 25-9)	undiluted 1% 0.1%	severe - extreme non-irritating non-irritating	Shell Chemical Company, 1984
$C_{12-15}AE_{12}$ (Neodol 25-12)	undiluted	severe	Shell Chemical Company, 1984
$C_{14-15}AE_7$ (Neodol 45-7)	undiluted 10% 1% 0.1%	moderate - severe mild non-irritating non-irritating	Shell Chemical Company, 1984
$C_{14-15}AE_{11}$ (Neodol 45-11)	undiluted 10% 1% 0.1%	severe severe slight - mild non-irritating	Shell Chemical Company, 1984; Shell Research Limited, 1984b
$C_{14-15}AE_{13}$ (Neodol 45-13)	undiluted	severe	Shell Chemical Company, 1984
$C_{14-15}AE_{18}$ (Neodol 45-18)	undiluted 10% 1% 0.1%	minimal - mild practically non-irritating non-irritating non-irritating	Shell Chemical Company, 1984
$C_{12,14,16,18}AE_9$ (Coconut alcohol EO_9)	undiluted	extreme	Shell Chemical Company, 1984
n-pri-$C_{12-14}AE_{6.3}$ (Tergitol 24-L-45)	undiluted (0.01mL) undiluted (0.005 mL)	moderate moderate	Union Carbide Corporation, 1987c
n-pri-$C_{12-14}AE_{7.2}$ (Tergitol 24-L-60)	undiluted (0.005mL)	persistent, moderate - severe	Union Carbide Corporation, 1987d
n-pri-$C_{12-14}AE_{3.5}$ (Tergitol 24-L-3N)	undiluted (0.1mL) undiluted (0.005 mL)	persistent, severe minor - moderate	Union Carbide Corporation, 1988a
n-pri-$C_{12-14}AE_{5.2}$ (Tergitol 24-L-25N)	undiluted (0.005 mL)	moderate - severe	Union Carbide Corporation, 1987f
n-pri-$C_{12-14}AE_{6.4}$ (Tergitol 24-L-50N)	undiluted (0.005mL)	persistent, moderate - severe	Union Carbide Corporation, 1987h
n-pri-$C_{12-14}AE_7$ (Tergitol 24-L-60N)	undiluted (0.01mL)	minor - severe	Union Carbide Corporation, 1986d
n-pri-$C_{12-14}AE_8$ (Tergitol 24-L-75N)	undiluted (0.005 mL)	persistent, moderate - severe	Union Carbide Corporation, 1987i
n-pri-$C_{12-14}AE_{11.6}$ (Tergitol 24-L-98N)	undiluted (0.005 mL)	persistent, moderate - severe	Union Carbide Corporation, 1987g
n-pri-$C_{12-16}AE_{1.6}$ (Tergitol 26-L-1.6)	undiluted	minor	Union Carbide Corporation, 1987e

Table 5-5. (cont.)

Surfactant/Trade Name	Concentration[a,b]	Irritation Category	Reference
n-pri-$C_{12\text{-}16}AE_3$ (Tergitol 26-L-3)	undiluted (0.1 mL) undiluted (0.001 mL)	moderate - severe minor	Union Carbide Corporation, 1986e
n-pri-$C_{12\text{-}16}AE_5$ (Tergitol 26-L-5)	undiluted (0.005 mL)	moderate - severe	Union Carbide Corporation, 1987a
n-sec-$C_{11\text{-}15}AE_3$ (Tergitol 15-S-3)	undiluted	minor - moderate	Union Carbide Corporation, 1986b
n-sec-$C_{11\text{-}15}AE_9$ (Tergitol 15-S-9)	undiluted (0.005 ml)	minor - severe	Union Carbide Corporation, 1986c
$C_4H_9(CH_2CHCH_3O)_x(CH_2CH_2O)_yH$[c] (Tergitol XD)	undiluted	minor - moderate	Union Carbide Corporation, 1986g
$C_4H_9O(CH_2CHCH_3)_x(CH_2CH_2O)_yH$[c] (Tergitol XH)	undiluted	minor - moderate	Union Carbide Corporation, 1986f
$C_4H_9O(CH_2CHCH_3O)_x(CH_2CH_2O)_yH$[c] (Tergitol XJ)	undiluted (0.1 mL) undiluted (0.01mL)	minor - moderate minor	Union Carbide Corporation, 1986h
$C_{8\text{-}10}H_{17\text{-}21}O(CH_2CH_2O)_x(CH_2CHCH_3O)_yH$[c] (Tergitol 80-L-50N)	undiluted (0.005 mL)	moderate	Union Carbide Corporation, 1987b
$C_{12\text{-}14}O(CH_2CH_2O)_x(CH_2CHCH_3O)_yH$[c] (Tergitol Mechanical Dish Surfactant)	undiluted	minor - moderate	Union Carbide Corporation, 1986a
$C_{12\text{-}14}AE_3$ (Surfonic L24-3)	undiluted	moderate	Texaco Chemical Company, 1991d
$C_{12\text{-}14}AE_9$ (Surfonic L24-9)	undiluted	moderate	Texaco Chemical Company, 1991e
$C_{12}AE_4$ (Laureth-4)	undiluted 20% 10%	moderate minimal minimal	CIR Expert Panel, 1983
$C_{12}AE_{23}$ (Laureth-23)	undiluted	practically non-irritating	CIR Expert Panel, 1983
$C_{18}AE_2$ (Steareth-2)	60% 40% 10%	minimal non-irritating non-irritating	CIR Expert Panel, 1988
$C_{18}AE_{10}$ (Steareth-10)	60% 10%	practically non-irritating minimal	CIR Expert Panel, 1988
$C_{18}AE_{20}$ (Steareth-20)	60%	minimal	CIR Expert Panel, 1988
$C_{12}AE_3$	1%	non-irritating	Grubb et al., 1960
n-pri-$C_{12\text{-}13}AE_{6.5}$ (Neodol 23-6.5)	undiluted	severe	Benke et al., 1977

Table 5-5. (cont.)

Surfactant/ Trade Name	Concentration[a,b]	Irritation Category	Reference
n-pri-$C_{14-15}AE_7$ (Neodol 45-7)	undiluted	severe	Benke et al., 1977

[a] Volume of 0.1 mL unless otherwise stated.
[b] Dilutions are aqueous solutions.
[c] Contains propylene oxide.

after an interval of time, by a topically-applied challenge dose of the test substance. Concentrations of the test substance are those which are slightly or non-irritant. The skin irritation response of the animal is scored and compared to that of control animals. Most of the following studies utilized the repeated topical (patch) induction test of Buehler (1965) or the more sensitive intradermal plus topical induction method of Magnusson and Kligman (1969).

Many Neodol products ($C_{9-11}AE_{2.5}$, $C_{9-11}AE_5$, $C_{9-11}AE_6$, $C_{9-11}AE_8$, $C_{12-13}AE_3$, $C_{12-13}AE_{6.5}$, $C_{12-13}AE_7$, $C_{12-15}AE_3$, $C_{12-15}AE_7$, $C_{12-15}AE_9$, $C_{14-15}AE_7$, $C_{14-15}AE_{11}$, $C_{14-15}AE_{13}$, $C_{14-15}AE_{18}$) have been tested for their skin sensitizing reactions and scored negative (Shell Chemical Company, 1984; Shell Development Company, 1982; Shell Oil Company, 1989d; Shell Research Limited, 1981; 1983b; 1984a; 1984b; Gingell and Lu, 1991). Only $C_{12-13}AE_3$ (Neodol 25-3) showed a very weak response in one test (Shell Chemical Company, 1984) but in another test, another sample of the same product (Dobanol 25-3) scored negative (Shell Research Limited, 1983a).

When injected intracutaneously three times a week for three weeks and challenged two weeks later, $C_{12}AE_9$ (laureth-9) did not produce immediate or delayed sensitivity reactions in male guinea pigs (Berberian et al., 1965a). The injections contained 0.05 or 0.1 mL of a 0.02% aqueous solution. Topical application of 5 mL/day of a 1% solution of n-pri-$C_{12}AE_7$ for 10 days to the intact and abraded skin of rabbits followed by a challenge dose 10 days later produced no sensitization reaction (Grubb et al., 1960).

6. Carcinogenicity

Chemicals administered orally and dermally in long-term studies include $C_{12-13}AE_{6.5}$ and $C_{14-15}AE_7$. No carcinogenic effects were noted in chronic (two year) studies in which rats were fed $C_{12-13}AE_{6.5}$ or $C_{14-15}AE_7$ at doses up to 500 mg/kg/day (Procter & Gamble Co., 1981). No treatment-related carcinogenic lesions, either on the skin or systemic, were observed in rats treated topically with 0.1 mL of a 5.0% aqueous solution of $C_{12-13}AE_{6.5}$ for 18 months (Procter & Gamble Co., 1981). (See Chronic Exposures for details of the studies).

HUMAN SAFETY

7. Genotoxicity

AE surfactants have not been found to cause genetic damage when tested in a variety of *in vitro* and *in vivo* systems. These short-term genotoxicity tests include reverse mutations in bacterial (*Salmonella typhimurium* and *Escherichia coli*) and eukaryotic (*Saccharomyces cerevisiae*) systems; DNA repair in a bacterial system (*Bacillus subtilis*); chromosomal aberrations, polyploidy, and endoreduplication in mammalian cells *in vitro*; and a dominant lethal assay with the rat and cytogenic studies with rat and hamster bone marrow cells *in vivo* (Table 5-6). In the *in vitro* studies, the use of a microsomal enzyme activating system is indicated by the presence (+) or absence (-) of rat liver S9 fraction. A metabolic activation system was not used for mammalian cell test systems. For n-pri-$C_{12-15}AE_3$, DMSO was used as the solvent. The $C_{14-15}AE_7$ was dissolved in distilled water. Further experimental details were not provided in the unpublished reports or summary secondary sources such as Yam et al. (1984).

The results of a variety of short-term genotoxicity assays of 15 AE surfactants were all negative. Alcohol ethoxylates did not induce reverse gene mutations in five strains of *Salmonella typhimurium* or two strains of *Escherichia coli* or mitotic gene conversion in the yeast *Saccharomyces cerevisiae*. Chromosome damage was not observed in cultured mammalian cells or when AE were administered orally to rats and hamsters.

8. Developmental/Reproductive Toxicity

AE surfactants have not been found to cause reproductive or developmental effects when males were treated prior to mating and females were treated continuously through the weaning period. In a two-generation reproduction study, 25 rats of each sex were fed $C_{14-15}AE_7$ in the diet at levels of 0, 0.05, 0.1, or 0.5% (0, 25, 50, or 250 mg/kg/day) (Procter & Gamble Co., 1977). Each dose level was divided into two treatment groups: (1) females received the compound during the 6th through the 15th day of gestation and the males were untreated, and (2) both sexes received the compound continuously during the study. There was one control group.

No treatment-related changes for the parental rats or pups with respect to general behavior, appearance, or survival were observed. Fertility, gestation, and viability indices were comparable for control and treated groups. At the highest dose, parental female rats and pups continuously treated did not gain as much body weight as did the control groups. Hematological parameters did not show any consistent significant differences between treated and control groups.

No treatment related teratogenesis was observed in either generation. Gross and microscopic pathologic examinations of parental rats were normal. Treatment-related effects in the 0.5% continuous feeding group included increased group mean relative liver weights of the P_1 generation rats at the 91 day sacrifice and increases in group mean relative liver weights of males of the P_2 generation.

Table 5-6. Genotoxicity Assays

Chemical	Assay	Indicator organism	Application/Activating system	Concentration/dose	Response[a]	Reference
Bacterial Assays						
n-pri-$C_{9-11}AE_6$ (Neodol 91-6)	Reverse gene mutation, histidine locus	Salmonella typhimurium TA98, TA100, TA1535, TA1537, TA1538	Preincubation plate incorporation/±S9	3×10^{-5} mg/plate	–	Shell Development Co, 1981c; Gingell and Lu, 1991
n-pri-$C_{12-15}AE_3$ (Dobanol 23-3)	Reverse gene mutation, histidine locus	Salmonella typhimurium TA98, TA100, TA1535, TA1537, TA1538	Plate incorporation/±S9	2 mg/plate	–	Shell Toxicology Laboratory, 1981
n-pri-$C_{12-15}AE_3$ (Dobanol 25-3)	Reverse gene mutation, histidine locus	Salmonella typhimurium TA98, TA100, TA1535, TA1537, TA1538	Plate incorporation/±S9	0.0002-2 mg/plate	–	Dean et al, 1985
n-pri-$C_{14-15}AE_7$ (Dobanol 45-7)	Reverse gene mutation, histidine locus	Salmonella typhimurium TA98, TA100, TA1535, TA1537, TA1538	Plate incorporation/±S9	1-4 mg/plate	–	Shell Research Limited, 1982b
$C_{12-14}AE_5$ (Surfonic L24-3)	Reverse gene mutation, histidine locus	Salmonella typhimurium TA98, TA100, TA1535, TA1537, TA1538	Plate incorporation/±S9	0.5-100 µg/plate	–	Texaco Chemical Company, 1990c
$C_{12-14}AE_9$ (Surfonic L24-9)	Reverse gene mutation, histidine locus	Salmonella typhimurium TA98, TA100, TA1535, TA1537, TA1538	Plate incorporation/±S9	0.5-100 µg/plate	–	Texaco Chemical Company, 1990d
$C_{12}AE_n$	Reverse gene mutation, histidine locus	Salmonella typhimurium TA98, TA100, TA1535, TA1537	Preincubation plate incorporation/±S9	0-333 µg/plate	–	Zeiger et al, 1987
$C_{13-15}AE_7$, $C_{13-15}AE_{11}$, $C_{13-15}AE_{20}$	Reverse gene mutation, histidine locus	Salmonella typhimurium TA98, TA100, TA1535, TA1538	Plate incorporation/±S9	Not reported	–	Imperial Chemical Industries, 1981
$C_{9-11}AE_8$	Reverse gene mutation, histidine locus	Salmonella typhimurium TA98, TA100, TA1535, TA1537, TA1538	Plate incorporation/±S9	Not reported	–	Procter & Gamble, Co, 1979

Table 5-6. (cont.)

Chemical	Assay	Indicator organism	Application/Activating system	Concentration/dose	Response[a]	Reference
$C_{12}AE_n$ (isolauryl) (Tergitol XD)	Reverse gene mutation, histidine locus	Salmonella typhimurium TA98, TA100, TA1535, TA1537, TA1538	Plate incorporation/ ±S9	Not reported	−	Proctor & Gamble Co., 1979
$C_4EO_{27}PO_{24}$ (Tergitol XD)	Reverse gene mutation, histidine locus	Salmonella typhimurium TA98, TA100, TA1535, TA1537, TA1538	Plate incorporation/ ±S9	Not reported	−	Proctor & Gamble Co., 1979
Modified polyoxylated linear alcohol (Triton DF-12)	Reverse gene mutation, histidine locus	Salmonella typhimurium TA98, TA100, TA1535, TA1537, TA1538	Plate incorporation/ ±S9	Not reported	−	Proctor & Gamble Co., 1979
$C_{16}AE_{15}$	Reverse gene mutation, histidine locus	Salmonella typhimurium TA98, TA100	Plate incorporation/ ±S9	Not reported	−	Morita et al., 1981
n-pri-$C_{14-15}AE_7$ (Dobanol 45-7)	Reverse gene mutation	Escherichia coli WP_2, Wp_2, uvrA	Plate incorporation/±S9	1-4 mg/plate	−	Shell Research Limited, 1982b
n-pri-$C_{12-15}AE_7$ (Dobanol 25-3)	Reverse gene mutation	Escherichia coli WP_2, Wp_2, uvrA	Plate incorporation/±S9	0.0002-2 mg/plate	−	Dean et al., 1985
$C_{16}AE_{45}$	DNA repair	Bacillus subtilis	Not reported	Not reported	−	Morita et al., 1981
Yeast Cells						
n-pri-$C_{12-15}AE_3$ (Dobanol 25-3)	Mitotic gene conversion histidine, tryptophan loci	Saccharomyces cerevisiae JD1	Cells in culture/±S9	0.05 mL/mL	−	Dean et al., 1985
n-pri-$C_{14-15}AE_7$ (Dobanol 45-7)	Mitotic gene conversion histidine, tryptophan loci	Saccharomyces cerevisiae JD1	Cells in culture/±S9	Not reported	−	Shell Research Limited, 1982b

Table 5-6. (cont.)

Chemical	Assay	Indicator organism	Application/ Activating system	Concentration/dose	Response[a]	Reference
Mammalian cells						
n-pri-$C_{14-15}AE_7$ (Dobanol 45-7)	Chromosome aberrations, polyploidy, endoredupliction	Rat liver cells	Cells in culture	10,15,20, or 25 $\mu g/mL$	−	Shell Research Limited, 1982b
n-pri-$C_{12-15}AE_3$ (Dobanol 25-3)	Chromosome aberrations	Rat liver cells	Cells in culture	Not reported	−	Dean et al., 1985
$C_{12-14}AE_3$ (Surfonic L24-3)	Unscheduled DNA synthesis	Rat primary hepatocytes	Cells in culture	0.25-100 $\mu g/mL$	−	Texaco Chemical Company, 1991f
$C_{12-14}AE_9$ (Surfonic L24-9)	Unscheduled DNA synthesis	Rat primary hepatocytes	Cells in culture	0.025-5.0 $\mu g/mL$	−	Texaco Chemical Company, 1991g
AE_6	Chromosome anomalies	Human leucocytes	Cells in culture	2-100 $\mu g/mL$	−	Procter & Gamble Co., 1977
AE_6	Dominant lethal assay	Male mice	in vivo	20-1000 mg/kg, oral	−	Procter & Gamble Co., 1977
AE_6	Chromosome anomalies	Hamster bone marrow cells	in vivo	80, 400 or 800 mg/kg, oral	−	Procter & Gamble Co., 1977
n-pri-$C_{14-15}AE_7$ (Dobanol 45-7)	Chromosome damage, polyploidy, endoreduplication	Rat bone marrow cells	in vivo	250, 500, or 1000 mg/kg, oral	−	Shell Research Limited, 1982b
$C_{12-14}AE_3$ (Surfonic L24-3)	Micronucleus test	CD-1 mice bone marrow cells	in vivo	100 mg/kg, intraperitoneally	−	Texaco Chemical Company, 1990e
$C_{12-14}AE_9$ (Surfonic L24-9)	Micronucleus test	CD-1 mice bone marrow cells	in vivo	50 mg/kg, intraperitoneally	−	Texaco Chemical Company, 1990f

[a] Results are indicated as negative (−) or positive (+).

Using the same protocol, $C_{12}AE_6$ was administered to rats in another study (Procter & Gamble Co., 1977). General behavior, appearance, and survival were not affected by treatment. Fertility, gestation and viability indices were similar to those of controls. At the highest dose (0.5% fed continuously), parental rats and pups gained less weight than control rats. Hematological parameters were unaffected.

Statistically significant increases in embryolethality and soft tissue anomalies in the F_2 generation in the group fed 0.5% continuously and a statistically significant decrease in mean fetal liver weights in the group fed 0.1% continuously were not considered compound related. No signs of toxicity, pathologic lesions in any generation, or changes in reproduction parameters were observed.

In another two-generation reproductive study, weanling Fischer 344 rats (30 males and 30 females/group) were treated dermally with 1.0 mL/kg of $C_{9-11}AE_6$ (Neodol 91-6) at concentrations of 0, 1, 10, or 25% (w/v), (doses of 0, 10, 100, or 250 mg/kg/day) three times a week (Shell Development Company, 1985b; Gingell and Lu, 1991). The F_o and F_1 males and females were treated for approximately four months prior to mating; females were treated through the gestation and weaning periods. Results of clinical and pathological observations and observations of systemic and reproductive toxicity showed that there were no adverse effects on reproductive performance or on the growth and development of the F_1 and F_2 offspring. In the highest dose group, body weights of both males and females in both treated generations were sporadically decreased compared to controls.

Administration of $C_{12}AE_6$ to pregnant rabbits at doses of 0, 50, 100, or 200 mg/kg/day from day 2 to day 16 of gestation resulted in no adverse effects on fetal mean body weight, sex, or external, visceral, or skeletal morphology (Procter & Gamble Co., 1977). At the two higher dose levels, an increase in maternal toxicity as evidenced by ataxia and a slight loss in body weight were observed, but no effects on corpora lutea, implantations, live fetuses, and spontaneous abortions were reported. Early deliveries were noted for two control rabbits and seven treated rabbits.

In two unpublished studies, rats and rabbits were treated topically with a 6% solution of $C_{12}AE_4$ dissolved in 52% ethanol and water at the rate of 0.4 mL/kg/day. Rats were treated on days 6 through 15 of gestation for a total dose of 0.24 mL/kg (240 mg/kg). Rabbits were treated on days 6 through 18 of gestation for a total dose of 0.31 mL/kg (310 mg/kg). No teratogenic or embryotoxic effects were observed. Details of the studies were not provided. In a third study, 26 pregnant rats treated with the same dose from day 15 of pregnancy through the weaning period, showed no mortalities or toxic effects. Peri-natal and post-natal development of the pups were normal. In addition, male rats treated with the above regimen for 60 days prior to mating and females treated 14 days before mating through weaning at 21 days showed no effects on fertility (CIR Expert Panel, 1983).

9. Absorption, Metabolism, and Disposition

Metabolism studies with rats administered ^{14}C-labeled AE showed that orally-administered compounds are rapidly and extensively absorbed from the gastrointestinal tract and rapidly eliminated. Following oral administration to rats, greater than 70% of the label was excreted within 24 hours. The major portion of the administered radioactivity appeared in the urine (54%) and feces (26%), with only 3% in the expired carbon dioxide and less than 4% in the body (Procter & Gamble Co., 1977). Chemical structure, position of radiolabel, and doses were not available in this study.

No studies addressed the pathway of AE metabolism, but studies with radiolabeled compounds showed that both the alkyl chain and the EO groups are sites of attack. Metabolism studies with a series of AE surfactants labeled with ^{14}C in either the α-carbon of the alkyl group or the hydroxyl-bearing position of the ethoxylate moiety showed that distribution and excretion of ethoxylate groups of varying lengths was similar but the metabolism of their alkyl chains was a function of chain length (Drotman, 1980). Absorption following oral administration to rats was extensive (>75% of the dose) and rapid as shown by the appearance of the label in the urine within 5-10 minutes after administration. When rats were administered $C_{12}AE_6$, $C_{13}AE_6$, or $C_{14}AE_7$ labeled in the ethoxylate portion, distribution and metabolism were similar with the major portion of the radioactivity appearing in the urine (52-55%) and smaller amounts in the feces (23-27%) and expired CO_2 (2-3%). When the ^{14}C label was in the α-position of the alkyl chain ($C_{12}AE_6$, $C_{13}AE_6$, and $C_{15}AE_7$), the disposition of radioactivity was influenced by the length of the alkyl chain. The distribution and excretion of $C_{12}AE_6$ closely resembled that of the ethoxylate-labeled AE. However, increasing the chain length from 12 to 15 carbons resulted in less excretion in the urine and feces and more (up to 54% of the label) in the expired CO_2.

In a study of the absorption of AE surfactants by neonatal animals, Benke (1976) bathed weanling and adult rats and guinea pigs in solutions of ^{14}C-labeled C_nAE_n. Penetration of the surfactant, as measured by blood concentration, was the same in weanling and adult rats but was greater in weanling guinea pigs than in adults. No data were provided.

Skin penetration of AE was studied by applying aqueous solutions of pure 1-^{14}C-labelled compounds made up in a solution of linear alkyl benzene sulphonate to the skin of rats. The excess solution was rinsed off and a patch was applied. The excreta and expired air were monitored for 48 hours. Most of the dose was recovered in the rinse water (86-102%), treated skin (1.5-7.3%), or patch (0.2-8.2%). Penetration was low based on excreted radioactivity during 48 hours and was greater for $C_{15}AE_3$ (8.3 ug/cm^2), than for $C_{12}AE_3$ (4.38 μg/cm^2), $C_{12}AE_6$ (4.88 μg/cm^2), or $C_{12}AE_{10}$ (0.83 /cm^2) (Black and Howes, 1979).

Following cutaneous administration of the labeled compounds to the shaved backs of rats, absorption was slow and incomplete compared with oral administration. Approximately half of the dose was still at the application site after 72 hours. When the ethoxylate-labeled AE were applied to the skin of rats, the patterns of distribution and elimination were similar and resembled that of the orally-administered compounds. When the alkyl-labeled compounds were applied, the patterns of disposition were similar to those observed after oral administration; increasing the chain length from 12 to 15 carbons increased the amount of radioactivity appearing in the expired air from 4 to 22% (Drotman, 1980)

10. Anesthetic and Analgesic Effects.

Orally or intraperitoneally administered AE surfactants produce neuropharmacologic effects in animals including anesthesia, ataxia, and loss of righting only at doses that far exceed those expected from accidental ingestion by humans. Dilute solutions produce anesthesia of the eye.

Arthur D. Little (1977) reviewed the anesthetic/analgesic activity of AE surfactants. They reported the research of several investigators in which effects were studied under a variety of experimental conditions using several laboratory species.

Aqueous solutions of 0.5% n-pri-$C_{12}AE_9$, 1% n-pri-$C_{12}AE_7$ and 1% n-pri-$C_{12}AE_{12}$ were effective surface anesthetics to the cornea of the rabbit eye (Soehring et al. 1952). Using a series of AE in which the alkyl chain ranged from C_1 to C_{14}, Zipf and Dittmann (1964) found that n-pri-C_4AE_9 and lower homologs had no surface anesthetic activity while n-pri-C_8AE_9 and higher homologs exhibited anesthetic effects, with activity increasing as the alkyl chain lengthened. No irritating effects were seen in either study.

Three nonionic surfactants were tested for corneal anesthesia in rabbits (five of each sex/group) by the Colgate-Palmolive Company (1983). A 0.1 ml aliquot of a 25% solution in distilled water was placed in the test eye and the eye was either left unrinsed or rinsed with 50 ml of water immediately after instillation. Anesthesia as indicated by the wink reflex was tested at 30 and 90 minutes. Two surfactants, $C_{12-14}AE_n$ (Tergitol 24-L-50) and $C_{12-15}AE_7$ (Neodol 25-7) tested positive for anesthesia in both rinsed and unrinsed eyes at both times. The n-sec-$C_{11-15}AE_{20}$ (Tergitol 15-S-20) tested negative in this test.

Subcutaneous injections to rats produced local anaesthetic effects. Doses ≥ 100 mg n-pri-$C_{12-13}AE_{6.5}$/kg delayed or prevented pain-induced responses. Intraperitoneal injections produced some analgesic action (Benke et al., 1977).

More recently, Zerkle et al. (1987) investigated the effects of systemic and oral exposure to AE. Intraperitoneal administration of doses of 0.11 to 0.15 g/kg (8% w/v aqueous solutions) of $C_{12-13}AE_{6.5}$ to rats produced a dose-dependent temporal progression of increasingly severe effects: ataxia, loss of righting reflex, respiratory depression, and death. At less than lethal doses, the effects were completely reversible.

Oral administration of aqueous solutions of $C_{12-13}AE_{6.5}$ at several dosing volumes and concentrations to rats and mice in the same study produced variable and inconsistent effects; effects were observed only at doses considered higher than would be encountered under accidental ingestion conditions (i.e., 10 mL/kg for mice and 30 mL/kg for rats). Although results were variable, effects were correlated with plasma concentrations following administration of a pure AE homologue. A 25% aqueous solution of $C_{13}AE_9$ (nonaethylene glycol mono-n-tridecyl ether) administered orally to rats at a dose of 10 mL/kg produced ataxia and loss of righting reflex starting at 1.5 hours post administration. Ataxia was present in 1/3 animals at a plasma level of 14.5 µg/mL and loss of righting reflex was observed in 3/3 animals at 25.3 µg/mL (Zerkle et al. 1987).

B. HUMAN STUDIES

1. Dermal Irritation and Sensitization

Using the Draize patch test, AE surfactants at dilutions of ≤60% produced no to slight skin irritation in human subjects. AE surfactants were not skin sensitizers.

A sensitization test with $C_{12}AE_9$ using 10, 15, and 20% aerosol cream preparations was performed by Berberian et al. (1965a). Repeated skin patch tests (usually nine) and a challenge test were tolerated well, producing only a mild erythema in most subjects. The day after placement of the challenge patch, 21 of the 51 subjects had a mild erythema reaction, not indicative of sensitization; none showed an edematous reaction or eczematous flare.

Skin patch tests with n-pri-$C_{12-13}AE_{6.5}$ and $C_{14-15}AE_7$ (0.4 mL) at concentrations of 10 or 25% for four hours produced only minor irritation (Benke et al., 1977). Brown and Benke (1977) conducted repeated insult patch tests on human subjects according to the method of Griffith (1969). Subjects wore occlusive patches containing 2.5% aqueous solutions of $C_{12-13}AE_{6.5}$ (176 subjects) or $C_{14-15}AE_7$ (144 subjects) for 24-hour periods on alternate days, three times/week, for three weeks. Seventeen days after the final induction patch, challenge applications were made on the induction site and another site. Sites were examined 48 and 96 hours later for evidence of skin sensitization. Skin hyper-reactivity to $C_{12-13}AE_{6.5}$ occurred in one only subject. In addition, the results of home usage tests with formulations containing these surfactants indicated no significant skin irritation.

In another repeated patch test for skin sensitization, a small percentage of subjects exposed to 1-25% solutions of Neodol products showed very slight skin irritation; none exhibited sensitization (Shell Chemical Company, 1984). No details of the studies were provided in this summary report.

Two lauryl alcohol ethoxylates were tested for primary irritation and sensitization on human subjects. $C_{12}AE_4$ (laureth-4) was tested for primary skin irritation and sensitization on 50 subjects. A 0.1 mL sample of the undiluted liquid was applied to one-inch square patches which were taped to the skin for 72 hours. Ten days after patch removal, a second patch was applied for 72 hours. No skin reaction or sensitization occurred to the induction or challenge patch (CIR Expert Panel, 1983).

$C_{12}AE_{23}$ (laureth-23) was tested undiluted and in aqueous solutions on human subjects. Undiluted laureth-23 was tested on 10 subjects for a 48-hour period, followed by a seven day nontreatment period, and an additional 48-hour treatment period. Erythema was present in one subject following both applications. In a similar test, 50 subjects were treated for 72-hour periods with a 60% weight/volume preparation. No primary cutaneous irritation was observed (CIR Expert Panel, 1983).

HUMAN SAFETY

In a repeated insult patch test, 0.1 mL of a 25% solution of $C_{12}AE_{23}$ (laureth-23) was placed on the backs of 168 male and female subjects. The test material was applied at 48-hour intervals, three times a week for three weeks. After a three week nontreatment period, the test area and a nontreated site were challenged using the same procedure. Sites were scored after the insult test as well as after the challenge test. No irritation or sensitization was reported. Repeated insult patch tests with 3% and 5% aqueous solutions of laureth-23 performed on 103 and 150 subjects, respectively, also resulted in no sensitization reaction (CIR Expert Panel, 1983).

Cotton dressings saturated with 60% aqueous solutions of $C_{18}AE_2$, $C_{18}AE_{10}$, or $C_{18}AE_{20}$ were applied to the skin of 200 subjects by means of an elastic adhesive patch. The patch was left in contact with the skin for 72 hours. Ten days later, a second occlusive dressing was applied to the original exposure site. None of the solutions produced skin irritation. Two additional tests with $C_{18}AE_2$ and $C_{18}AE_{20}$, using the same protocol, also produced no dermal reaction. In these tests, the 60% solutions of stearyl alcohol ethoxylates were neither Primary Irritants nor sensitizers (CIR Expert Panel, 1988).

2. Absorption, Metabolism, and Disposition

Studies with radiolabeled compounds show that orally-administered AE surfactants are rapidly absorbed and eliminated. Adult males (60-90 kg) ingested 50 mg of $C_{12}AE_6$ or $C_{13}AE_6$ labeled with ^{14}C in either the ethoxylate moiety or the α-carbon of the alkyl group (Drotman, 1980). The pattern of disposition was similar to that seen in the study with rats, with the ethoxylate portion rapidly eliminated in the urine. Lengthening the alkyl chain from 12 to 13 carbons increased the fraction that was expired as CO_2 from 3 to 13%.

The disposition following dermal application was likewise similar to that observed in the rat (Drotman, 1980). Absorption was slow and incomplete; 74 to 88% remained on the skin after 144 hours. Elimination was primarily in the urine.

3. Therapeutic/Contraceptive Uses

In early studies, AE were tested in contraceptive preparations and as analgesics and anesthetics. No untoward effects were reported.

Berberian (1965b) tested the irritancy of an aerosol cream formulation containing $C_{12}AE_9$ intended for use as a contraceptive cream. The formulation was applied to the penile surfaces of 13 male subjects for 8 hours or longer, four times a week, for two weeks without recurring irritant effects. A challenge application two weeks later did not elicit sensitization. $C_{12}AE_9$ was spermicidal (sperm immobilization) to human spermatozoa within 20 seconds at a dilution of 1:1200 to 1:3000 (Berberian, 1965a). The following were also spermicidal in 5-30 minutes: a 0.25% solution of unspecified AE (Holzaepfel et al., 1959), a 0.06-0.125% aqueous solution of pri-C_8AE_7, a 0.03-0.06% aqueous solution of pri-$C_{12}AE_{19-23}$, and a 0.03-0.06%

aqueous solution of pri-$C_{16}AE_{19-23}$ (Harvey and Stucky, 1962). A 0.05% aqueous solution of pri-C_6AE_4 was not lethal to human sperm cells (Harvey and Stucky, 1962).

In the following studies formulations containing AE were tested for local anesthetic effects which were achieved, in most cases, without irritant side effects. The n-pri-$C_{12}AE_9$ was tested as a skin anesthetic in a variety of oils, lotions, and pastes with only drying effects (Lutzenkirchen, 1952). Blasiu (1953) applied n-pri-$C_{12}AE_{7.13}$ and n-pri-$C_{12}AE_{11.9}$ (Thesit) to patients experiencing X-ray dermatitis. Schoog (1953) reported the use of Thesit as an antipruritic and Heyman (1954) treated burns with aqueous solutions of Thesit. Thesit (1% or 2% solutions) was used as an analgesic for a variety of skin and oral lesions (Schulz 1952). Thirty-eight of 2557 eczema patients treated with a Thesit preparation experienced contact dermatitis (Hartung and Rudolph, 1970).

Two studies reported the use of AE as mucosal analgesics. Both Strack (1950) and Hochrein and Schleicher (1951) used n-pri-$C_{12}AE_9$, n-pri-$C_{12}AE_{7.13}$, and $C_{12}AE_{11.9}$ as a treatment for gastric ulcers and gastritis without side effects.

Sixteen adult test subjects continuously exposed to steam from a water vaporizer containing up to 20 tablets/quart of a mixture of n-pri-$C_{12}AE_7$, menthol, camphor, eucalyptol, and benzoin (dose not given) for eight hours showed no effects (Larkin, 1957). In a clinical test designed to treat respiratory infections, the author treated 92 infants and children with vapors from a 20% solution of n-pri-$C_{12}AE_7$ by inhalation. No harmful effects were noted.

4. Cosmetic Uses

The Cosmetic Ingredient Review (CIR) Expert Panel of the American College of Toxicology assessed the safety of $C_{12}AE_4$ (laureth-4) and $C_{12}AE_{23}$ (laureth-23) which are used in cosmetic and pharmaceutical products (CIR Expert Panel, 1983). Based on published data from laboratory animal studies and unpublished data from clinical studies in which subjects were treated dermally with laureth (up to 100%) and laureth formulations, the Panel concluded that both AE are safe as cosmetic ingredients as presently used and at the concentrations used.

The CIR Expert Panel also assessed the safety of $C_{18}AE_2$ - $C_{18}AE_{20}$ (steareth-2, -4, -6, -7, -10, -13, -15, and -20) which are used as emulsifiers in cosmetic products as well as wetting agents, solubilizers, and surfactants (CIR Expert Panel, 1988). The majority of the data were industry submissions of unpublished data. On the basis of acute oral and subchronic dermal toxicity tests with laboratory animals and dermal irritation and sensitization tests with humans, the Panel concluded that the steareths are safe as cosmetic ingredients in the present practices of use and concentration.

5. Epidemiology

No reports on human exposure to AE during manufacture were located. No epidemiology studies were reported.

REFERENCES

Arthur D. Little. 1977. Human Safety and Environmental Aspects of Major Surfactants. A Report to the Soap and Detergent Association. Arthur D. Little, Cambridge, MA.

Balls, M., S. Reader, K. Atkinson, J. Tarrent and R. Clothier. 1991. Non-animal alternative toxicity tests for detergents: genuine replacements or mere prescreens? J. Chem. Tech. Biotechnol. 50:423-433.

Benke, G.M. 1976. An approach to safety factor estimation for infant bathing. Toxicol. Appl. Pharmacol. 37:96. (Abstract).

Benke, G.M., N.M. Brown, M.J. Walsh and R.D. Drotman. 1977. Safety testing of alkyl polyethoxylate nonionic surfactants. I. Acute effects. Fd. Cosmet. Toxicol. 15:309-318.

Berberian, D.A., W.G. Gorman, H.P. Drobeck, F. Coulston and R.G. Slighter, Jr. 1965a. The toxicology and biological properties of laureth 9 (a polyoxyethylene lauryl ether), a new spermicidal agent. Toxicol. Appl. Pharmacol. 7:206-214.

Berberian, D.A. W.G. Gorman, H.P. Drobeck, F. Coulston and R.G. Slighter, Jr. 1965b. Toxicology and spermicidal activity of a new contraceptive cream containing chlorindanol and Laureth 9. Toxicol. Appl. Pharmacol. 7:215-226.

Black, J.G. and D. Howes. 1979. Skin penetration of chemically related detergents. J. Soc. Cosmet. Chem. 30:157-165.

Blasiu, A.P. 1953. Experiences with Thesit in X-ray dermatitis. Stralentherapie 92:75-79.

Brown, N.M. and G.M. Benke. 1977. Safety testing of alkyl polyethoxylate nonionic surfactants. II. Subchronic studies. Fd. Cosmet. Toxicol. 15: 319-324.

Buehler, E.V. 1965. Delayed contact hypersensitivity in the guinea pig. Arch. Derm. 91:171-177.

CIR (Cosmetic Ingredient Review) Expert Panel of the American College of Toxicology. 1983. Final report on the safety assessment of Laureths -4 and -23. J. Am. Coll. Toxicol. 2:1-15.

CIR (Cosmetic Ingredient Review) Expert Panel of the American College of Toxicology. 1988. Final report on the safety assessment of steareth-2, -4, -6, -7, -10, -11, -13, -15, and -20. J. Am. Coll. Toxicol. 7:881-910.

Colgate-Palmolive Company. 1979. Product safety evaluation data sheets.

Colgate-Palmolive Company. 1983. Product safety evaluation data sheets.

Dean, B.J., T.M. Brooks, G. Hodson-Walker and D.H. Hutson. 1985. Genetic toxicology testing of 41 industrial chemicals. Mutat. Res. 153:57-77.

Draize, J.H., G. Woodward and H.O. Calvery. 1944. Methods for the study of irritation and toxicity of substances applied topically to the skin and mucous membrane. J. Pharmacol. Exptl. Therap. 82:377-390.

Draize, J.H. 1959. Dermal toxicity. In: Appraisal of the Safety of Chemicals in Foods, Drugs and Cosmetics. Association of Food and Drug Officials of the U.S., Baltimore, MD. The Editorial Committee Publication, pp. 48-59.

Drotman, R.B. 1980. The absorption, distribution, and excretion of alkylpolyethoxylates by rats and humans. Toxicol. Appl. Pharmacol. 52:38-44.

Duprey, L. and J.O. Hoppe. 1960. Acute toxicity of laureth 9. Report: Esta Medical Laboratories, Rensselaer, New York. (Cited in Berberian, 1965a).

Galante, D.C. and K.W. Dillan. 1981. Heavy-duty detergents. J. Amer. Oil Chem. Soc. 58:356A-362A.

Gerarde, H.W. 1963. Toxicology studies on hydrocarbons. XI. Aspiration hazard and toxicity of hydrocarbons and hydrocarbon mixtures. Arch. Environ. Health 6:35-47.

Gingell, R. and C.C. Lu. 1991. Acute, subchronic, and reproductive toxicity of a linear alcohol ethoxylate surfactant in the rat. J. Am. Coll. Toxicol. 10:477-486.

Gosselin, R.E., R.P.Smith, H.C. Hodge and J.E. Braddock. 1984. *Clinical Toxicology of Commercial Products.* 5th ed. Baltimore: The Williams and Wilkins Co.

Goyer, M.M., J.H. Perwak, A. Sivak and P.S. Thayer. 1981. Human Safety and Environmental Aspects of Major Surfactants (Supplement). A Report to the Soap and Detergent Association. Arthur D. Little, Cambridge, MA.

Griffith, J.F. 1969. Predictive and diagnostic testing for contact sensitization. Toxicol. Appl. Pharmacol. Suppl. 3:90.

Grubb, T.C., L.C. Dick, and M. Oser. 1960. Studies on the toxicity of polyoxyethylene dodecanol. Toxicol. Appl. Pharmacol. 2:133-143.

Hartung, J. and Rudolph. 1970. Epidermale allergie gegen hydroxypolyaetoxydodekan. Z. Haut.-Geschl. Kr. 45:547-550.

Harvey, C. and R.E. Stucky. 1962. Spermicidal activity of surface-active agents. J. Reprod. Fertil. 3:124-131.

Heyman, J. 1954. Local pain alleviation in burns. Med. Klinik. 49:559.

Hochrein, M. and I. Schleicher. 1951. Circulatory disturbances as the basis for pathogenesis and therapy of peptic ulcer. Deut. Med. Wschr. 76:735-739.

Holzaepfel, J.H., R.W. Greenlee, R.E. Wyant and W.C. Ellis, Jr. 1959. Screening of organic compounds for spermicidal activity. Fertility Sterility 10:272-284.

Imperial Chemical Industries, Unpublished data, 1981. (Cited in Yam et al., 1984).

Kay, J.H. and J.C. Calandra. 1962. Interpretation of eye irritation tests. J. Soc. Cosmet. Chem. 13:281-289.

Larkin, V. De P. 1957. Polyoxyethylene dodecanol vaporization in the treatment of respiratory infections of infants and children. N. Y. State J. Med. 57:2667-2672.

Lehman, A.J., et al. 1965. Appraisal of the Safety of Chemicals in Food, Drugs and Cosmetics Association, Food and Drug Officials of the U.S., Topeka, KS.

Lutzenkirchen, A.L. 1952. Clinical experiences with a new surface anesthetic in dermatology. Med. Klin. 47:618-620.

Magnusson, B. and A.M. Kligman. 1969. The identification of contact allergens by animal assay. The guinea pig maximization test. J. Inves. Dermatol. 52:268-276.

Morita, K., M. Ishigake and T. Abe. 1981. Mutagenicity of materials related with cosmetics. J. Soc. Cosmet. Chem. Jap. 15:243. (Cited in Yam et al., 1984).

Osterberg, R.E., S.P. Bayard and A.G. Ulsamer. 1976. Appraisal of existing methodology in aspiration toxicity testing. J. Assoc. Off. Anal. Chem. 59:516-525.

Procter & Gamble Co. 1977. Unpublished data. (Cited in Arthur D. Little, 1977).

Procter & Gamble Co. 1979. Unpublished data. (Cited in Yam et al., 1984).

Procter & Gamble Co. 1981. Unpublished data. (Cited in Goyer et al., 1981).

Procter & Gamble Co. 1985. Unpublished data.

Schick, M.J. 1967. *Nonionic Surfactants*. Marcel Dekker, Inc., New York. pp. 926-928.

Schoog, M. 1953. Der pruritus and seine behandlung. Fortschr. Medizin 71:31-33.

Schulz, K.H. 1952. On the use of alkyl-polyethylene oxide derivatives as surface anesthetics. Dermatolog. Wocheschr. 126:657-662.

Shell Chemical Company. 1984. Human Safety of Neodol Products. SC:793-84.

Shell Chemical Company. 1991. NEODOL Product Guide for Alcohols, Ethoxylates and Ethoxysulfates, Technical Bulletin SC:7-91.

Shell Development Company. 1981a. Acute dermal toxicity of Neodol 91-6 ethoxylate in the rabbit. Shell Internal Report WRC RIR-164. Unpublished data.

Shell Development Company. 1981b. Acute oral toxicity of Neodol 91-6 ethoxylate in the rat. Shell Internal Report WRC RIR-165. Unpublished data.

Shell Development Company. 1981c. Gene mutation in *Salmonella typhimurium* assay of Neodol 91-6 alcohol ethoxylate. Shell Internal Report WRC RIR-117. Unpublished data.

Shell Development Company. 1981d. Primary skin irritation of Neodol 91-6 ethoxylate. Shell Internal Report WRC RIR-166. Unpublished data.

Shell Development Company. 1982. Guinea pig skin sensitization of Neodol 91-6 ethoxylate. Shell Internal Report WRC RIR-210. Unpublished data.

Shell Development Company. 1985a. 90-Day dermal toxicity study in rats of Neodol® 91-6 ethoxylate. Shell Internal Report WRC RIR-342. Unpublished data.

Shell Development Company. 1985b. Two generation reproduction study of Neodol® 91-6 ethoxylate in Fischer 344 Rats. Shell Internal Report WRC RIR-363. Unpublished data.

Shell Oil Company. 1989a. Acute dermal toxicity (LD_{50}) study in albino rabbits with Neodol 23-7P7. Report HSE-89-0026. Unpublished data.

Shell Oil Company. 1989b. Acute oral toxicity (LD_{50}) study in albino rats with Neodol 23-7P7. Report HSE-89-0025. Unpublished data.

Shell Oil Company. 1989c. Primary dermal irritation study in albino rabbits with Neodol 23-7P7. Report HSE-89-0027. Unpublished data.

Shell Oil Company. 1989d. Skin sensitization study in albino guinea pigs with Neodol 23-7P7. Report HSE-89-0028. Unpublished data.

Shell Oil Company. 1990. Acute oral toxicity study in rats (Neodol 23-6.5): EPA Guidelines No. 81-1. Unpublished data.

Shell Research Limited. 1980a. Acute inhalation toxicity of Dobanol 91-5. Shell Toxicology Laboratory, Tunstall, England: Group Research Report TLGR.80.053. Unpublished data.

Shell Research Limited. 1980b. Toxicology of detergents: the acute toxicity of Dobanol 91 ethoxylates: 91-2.5, 91-5, 91-6, 91-8. Shell Toxicology Laboratory, Tunstall, England: Group Research Report TLGR.80.088. Unpublished data.

Shell Research Limited. 1981. Toxicology of detergents: the skin sensitizing potential of Dobanol 45-18. Shell Toxicology Laboratory, Tunstall, England: Group Research Report TLGR.80.133. Unpublished data.

Shell Research Limited. 1982a. A subchronic (90-day) feeding study of Dobanol 45-7 in rats. Shell Internal Report SBGR.81.330. Unpublished data.

Shell Research Limited. 1982b. Toxicity studies with detergents: Short-term tests for genotoxic activity using Dobanol 45-7. Shell Internal Report SBGR.82.252. Unpublished data.

Shell Research Limited. 1983a. Toxicology of detergents: The skin sensitizing potential of Dobanol 25-3 (shop sample). Report SBGR.83.313. Unpublished data.

Shell Research Limited. 1983b. Toxicology of detergents: The skin sensitizing potential of Dobanol 25-7). Report SBGR.83.329. Unpublished data.

Shell Research Limited. 1984a. Toxicology of detergents: the acute oral and percutaneous toxicity, skin and eye irritancy and skin sensitizing potential of Dobanol 25-7. Report SBGR.84.263. Unpublished data.

Shell Research Limited. 1984b. Toxicology of detergents: the acute oral and percutaneous toxicity, skin and eye irritancy and skin sensitizing potential of Dobanol 45-11. Report SBGR.84.296. Unpublished data.

Shell Research Limited. 1984c. Toxicology of detergents: the comparative eye irritancy of various Dobanol based nonionic and anionic and nonionic detergents. Report SBGR.84.092. Unpublished data.

Shell Research Limited. 1986. Toxicology of industrial chemicals (detergents): the acute oral toxicity of Dobanol 45-11 and Dobanol 23-6.5 (administered as 50% (m/v) solutions in corn oil). Report SBGR.86.124. Unpublished data.

Shell Research Limited. 1990. Dobanol 23-6.5: acute oral toxicity. Report SBGR.90.093, Sittingbourne Research Centre, Sittingbourne, Kent, England. Unpublished data.

Shell Toxicology Laboratory. 1981. Unpublished data cited in Goyer et al., 1981.

Smyth, H.F., Jr., C.P. Carpenter and C.S. Weil. 1950. The toxicology of the polyethylene glycols. (Sci. ed.) J. Am. Pharm. Assoc. 39:349-354.

Smyth, H.F., Jr., C.P. Carpenter and C,S. Weil. 1955. The chronic oral toxicology of the polyethylene glycols. (Sci. ed.) J. Am. Pharm. Assoc. 44:27-30.

Soehring, K., M. Frahm and K. Mletzko. 1952. Contributions to the pharmacology of the alkyl derivatives of poly(ethylene oxide). IV. Local and analgesic effects and comments on the comparative evaluation of conductivity-inhibiting substances. Arch. Int. Pharmacodyn. 91:112-130. (Cited in Arthur D. Little, 1977).

Strack, K. 1950. On the treatment of peptic ulcers and gastritis with mucous membrane-anesthesizing substances. Muench. Med. Wschr. No. 21-22.

Texaco Chemical Company. 1990a. Acute dermal toxicity: Surfonic L24-3. Unpublished report.

Texaco Chemical Company. 1990b. Primary dermal irritation study: Surfonic L24-3. Unpublished report.

Texaco Chemical Company. 1990c. Ames/*Salmonella* plate incorporation assay: Surfonic L24-3. Unpublished report.

Texaco Chemical Company. 1990d. Ames/*Salmonella* plate incorporation assay: Surfonic L24-9. Unpublished report.

Texaco Chemical Company. 1990e. Micronucleus test (MNT) EPA: Surfonic L24-3. Unpublished report.

Texaco Chemical Company. 1990f. Micronucleus test (MNT) EPA: Surfonic L24-9. Unpublished report.

Texaco Chemical Company. 1991a. Acute exposure oral toxicity: Surfonic L24-9. Unpublished report.

Texaco Chemical Company. 1991b. Acute dermal toxicity: Surfonic L24-9. Unpublished report.

Texaco Chemical Company. 1991c. Primary dermal irritation study: Surfonic L24-9. Unpublished report.

Texaco Chemical Company. 1991d. Primary eye irritation: Surfonic L24-3. Unpublished report.

Texaco Chemical Company. 1991e. Primary eye irritation: Surfonic L24-9. Unpublished report.

Texaco Chemical Company. 1991f. Rat hepatocyte primary culture/DNA repair test: Surfonic L24-3. Unpublished report.

Texaco Chemical Company. 1991g. Rat hepatocyte primary culture/DNA repair test: Surfonic L24-9. Unpublished report.

Texaco Chemical Company. 1992. Acute exposure oral toxicity: Surfonic L24-3. Unpublished report.

Tyler, T.R., R.C. Myers, S.M. Christopher and E.H. Fowler. 1988. Pulmonary toxicity of ethoxylates given endotracheally to rats. Toxicologist 8:147. (Abstract).

Union Carbide Corporation. 1981. Toxicity of Tergitol materials to the lung: A retrospective examination of experimental evidence. Project Report 44-71. Unpublished data.

Union Carbide Corporation. 1986a. Tergitol mechanical dish surfactant: acute toxicity and primary irritancy studies. Project Report 49-84. Unpublished data.

Union Carbide Corporation. 1986b. Tergitol nonionic surfactant 15-S-3: acute toxicity and primary irritancy studies. Project Report 49-142. Unpublished data.

Union Carbide Corporation. 1986c. Tergitol nonionic surfactant 15-S-9: acute toxicity and primary irritancy studies. Project Report 49-143. Unpublished data.

Union Carbide Corporation. 1986d. Tergitol nonionic surfactant 24-L-60N: acute toxicity and primary irritancy studies. Project Report 49-153. Unpublished data.
Union Carbide Corporation. 1986e. Tergitol nonionic surfactant 26-L-3: acute toxicity and primary irritancy studies. Project Report 49-156. Unpublished data.
Union Carbide Corporation. 1986f. Tergitol nonionic surfactant XH: acute toxicity and primary irritancy studies. Project Report 49-160. Unpublished data.
Union Carbide Corporation. 1986g. Tergitol nonionic surfactant XD: acute toxicity and primary irritancy studies. Project Report 49-161. Unpublished data.
Union Carbide Corporation. 1986h. Tergitol nonionic surfactant XJ: acute toxicity and primary irritancy studies. Project Report 49-162. Unpublished data.
Union Carbide Corporation. 1987a. Tergitol nonionic surfactant 26-L-5: acute toxicity and primary irritancy studies. Project Report 50-44. Unpublished data.
Union Carbide Corporation. 1987b. Tergitol nonionic surfactant 80-L-50N: acute toxicity and primary irritancy studies. Project Report 50-49. Unpublished data.
Union Carbide Corporation. 1987c. Tergitol nonionic surfactant 24-L-45: acute toxicity and primary irritancy studies. Project Report 50-73. Unpublished data.
Union Carbide Corporation. 1987d. Tergitol nonionic surfactant 24-L-60: acute toxicity and primary irritancy studies. Project Report 50-142. Unpublished data.
Union Carbide Corporation. 1987e. Tergitol nonionic surfactant 26-L-1.6: acute toxicity and primary irritancy studies. Project Report 50-148. Unpublished data.
Union Carbide Corporation. 1987f. Tergitol nonionic surfactant 24-L-25N: acute toxicity and primary irritancy studies. Project Report 50-154. Unpublished data.
Union Carbide Corporation. 1987g. Tergitol nonionic surfactant 24-L-98N: acute toxicity and primary irritancy studies. Project Report 50-157. Unpublished data.
Union Carbide Corporation. 1987h. Tergitol nonionic surfactant 24-L-50N: acute toxicity and primary irritancy studies. Project Report 50-155. Unpublished data.
Union Carbide Corporation. 1987i. Tergitol nonionic surfactant 24-L-75N: acute toxicity and primary irritancy studies. Project Report 50-156. Unpublished data.
Union Carbide Corporation. 1988a. Tergitol nonionic surfactant 24-L-3N: acute toxicity and primary irritancy studies. Project Report 50-153. Unpublished data.
Union Carbide Corporation. 1988b. Tergitol and Carbowax samples: assessment of toxicity and pulmonary effects in the rat following single endotracheal injection, revised. Project Report 50-128. Unpublished data.
Union Carbide Corporation. 1988c. Carbowax PEG 600, PEG 3350, MPEG 2000: assessment of pulmonary effects in the rat following single endotracheal injection. Project Report 51-38. Unpublished data.
Union Carbide Corporation. 1988d. Tergitol nonionic surfactant 24-L-60N (25% and 10% dilutions): acute percutaneous toxicity studies. Project Report 51-98. Unpublished data.

Union Carbide Corporation. 1988e. Tergitol nonionic surfactant 24-L-60N: acute peroral (rabbit) and percutaneous (rat) toxicity studies. Project Report 51-101. Unpublished data.

Union Carbide Corporation. 1988f. Tergitol nonionic surfactant 15-S-9: acute peroral (rabbit) and percutaneous (rat) toxicity studies. Project Report 51-102. Unpublished data.

Union Carbide Corporation. 1988g. Tergitol nonionic surfactant 15-S-9: acute percutaneous toxicity studies in the rabbit (restricted diet). Project Report 51-112. Unpublished data.

Union Carbide Corporation. 1988h. Tergitol nonionic surfactant 24-L-60N: acute percutaneous toxicity studies in the rabbit (restricted diet). Project Report 51-109. Unpublished data.

Union Carbide Corporation. 1988i. Tergitol nonionic surfactant 24-L-60N and TERGITOL nonionic surfactant 15-S-9: fourteen-day temporal toxicity study following a single cutaneous dose in rats. Project Report 51-60. Unpublished data.

Union Carbide Corporation. 1988j. Tergitol nonionic surfactant 24-L-60N and Tergitol nonionic surfactant 15-S-9: fourteen-day temporal toxicity study following a single cutaneous dose in rabbits. Project Report 51-79. Unpublished data.

Union Carbide Corporation. 1988k. Tergitol nonionic surfactant 24-L-60N: nine-day repeated cutaneous dose toxicity study in albino rabbits. Project Report 51-2. Unpublished data.

Union Carbide Corporation. 1989. Tergitol nonionic surfactant 24-L-60N: acute peroral, percutaneous, and subcutaneous toxicity studies. Project Report 52-90. Unpublished data.

Union Carbide Corporation. 1990. Tergitol nonionic surfactant 24-L-60N: nine-day repeated cutaneous dose toxicity study with neurotoxicity evaluation in albino rats. Project Report 52-42. Unpublished data.

U.S. EPA. 1985. Reference values for risk assessment. SRC TR 85-300. U.S. Environmental Protection Agency, Cincinnati, OH.

Vista Chemical Company. 1964a. Toxicity tests on Alfonic 1014-40. Unpublished report.

Vista Chemical Company. 1964b. Toxicity tests on Alfonic 1012-60. Unpublished report.

Vista Chemical Company. 1964c. Toxicity studies on Alfonic 1218-40 ethoxylate. Unpublished report.

Vista Chemical Company. 1967. Alfonic 1014-60: acute oral toxicity (LD_{50}) study in rats. acute dermal toxicity (LD_{50}) in rabbits, dermal irritation in rabbits, eye irritation in rabbits, and inhalation toxicity in rats. Unpublished report.

Vista Chemical Company. 1977. Alfonic 610-50: acute oral toxicity (LD_{50}) in rats, acute dermal toxicity (LD_{50}) in rabbits, dermal irritation test in rabbits, eye irritation test in rabbits, inhalation toxicity test in rats. Unpublished report.

Vista Chemical Company. 1978a. Alfonic 1012-60: acute dermal toxicity study in rabbits. Unpublished report.

Vista Chemical Company. 1978b. Inhalation toxicity study of Alfonic 1012-60. Unpublished report.

Vista Chemical Company. 1979a. Alfonic 1214-HB-58. Acute dermal toxicity (LD_{50}) test in rabbits. Unpublished report.

Vista Chemical Company. 1979b. Alfonic 1214-HB-58. Acute dermal toxicity (LD_{50}) test in rats. Unpublished report.

Vista Chemical Company. 1979c. Alfonic 1214-HB-58. Acute oral toxicity (LD_{50}) test in rabbits. Unpublished report.

Vista Chemical Company. 1979d. Alfonic 1214-HB-58. Acute oral toxicity test (LD_{50}) in rats. Unpublished report.

Vista Chemical Company. 1979e. Alfonic 1214-HB-58. Dermal irritation test in rabbits. Unpublished report.

Vista Chemical Company. 1979f. Alfonic 1214-HB-58. Eye irritation test in rabbits (Draize and Kelly modified). Unpublished report.

Vista Chemical Company. 1979g. Alfonic 1214-HB-58. Inhalation toxicity in rats. Unpublished report.

Vista Chemical Company. 1979h. Alfonic 1216-22: eye irritation test in rabbits. Unpublished report.

Vista Chemical Company. 1980a. Alfonic 1216-22: two level acute inhalation toxicity test in rats. Unpublished report.

Vista Chemical Company. 1980b. Alfonic 1216-22: two level acute oral toxicity test in rats. Unpublished report.

Vista Chemical Company. 1981. Alfonic 610-50: single level inhalation toxicity test in rats. Unpublished report.

Vista Chemical Company. 1982. Acute inhalation toxicity study in rats: Alfonic 1214-HA-30, final report. Unpublished report.

Vista Chemical Company. 1984. Material Safety Data Sheets.

Vista Chemical Company. 1985. Material Safety Data Sheets.

Vista Chemical Company. n.d. Alfonic 1216-22: dermal irritation test in rabbits. Unpublished report.

Weiss, G. 1980. *Hazardous Chemicals Data Book*, Part 2. Noyes Data Corporation, Park Ridge, NJ. p. 945.

Yam, J., K.A. Booman, W. Broddle, L. Geiger, J.E. Heinze, Y.J. Lin, K. McCarthy, S. Reiss, V. Sawin, R.I. Sedlak, R.S. Slesinski and G.A. Wright. 1984. Surfactants: a survey of short-term genotoxicity testing. Fd. Chem. Toxicol. 22: 761-769.

Zeiger, E., B. Anderson, S. Haworth, T. Lawlor, K. Mortelmans and W. Speck. 1987. *Salmonella* mutagenicity tests. III. Results from the testing of 255 chemicals. Environ. Mutagen. 9:1-110.

Zerkle, T.B., J.F. Ross and B.E. Domeyer. 1987. Alkyl ethoxylates: an assessment of their oral safety alone and in mixtures. J. Am. Oil Chem. Soc. 64:269-272.

Zipf, H.F. and E.C. Dittmann. 1964. Relationships between alkyl chain length, lipophilicity, and local anesthetic and endoanesthetic activity in homologous alkyl polyglycol ethers. Naun. Schmied. Arch. Exp. Pathol. Paramakol. 247:544-557. (Cited in Arthur D. Little, 1977).

Zipf, H.F., E. Wetzels, H. Ludwig and M. Friedrich. 1957. General and local toxic effects of dodecylpolyethyleneoxide ethers. Arzneimittel-Forschung. 7:162-166.

APPENDIX

Laboratory Studies of Biodegradation

Appendix. Laboratory Studies of Biodegradation

Chemical[a]	Test System/Protocol	Extent of Biodegradation[b]	Measurement Method	Reference
C_9AE_4	River-water dieaway	100% in 7 days	Phosphomolybdate	Huyser, 1960
C_8AE_{14}	River-water dieaway	84% in 34 days	Phosphomolybdate	
C_8AE_{25}	River-water dieaway	35% in 34 days	Phosphomolybdate	
$C_{18}AE_4$	River-water dieaway	100% in 27 days	Phosphomolybdate	
$C_{18}AE_8$	River-water dieaway	100% in 27 days	Phosphomolybdate	
$C_{18}AE_{14}$	River-water dieaway	56% in 34 days	Phosphomolybdate	
$C_{10}AE_7$	River-water dieaway	97% in 6 days	Surface tension	Blankenship and Piccolini, 1963
$C_{12}AE_6$	River-water dieaway	99% in 6 days	Surface tension	
$C_{12}AE_8$	River-water dieaway	96-99% in 5-8 days	Surface tension	
	Warburg respirometer	48% in >3 hours	Oxygen consumption	
$C_{12}AE_{10}$	River-water dieaway	99% in 6 days	Surface tension	
$C_{12}AE_{20}$	River-water dieaway	99% in 23 days	Surface tension	
$C_{12}AE_{30}$	River-water dieaway	95% in 20 days	Surface tension	
$C_{14}AE_{9.3}$	River-water dieaway	97% in 6 days	Surface tension	
$C_{16}AE_{10.4}$	River-water dieaway	97% in 7 days	Surface tension	
n-sec-$C_{12}AE_8$ (4-OH)	River-water dieaway	99% in 10 days	Surface tension	
n-sec-$C_{12}AE_8$ (6-OH)	River-water dieaway	70% in 13 days	Surface tension	
$C_{12}AE_4$ (lauryl)	Continuous-flow activated sludge	100% in 10 hours	Phosphotungstate	Pitter and Trauc, 1963
$C_{16}C_{18}$–AE_{20} (cetyl/oleyl)	Continuous-flow activated sludge	0% in 10 hours	Phosphotungstate	
$C_{12}AE_5$ (Alfonic)	Inoculated medium	65% in 20 days	Oxygen consumption	Ruschenberg, 1963
$C_{12}AE_{10}$ (Alfonic)	Inoculated medium	53% in 20 days	Oxygen consumption	
$C_{12}AE_{20}$ (Alfonic)	Inoculated medium	33% in 20 days	Oxygen consumption	
$C_{12}AE_5$ (lauryl)	Inoculated medium	25-100% in 20 days	Mercuric iodide	Cuta and Hanusova, 1964
$C_{12\text{-}16}AE_5 + C_{12\text{-}16}AE_{14}$	Inoculated medium	80-90% in 20 days	Mercuric iodide	
$C_{12\text{-}16}AE_{13}$	Inoculated medium	100% in 20 days	Mercuric iodide	
$C_{16}C_{18}$–AE_8 (cetyl/oleyl)	Inoculated medium	60-85% in 20 days	Mercuric iodide	
$C_{16}C_{18}$–AE_{20} (cetyl/oleyl)	Inoculated medium	80% in 20 days	Mercuric iodide	
$C_{14}AE_8$	River-water dieaway	100% in 5 days	Infrared spectroscopy	Frazee et al., 1964

APPENDIX

Appendix (cont.)

Chemical[a]	Test System/Protocol	Extent of Biodegradation[b]	Measurement Method	Reference
$C_{12}AE_3$ (lauryl)	Warburg respirometer	—	Oxygen consumption	Garrison and Matson, 1964
$C_{12-18}AE_8$	Inoculated medium	99-100% in 9 days	Cobalt thiocyanate, surface tension, foam	
	Shake flask	98-100% in 7 days	Cobalt thiocyanate, surface tension, foam	
br-pri-$C_{13}AE_9$	Warburg respirometer	Oxygen uptake	Oxygen consumption	
	Shake flask	11% in 7 days	Cobalt thiocyanate	
$C_{12-18}AE$ (60%) (Alfonic 1218-6)	Continuous-flow activated sludge	100% in 4 hours	Cobalt thiocyanate	Huddleston and Allred, 1964a
C_8AE_5	Continuous-flow activated sludge	100% in 4 hours	Cobalt thiocyanate	Huddleston and Allred, 1964b; Allred and Huddleston, 1967
	Shake flask	100% in 3 days	Cobalt thiocyanate	
	River-water dieaway	100% in 8 days	Cobalt thiocyanate	
$C_{10}AE_7$	Continuous-flow activated sludge	100% in 4 hours	Cobalt thiocyanate	
	Shake flask	100% in 3 days	Cobalt thiocyanate	
	River-water dieaway	100% in 5 days	Cobalt thiocyanate	
$C_{12}AE_{4.4}$	Continuous-flow activated sludge	100% in 4 hours	Cobalt thiocyanate	
	Shake flask	100% in 3 days	Cobalt thiocyanate	
	River-water dieaway	100% in 4 days	Cobalt thiocyanate	
$C_{12}AE_6$	Continuous-flow activated sludge	100% in 4 hours	Cobalt thiocyanate	
	Shake flask	100% in 3 days	Cobalt thiocyanate	
	River-water dieaway	100% in 4 days	Cobalt thiocyanate	
$C_{12}AE_9$	Continuous-flow activated sludge	100% in 4 hours	Cobalt thiocyanate	
	Shake flask	100% in 2 days	Cobalt thiocyanate	
	River-water dieaway	100% in 3 days	Cobalt thiocyanate	

Appendix (cont.)

Chemical[a]	Test System/Protocol	Extent of Biodegradation[b]	Measurement Method	Reference
$C_{16}AE_{10}$	Continuous-flow activated sludge	100% in 4 hours	Cobalt thiocyanate	Huddleston and Allred, 1964b; Allred and Huddleston, 1967 (cont.)
	Shake flask	100% in 2 days	Cobalt thiocyanate	
	River-water dieaway	100% in 5 days	Cobalt thiocyanate	
$C_{18}AE_{10.5}$	Continuous-flow activated sludge	100% in 4 hours	Cobalt thiocyanate	
	Shake flask	100% in 2 days	Cobalt thiocyanate	
	River-water dieaway	100% in 13 days	Cobalt thiocyanate	
$C_{10-12}AE$ (58%)	Continuous-flow activated sludge	100% in 4 hours	Cobalt thiocyanate	
	Shake flask	100% in 2 days	Cobalt thiocyanate	
	River-water dieaway	100% in 5 days	Cobalt thiocyanate	
$C_{12-18}AE$ (62%)	Continuous-flow activated sludge	100% in 4 hours	Cobalt thiocyanate	
	Shake flask	100% in 2 days	Cobalt thiocyanate	
	River-water dieaway	100% in 5 days	Cobalt thiocyanate	
$C_{16-18}AE$ (63%)	Continuous-flow activated sludge	100% in 4 hours	Cobalt thiocyanate	
	Shake-flask	100% in 2 days	Cobalt thiocyanate	
	River-water dieaway	100% in 5 days	Cobalt thiocyanate	
$br-C_{15}AE_{14.5}$	Shake flask culture	33% in 4 days	Cobalt thiocyanate	
	River-water dieaway	86% in 26 days	Cobalt thiocyanate	
$C_{12}AE_3$ (lauryl)	Warburg respirometer	66% in 10 days	Oxygen uptake	Hunter and Heukelekian, 1964
$C_{12}AE_9$ (lauryl)	Warburg respirometer	56% in 10 days	Oxygen uptake	
$C_{12}AE_9$ (lauryl)	River-water dieaway	100% in 12 days	Surface tension	Knaggs, 1964
	Warburg respirometer	72% in 15 days	Oxygen uptake	
$C_{12-14}AE_9$	River-water dieaway	97% in 6 days	Cobalt thiocyanate	Myerly et al., 1964
	Warburg respirometer	50% in 3 days	Oxygen uptake	
$br-C_{12-14}AE_9$	River-water dieaway	99% in 10 days	Cobalt thiocyanate	
$sec-C_{11-15}AE_9$	River-water dieaway	97% in 6 days	Cobalt thiocyanate	
	Warburg respirometer	55% in 3 days	Oxygen uptake	

Appendix (cont.)

Chemical[a]	Test System/ Protocol	Extent of Biodegradation[b]	Measurement Method	Reference
$C_{12}AE_4$ (lauryl)	Inoculated medium	84% in 20 days	Oxygen uptake	Pitter, 1964
$C_{16}C_{18}$–$C_{18}AE_3$ (cetyl, oleyl, stearyl)	Inoculated medium	88% in 20 days	Oxygen uptake	
C_nAE_8	River-water dieaway	90% in 4 days	Cobalt thiocyanate	Steinle et al., 1964
		90% in 5 days	Surface tension	
	Warburg respirometer	50% in 2 days	Oxygen uptake	
C_nAE_9	River-water dieaway	90% in 6 days	Cobalt thiocyanate	
		90% in 4 days	Surface tension	
	Warburg respirometer	60% in 2 days	Oxygen uptake	
C_nAE_8	River-water dieaway	90% in 8 days	Cobalt thiocyanate	
		90% in 6 days	Surface tension	
n-sec-C_nAE_6	Warburg respirometer	50%	Oxygen consumption	
n-sec-C_nAE_8	River-water dieaway	90% in 6 days	Cobalt thiocyanate	
n-sec-C_nAE_9	Warburg respirometer	35%	Oxygen consumption	
n-sec-C_nAE_{10}	River-water dieaway	90% in 6 days	Cobalt thiocyanate	
n-sec-C_nAE_{11}	River-water dieaway	90% in 6 days	Cobalt thiocyanate, surface tension	
n-pri-$C_{16}AE_n$	River-water dieaway	97-98% in 28 days	Cobalt thiocyanate	Vath, 1964
			Surface tension	
			Foaming	
n-pri-$C_{12-14}AE_{7,4}$	River-water dieaway	100% in 16 days	Cobalt thiocyanate	
			Surface tension	
			Foaming	
	Warburg respirometer	50% in 30 hours	Oxygen consumption	
n-sec-$C_{16}AE_n$	River-water dieaway	93-98% in 28 days	Cobalt thiocyanate	
			Surface tension	
			Foaming	
n-sec-$C_{11-13}AE_6$	Warburg respirometer	39%; 26%	Oxygen consumption Organic ^{14}carbon	
n-sec-$C_{11-13}AE_8$	River-water dieaway	100% in 17 days	Cobalt thiocyanate	
			Surface tension	
			Foaming	

Appendix (cont.)

Chemical[a]	Test System/ Protocol	Extent of Biodegradation[b]	Measurement Method	Reference
n-sec-$C_{11-15}AE_9$	Anaerobic sewage dieaway	degradation in 14 days	Cobalt thiocyanate Foaming	Vath, 1964 (cont.)
n-sec-$C_{13-15}AE_6$	Warburg respirometer	39%; 26%	Oxygen consumption	
n-sec-$C_{13-15}AE_{9,5}$	Warburg respirometer	55% in 30 hours	Organic ^{14}carbon Oxygen consumption	
$C_{16}AE_{10}$	River-water dieaway	95% in 2 days	Surface tension, foam	Weil and Stirton, 1964
$C_{16}AE_{20}$	River-water dieaway	95% in 2 days 95% in >25 days	Surface tension Foaming	
$C_{18}AE_6$	River-water dieaway	100% in 2 days 80% in 3 days	Foaming Surface tension	
$C_{12}AE_3$ (lauryl)	Warburg respirometer	75%; 78%	Oxygen consumption	Barbaro and Hunter, 1965
sec-$C_{11-13}AE_9$	Activated sludge	95% in 1 day	Cobalt thiocyanate	Booman et al., 1965
sec-$C_{11-15}AE_9$ (Tergitol 15-S-9)	Activated sludge	93% in 2 days	Cobalt thiocyanate Surface tension	Conway et al., 1965
$C_{10-12}AE_6$	Shake flask	100% in 7 days	Cobalt thiocyanate Surface tension Foaming	Huddleston and Allred, 1965
	River-water dieaway	100% in 2 days	Cobalt thiocyanate Surface tension Foaming Soluble organics	
	Batch or semi-continuous activated sludge	91% in 2 days 100% in 1 day 95% in 1 day 93% in 1 day	Cobalt thiocyanate Surface tension Foaming Soluble organics	

Appendix (cont.)

Chemical[a]	Test System/ Protocol	Extent of Biodegradation[b]	Measurement Method	Reference
sec-$C_{11-15}AE_7$	River-water dieaway	98% in 14 days	Cobalt thiocyanate Surface tension	Conway and Waggy, 1966
	Trickling filter	65%	Cobalt thiocyanate	
	BOD bottle procedure	0.6% uptake	Oxygen uptake	
sec-$C_{11-15}AE_9$ (Tergitol 15-S-9)	River-water dieaway	98% in 21 days	Cobalt thiocyanate Surface tension	
	River-water dieaway	100% in 30 days	Cobalt thiocyanate Surface tension	
	Anaerobic river-water dieaway	90% in 8 hours	Cobalt thiocyanate Surface tension	
	Continuous-flow activated sludge	99% in 8 days	Cobalt thiocyanate Surface tension	
	Shake flask	93% in 8 days	Foaming	
		94% in 8 days	Cobalt thiocyanate	
	Trickling filter	65%	Oxygen uptake	
	BOD bottle procedure	0.4 (oxygen/substrate)	Carbon dioxide	
	Inoculated medium	Not given		
$C_{10-12}AE_{4,9}$ (Alfonic 1012-60)	Continuous-flow activated sludge	100% in 4 hours	Cobalt thiocyanate	Huddleston, 1966
	Shake flask	100% in 8 days	Cobalt thiocyanate	
	River-water dieaway	100% in 11 days	Cobalt thiocyanate Surface tension Foaming	
	BOD bottle procedure	0.7 (oxygen/substrate)	Oxygen consumption	
$C_{16-18}AE_9$	Inoculated medium	100% in 5 days	Thin-layer chromatography	Laboratory of the Government Chemist, 1966
$C_{16-18}AE_{20}$	Inoculated medium	100% in 8 days	Thin-layer chromatography	
sec-$C_{11-15}AE_9$ (Tergitol 15-S-9)	Continuous-flow activated sludge	92% in 3 hours	Cobalt thiocyanate Foaming	Lashen et al., 1966
		82% in 3 hours		
C_nAE_9	Sewage dieaway	98% in 4 days	Thin-layer chromatography	Patterson et al., 1966
C_nAE_{20}	Sewage dieaway	100% in 14 days	Thin-layer chromatography	

Appendix (cont.)

Chemical[a]	Test System/ Protocol	Extent of Biodegradation[b]	Measurement Method	Reference
sec-$C_{14}AE_9$, sec-$C_{14}AE_{12}$	Continuous-flow activated sludge	95% in 3 hours	Bismuth iodide	Wickbold, 1966
sec-$C_{11-15}AE_9$	River-water dieaway	95% in 21 days	Foaming	Booman et al., 1967
sec-$C_{11-15}AE_{12.5}$	River-water dieaway	70% in 21 days	Foaming	
sec-$C_{11-15}AE_{16.5}$	River-water dieaway	40% in 21 days	Foaming	
$C_{12}AE_9$ (lauryl)	Inoculated medium	100% in 28 days	Soluble organics	Borstlap and Kortland, 1967
$C_{12}C_{14}AE_6$ (lauryl, myristyl)	Inoculated medium	100% in 28 days	Soluble organics	
$C_{12}C_{14}AE_9$ (lauryl, myristyl)	Inoculated medium	100% in 28 days	Soluble organics	
$C_{12}C_{14}AE_{12}$ (lauryl, myristyl)	Inoculated medium	94% in 28 days	Soluble organics	
$C_{12}C_{14}AE_{15}$ (lauryl, myristyl)	Inoculated medium	89% in 28 days	Soluble organics	
$C_{12}C_{14}AE_{18}$ (lauryl, myristyl)	Inoculated medium	81% in 28 days	Soluble organics	
$C_{12}C_{14}AE_{30}$ (lauryl, myristyl)	Inoculated medium	64% in 28 days	Soluble organics	
$C_{12-15}AE_6$	Inoculated medium	100% in 28 days	Soluble organics	
$C_{12-15}AE_9$	Inoculated medium	95% in 28 days	Soluble organics	
$C_{12-15}AE_{12}$	Inoculated medium	95% in 28 days	Soluble organics	
$C_{12-15}AE_{15}$	Inoculated medium	89% in 28 days	Soluble organics	
$C_{12-15}AE_{30}$	Inoculated medium	77% in 28 days	Soluble organics	
$C_{12-15}AE_{30}$	Inoculated medium	68% in 28 days	Soluble organics	
n-pri-$C_{12}AE_9$	Inoculated medium	98-100% in 7 days	Cobalt thiocyanate	Bunch and Chambers, 1967
n-pri-$C_{14}AE_8$	Inoculated medium	95-100% in 7 days	Cobalt thiocyanate	
br-pri-$C_{13}AE_9$	Inoculated medium	0-10% in 7 days	Cobalt thiocyanate	

Appendix (cont.)

Chemical[a]	Test System/ Protocol	Extent of Biodegradation[b]	Measurement Method	Reference
$C_{12-16}AE_8$	Batch or semi-continuous activated sludge	100% in 1 day	Sulfation methylene blue	Han, 1967
$C_{12-16}AE_{16}$	Batch or semi-continuous activated sludge	86% in 1 day	Sulfation methylene blue	
$C_{16,18}AE_{40}$ (tallow)	Batch or semi-continuous activated sludge	12% in 1 day	Sulfation methylene blue	
$C_{11-14}AE_3$	Batch or semi-continuous activated sludge	100% in 1 day	Sulfation methylene blue	
$C_{11-14}AE_{13}$	Batch or semi-continuous activated sludge	92% in 1 day	Sulfation methylene blue	
sec-$C_{11-15}AE_8$	Batch or semi-continuous activated sludge	92% in 1 day	Sulfation methylene blue	
sec-$C_{11-15}AE_{15}$	Batch or semi-continuous activated sludge	57% in 1 day	Sulfation methylene blue	
$C_{12-18}AE_9$	Warburg respirometer	0.8-1.3 (oxygen/substrate)	Oxygen consumption	Hartmann et al., 1967
$C_{12}AE_4$ (lauryl)	Inoculated medium	97% in 30 days	Oxygen consumption	Heinz and Fischer, 1967
$C_{12}AE_6$	Inoculated medium	76% in 30 days	Oxygen consumption	
$C_{12}AE_8$	Inoculated medium	72-78% in 30 days	Oxygen consumption	
$C_{12}AE_{10}$	Inoculated medium	73% in 30 days	Oxygen consumption	
$C_{12}AE_{16}$	Inoculated medium	62% in 30 days	Oxygen consumption	
$C_{12}AE_{20}$	Inoculated medium	41% in 30 days	Oxygen consumption	
$C_{12}AE_{30}$	Inoculated medium	-3% in 30 days	Oxygen consumption	
br-$C_{13}AE_9$	Inoculated medium	15% in 30 days	Oxygen consumption	
$C_{12}AE_n$ (Lissapol DS 4229)	Recycle trickling filter	100% in 1 day	Cobalt thiocyanate Thin-layer chromatography	Jenkins et al., 1967

Appendix (cont.)

Chemical[a]	Test System/ Protocol	Extent of Biodegradation[b]	Measurement Method	Reference
$C_{18}AE_8$ (stearyl)	Sludge inoculated	100% in 10 days	Thin-layer chromatography	Patterson et al., 1967
$C_{10-12}AE$ (58%)	Sludge inoculated	100% in 3 days	Thin-layer chromatography	
$C_{12-15}AE_9$	Sludge inoculated	100% in 24 days	Thin-layer chromatography	
$C_{12-18}AE$ (62%)	Sludge inoculated	100% in 4 days	Thin-layer chromatography	
$C_{16-18}AE_6$	Sludge inoculated	100% in 7 days	Thin-layer chromatography	
$C_{12-18}AE_9$	Sludge inoculated	100% in 7 days	Thin-layer chromatography	
$C_{16-18}AE_{15}$	Sludge inoculated	100% in 7 days	Thin-layer chromatography	
$C_{16-18}AE_{20}$	Sludge inoculated	100% in 9 days	Thin-layer chromatography	
$C_{16-18}AE_{22}$	Sludge inoculated	98% in 28 days	Thin-layer chromatography	
$C_{16-18}AE_{30}$	Sludge inoculated	100% in 22 days	Thin-layer chromatography	
sec-$C_{11-15}AE_9$	Sludge inoculated	100% in 7 days	Thin-layer chromatography	
sec-$C_{11-15}AE_{13}$	Sludge inoculated	100% in 20 days	Thin-layer chromatography	
br-$C_{13}AE_8$	Sludge inoculated	70% in 49 days	Thin-layer chromatography	
$C_{16-18}AE_5$	Inoculated medium	91% in 20 days	Phosphotungstate	Pitter, 1968a
		87% in 20 days	Chemical oxygen demand	
$C_{16-18}AE_{10}$	Inoculated medium	74% in 20 days	Phosphotungstate	
		70% in 20 days	Chemical oxygen demand	
$C_{16-18}AE_{15}$	Inoculated medium	36% in 20 days	Phosphotungstate	
		46% in 20 days	Chemical oxygen demand	
$C_{16-18}AE_{20}$	Inoculated medium	7% in 20 days	Phosphotungstate	
		32% in 20 days	Chemical oxygen demand	
$C_{16-18}AE_{25}$	Inoculated medium	0% in 20 days	Phosphotungstate	
		13% in 20 days	Chemical oxygen demand	
$C_{10-16}AE_3$	Inoculated medium	90% in 20 days	Phosphotungstate	Pitter, 1968a; 1968b
		93% in 20 days	Chemical oxygen demand	
		66% in 20 days	Oxygen consumption	
$C_{10-16}AE_5$	Inoculated medium	75% in 20 days	Phosphotungstate	
		84% in 20 days	Chemical oxygen demand	
		62% in 20 days	Oxygen consumption	
$C_{10-16}AE_8$	Inoculated medium	65% in 20 days	Phosphotungstate	
		73% in 20 days	Chemical oxygen demand	
		52% in 20 days	Oxygen consumption	

Appendix (cont.)

Chemical[a]	Test System/Protocol	Extent of Biodegradation[b]	Measurement Method	Reference
$C_{10-16}AE_{10}$	Inoculated medium	50% in 20 days 64% in 20 days 48% in 20 days	Phosphotungstate Chemical oxygen demand Oxygen consumption	Pitter, 1968a; 1968b (cont.)
$C_{10-16}AE_{14}$	Inoculated medium	48% in 20 days 34% in 20 days	Chemical oxygen demand Oxygen consumption	
$C_{10-16}AE_{15}$	Inoculated medium	33% in 20 days 41% in 20 days 32% in 20 days	Phosphotungstate Chemical oxygen demand Oxygen consumption	
$C_{10-16}AE_{18}$	Inoculated medium	36% in 20 days 23% in 20 days	Chemical oxygen demand Oxygen consumption	
$C_{10-16}AE_{20}$	Inoculated medium	8% in 20 days 27% in 20 days 18% in 20 days	Phosphotungstate Chemical oxygen demand Oxygen consumption	
C_nAE_{3-6}	Inoculated medium	80-96% in 20 days 61-88% in 20 days	Chemical oxygen demand Oxygen consumption	Pitter, 1968c
C_nAE_{20-30}	Inoculated medium	16-50% in 20 days 12-27% in 20 days	Chemical oxygen demand Oxygen consumption	
C_nAE_9 (Empilan KM9)	Inoculated medium Batch or semi-continuous activated sludge Recycle trickling filter	98-99% in 21 days 98-99% in 21 days 98-99% in 21 days	— — —	Truesdale et al., 1968
C_nAE_{20} (Empilan KM20)	Inoculated medium Recycle trickling filter	98-99% in 1 day 98-99% in 7 days	— —	
$C_{16,18}AE_{10}$ (tallow) $C_{16,18}AE_{20}$ (tallow) $C_{16,18}AE_{25}$ (tallow)	Inoculated medium Inoculated medium Inoculated medium	100% in 2 days 100% in 2 days 98% in 4 days	Surface tension Surface tension Surface tension	Arpino, 1969
sec-$C_{14,7}AE_8$ sec-$C_{14,7}AE_9$ sec-$C_{14,7}AE_{18}$ sec-$C_{14,7}AE_{23}$	Inoculated medium Inoculated medium Inoculated medium Inoculated medium	96% in 2 days 95% in 2 days 95% in 4 days 94% in 5 days	Cobalt thiocyanate Cobalt thiocyanate Cobalt thiocyanate Cobalt thiocyanate	Gebril and Naim, 1969a

Appendix (cont.)

Chemical[a]	Test System/Protocol	Extent of Biodegradation[b]	Measurement Method	Reference
sec-$C_{14.7}AE_{36}$	Inoculated medium	94% in 6 days	Cobalt thiocyanate	Gebril and Naim, 1969a (cont.)
sec-$C_{17.8}AE_9$	Inoculated medium	98% in 2 days	Cobalt thiocyanate	
sec-$C_{17.8}AE_{21}$	Inoculated medium	98% in 3 days	Cobalt thiocyanate	
sec-$C_{17.8}AE_{29}$	Inoculated medium	96% in 4 days	Cobalt thiocyanate	
sec-$C_{17.8}AE_{36}$	Inoculated medium	94% in 5 days	Cobalt thiocyanate	
sec-$C_{11}AE_6$	Inoculated medium	93% in 2 days	Cobalt thiocyanate	Gebril and Naim, 1969b
sec-$C_{11}AE_8$	Inoculated medium	93% in 2 days	Cobalt thiocyanate	
sec-$C_{11}AE_{13}$	Inoculated medium	93% in 4 days	Cobalt thiocyanate	
sec-$C_{11}AE_{18}$	Inoculated medium	91% in 5 days	Cobalt thiocyanate	
sec-$C_{11}AE_{24}$	Inoculated medium	95% in 7 days	Cobalt thiocyanate	
$C_{12-14}AE_{10}$	Semi-continuous activated sludge	100% in 1 day	Foaming	Mausner et al., 1969
	River-water dieaway	100% in 14 days	Cobalt thiocyanate	
	Shake flask	100% in 14 days	Foaming	
$C_{12-15}AE_9$	Shake flask	100% in 7 days	Cobalt thiocyanate	
	Semi-continuous activated sludge	99% in 17 hours	Foaming	
sec-$C_{11-15}AE_9$	Semi-continuous activated sludge	85-96% in 1 day	Foaming	
		93% in 17 hours	Foaming	
	River-water dieaway	94% in 1 day	Cobalt thiocyanate	
		98% in 28 days	Foaming	
		100% in 28 days	Cobalt thiocyanate	
	Shake flask	62% in 14 days	Foaming	
		93% in 14 days	Cobalt thiocyanate	
br-$C_{13}AE_9$	Inoculated medium	85% in 7 days	Foaming	
	Semi-continuous activated sludge	75% in 1 day	Foaming	
		63% in 1 day	Cobalt thiocyanate	
	River-water dieaway	91% in 28 days	Foaming	
		100% in 28 days	Cobalt thiocyanate	
	Shake flask	0% in 14 days	Foaming	
		31% in 14 days	Cobalt thiocyanate	

APPENDIX

Appendix (cont.)

Chemical[a]	Test System/ Protocol	Extent of Biodegradation[b]	Measurement Method	Reference
*$C_{18}AE_6$	Inoculated medium	48% in 7 days	$^{14}CO_2$ formation	Nooi et al., 1970
		97% in 7 days	Organic ^{14}carbon	
*$C_{18}AE_{21}$	Inoculated medium	49% in 7 days	$^{14}CO_2$ formation	
C_{18}*AE_8	Inoculated medium	49% in 7 days	$^{14}CO_2$ formation	
C_{18}*AE_{21}	Inoculated medium	41% in 7 days	$^{14}CO_2$ formation	
$C_{12}AE_4$	Inoculated medium	74-96% in 30 days	Oxygen consumption	Fischer, 1971
$C_{12}AE_6$	Inoculated medium	79-85% in 30 days	Oxygen consumption	
$C_{12}AE_8$	Inoculated medium	77-92% in 30 days	Oxygen consumption	
$C_{12}AE_{10}$	Inoculated medium	73-75% in 30 days	Oxygen consumption	
$C_{12}AE_{12}$	Inoculated medium	76-78% in 30 days	Oxygen consumption	
$C_{12}AE_{16}$	Inoculated medium	60% in 30 days	Oxygen consumption	
$C_{12}AE_{18}$	Inoculated medium	56-57% in 30 days	Oxygen consumption	
$C_{12}AE_{20}$	Inoculated medium	43-50% in 30 days	Oxygen consumption	
$C_{12}AE_{30}$	Inoculated medium	4-5% in 30 days	Oxygen consumption	
br-$C_{13}AE_9$	Inoculated medium	15% in 30 days	Oxygen consumption	
$C_{12-15}AE_9$ (Dobanol 25-9)	Continuous-flow activated sludge	95-98%	Thin-layer chromatography	Mann and Reid, 1971
$C_{9-11}AE_8$ (Dobanol 91-8)	Continuous-flow activated sludge	97-98%	Thin-layer chromatography	
$C_{11-15}AE_9$	Inoculated medium	0.44 (oxygen/substrate)	Oxygen consumption	Zika, 1971
$C_{12}AE_8$	Inoculated medium	85% in 30 days	Oxygen consumption	Fischer, 1972
sec-$C_{11-15}AE_9$	Inoculated medium	63% in 30 days	Oxygen consumption	
$C_{18}AE_5$ (stearyl)	Inoculated medium	31% in 5 days	Oxygen consumption	Moller, 1972
$C_{18}AE_{7-8}$ (stearyl)	Inoculated medium	28% in 5 days	Oxygen consumption	
$C_{18}AE_{10}$ (stearyl)	Inoculated medium	23% in 5 days	Oxygen consumption	
$C_{18}AE_{15}$ (stearyl)	Inoculated medium	17% in 5 days	Oxygen consumption	
$C_{18}AE_{20}$ (stearyl)	Inoculated medium	12% in 5 days	Oxygen consumption	
$C_{18}AE_{25}$ (stearyl)	Inoculated medium	10% in 5 days	Oxygen consumption	
C_{18}=AE_5 (oleyl)	Inoculated medium	39% in 5 days	Oxygen consumption	
C_{18}=AE_{7-8} (oleyl)	Inoculated medium	28% in 5 days	Oxygen consumption	
$C_{12,14,16,18}AE_5$ (coconut)	Inoculated medium	44% in 5 days	Oxygen consumption	
$C_{12,14,16,18}AE_{7-8}$ (coconut)	Inoculated medium	22% in 5 days	Oxygen consumption	

Appendix (cont.)

Chemical[a]	Test System/ Protocol	Extent of Biodegradation[b]	Measurement Method	Reference
$C_{16-20}AE_{10}$	Inoculated medium	50% in 6 days >95% in 13-20 days >95% in 13-20 days	Oxygen consumption Gas chromatography Thin-layer chromatography	Rudling, 1972
$C_{12-15}AE_6$	Inoculated medium	99% in 12 days 99% in 13 days	Thin-layer chromatography Foaming	Stead et al., 1972
sec-$C_{11-15}AE_9$	Inoculated medium	99% in 14 days 96% in 19 days	Thin-layer chromatography Foaming	
$C_{16}AE_{10}$	Continuous-flow activated sludge	99% in 3 hours	Bismuth iodide	Gerike and Schmied, 1973
$C_{16}AE_{11}$	Inoculated medium	99% in 30 days 80% in 30 days	Bismuth iodide Oxygen consumption	
$C_{16}AE_{31}$	Continuous-flow activated sludge	98% in 3 hours	Bismuth iodide	
	Inoculated medium	98% in 30 days 57% in 30 days	Bismuth iodide Oxygen consumption	
$C_{16,18}AE_{14}$ (tallow)	Continuous-flow activated sludge	100% in 3 hours	Bismuth iodide	
	Inoculated medium	98% in 30 days 81% in 30 days	Bismuth iodide Oxygen consumption	
$C_{12}AE_6$	Inoculated medium	99% in 30 days 81% in 30 days	Bismuth iodide Oxygen consumption	
$C_{12}AE_8$	Inoculated medium	99% in 30 days 82% in 30 days	Bismuth iodide Oxygen consumption	
$C_{12}AE_{10}$	Inoculated medium	99% in 30 days 83% in 30 days	Bismuth iodide Oxygen consumption	
$C_{12}AE_{12}$	Inoculated medium	98% in 30 days 85% in 30 days	Bismuth iodide Oxygen consumption	
$C_{12}AE_{14}$	Inoculated medium	98% in 30 days 85% in 30 days	Bismuth iodide Oxygen consumption	
$C_{12}AE_{16}$	Inoculated medium	98% in 30 days 90% in 30 days	Bismuth iodide Oxygen consumption	
$C_{12}AE_{18}$	Inoculated medium	98% in 30 days 70% in 30 days	Bismuth iodide Oxygen consumption	
$C_{12}AE_{20}$				

Appendix (cont.)

Chemical[a]	Test System/Protocol	Extent of Biodegradation[b]	Measurement Method	Reference
$C_{9-11}AE_8$	Continuous-flow activated sludge	96% in 6 hours	Thin-layer chromatography	Stiff et al., 1973
$C_{12-15}AE_9$	Continuous-flow activated sludge	97% in 6 hours	Thin-layer chromatography	
C_8AE_3	Inoculated medium	77% in 28 days	Carbon dioxide formation	Sturm, 1973
$C_{10}AE_3$	Inoculated medium	75% in 28 days	Carbon dioxide formation	
$C_{12}AE_3$	Inoculated medium	63% in 28 days	Carbon dioxide formation	
$C_{14}AE_3$	Inoculated medium	66% in 28 days	Carbon dioxide formation	
$C_{16}AE_3$	Inoculated medium	80% in 28 days	Carbon dioxide formation	
$C_{18}AE_3$	Inoculated medium	85% in 28 days	Carbon dioxide formation	
$C_{20}AE_3$	Inoculated medium	77% in 28 days	Carbon dioxide formation	
$C_{14-18}AE_3$	Inoculated medium	80% in 28 days	Carbon dioxide formation	
$C_{14-18}AE_6$	Inoculated medium	79% in 28 days	Carbon dioxide formation	
$C_{14-18}AE_9$	Inoculated medium	69% in 28 days	Carbon dioxide formation	
$C_{14-18}AE_{10.6}$	Inoculated medium	74% in 28 days	Carbon dioxide formation	
$C_{41-18}AE_{11}$	Inoculated medium	76% in 28 days	Carbon dioxide formation	
$C_{14-18}AE_{20}$	Inoculated medium	43% in 28 days	Carbon dioxide formation	
$C_{14-18}AE_{30}$	Inoculated medium	33% in 28 days	Carbon dioxide formation	
$C_{16,18}AE_9$ (tallow)	Inoculated medium	100% in 3 days	Cobalt thiocyanate Foaming	
		100% in 3 days		
		70% in 28 days	Carbon dioxide formation	
$C_{16,18}AE_{30}$ (tallow)	Inoculated medium	100% in 4 days	Cobalt thiocyanate Foaming	
		100% in 6 days		
		32% in 28 days	Carbon dioxide formation	
$C_{10-14}AE_6$	Inoculated medium	66% in 28 days	Carbon dioxide formation	
$C_{10-14}AE_{12}$	Inoculated medium	47% in 28 days	Carbon dioxide formation	
sec-$C_{10-14}AE_9$	Inoculated medium	54% in 28 days	Carbon dioxide formation	
$C_{16,18}AE_{10}$ (tallow)	Inoculated medium	96% in 2 days	Bismuth iodide	Treccani et al., 1973
$C_{12-15}AE_9$	Inoculated medium	100% in 2 days	Bismuth iodide	

Appendix (cont.)

Chemical[a]	Test System/Protocol	Extent of Biodegradation[b]	Measurement Method	Reference
$C_{14}AE_7$	Inoculated medium	100% in 2 days	Surface tension	Albanese and Capuci, 1974
		100% in 3 days	Foaming	
		100% in 4 days	Bismuth iodide	
		100% in 4 days	Thin-layer chromatography	
$C_{14}AE_{27}$	Inoculated medium	87% in 8 days	Surface tension	
		60% in 8 days	Foaming	
		96% in 7 days	Bismuth iodide	
		80% in 8 days	Thin-layer chromatography	
$C_{12,14,16}AE_7$ (Alfonic)	Inoculated medium	98% in 8 days	Surface tension	
		100% in 8 days	Foaming	
$C_{12,14,16}AE_{27}$ (Alfonic)	Inoculated medium	78% in 8 days	Surface tension	
		60% in 8 days	Foaming	
		60% in 8 days	Thin-layer chromatography	
$C_{16,18}AE_{11}$ (Alfonic)	Inoculated medium	86% in 6 days	Surface tension	
		91% in 8 days	Foaming	
		100% in 6 days	Thin-layer chromatography	
$C_{16,18}AE_{30}$ (Alfonic)	Inoculated medium	89% in 8 days	Surface tension	
		81% in 8 days	Foaming	
$C_{12-15}AE_7$ (80% linear)	Inoculated medium	98% in 8 days	Surface tension	
		98% in 8 days	Foaming	
$C_{12-15}AE_{27}$ (80% linear)	Inoculated medium	56% in 8 days	Surface tension	
		25% in 8 days	Foaming	
		60% in 8 days	Thin-layer chromatography	
$C_{11-15}AE_7$ (50% linear)	Inoculated medium	96% in 6 days	Surface tension	
		96% in 6 days	Foaming	
$C_{11-15}AE_{11}$ (50% linear)	Inoculated medium	92% in 8 days	Surface tension	
		92% in 8 days	Foaming	
		93% in 6 days	Thin-layer chromatography	
$C_{11-15}AE_{26}$ (50% linear)	Inoculated medium	73% in 8 days	Surface tension	
		38% in 8 days	Foaming	
		70% in 8 days	Thin-layer chromatography	

Appendix (cont.)

Chemical[a]	Test System/ Protocol	Extent of Biodegradation[b]	Measurement Method	Reference
$C_{12,14,16}AE_5$ (coconut)	Inoculated medium	100% in 4 days	Bismuth iodide	Arpino et al., 1974
		100% in 2 days	Surface tension	
$C_{12,14,16}AE_{15}$ (coconut)	Inoculated medium	99% in 7 days	Bismuth iodide	
		95% in 7 days	Surface tension	
$C_{12,14}AE_6$ (Alfonic)	Inoculated medium	100% in 2 days	Bismuth iodide	
		100% in 2 days	Surface tension	
$C_{12,14}AE_3$ (Alfonic)	Inoculated medium	98% in 9 days	Bismuth iodide	
		98% in 9 days	Surface tension	
$C_{12-15}AE_5$ (75% linear)	Inoculated medium	100% in 4 days	Bismuth iodide	
		100% in 2 days	Surface tension	
$C_{12-15}AE_{15}$	Inoculated medium	98% in 4 days	Bismuth iodide	
		98% in 9 days	Surface tension	
$C_{12-15}AE_5$ (61% linear)	Inoculated medium	100% in 9 days	Surface tension	
		100% in 7 days	Foaming	
$C_{12-15}AE_{15}$	Inoculated medium	100% in 9 days	Surface tension	
		100% in 9 days	Foaming	
$C_{12-15}AE_6$ (49% linear)	Inoculated medium	100% in 4 days	Surface tension	
		100% in 4 days	Foaming	
$C_{12-15}AE_{16}$	Inoculated medium	100% in 7 days	Surface tension	
		100% in 7 days	Foaming	
$C_{12-15}AE_5$ (41% linear)	Inoculated medium	100% in 9 days	Surface tension	
		100% in 9 days	Foaming	
$C_{12-15}AE_{15}$	Inoculated medium	99% in 11 days	Surface tension	
		99% in 11 days	Foaming	
br-$C_{13}AE_5$	Inoculated medium	66% in 14 days	Bismuth iodide	
		78% in 14 days	Surface tension	
br-$C_{13}AE_{15}$	Inoculated medium	26% in 14 days	Bismuth iodide	
		50% in 14 days	Surface tension	

Appendix (cont.)

Chemical[a]	Test System/ Protocol	Extent of Biodegradation[b]	Measurement Method	Reference
$C_{12}AE_8$	Sewage dieaway	15% in 5 days	Iodine method	Baleux and Caumette, 1974
	Recycle trickling filter	12% in 5 days	Iodine method	
C_nAE_{12}	Sewage dieaway	75% in 5 days	Iodine method	
	Recycle trickling filter	62% in 5 days	Iodine method	
$C_{9-11}AE_6$	Sewage dieaway	100% in 3 days	Iodine method	
	Recycle trickling filter	199% in 3 days	Iodine method	
$C_{12,14,16}AE_5$ (coconut)	Inoculated medium	98% in 2 days	Surface tension	Borsari et al., 1974
$C_{12,14,16}AE_{15}$ (coconut)	Inoculated medium	97% in 2 days	Surface tension	
$C_{12,14}AE_5$ (Alfonic)	Inoculated medium	97% in 2 days	Surface tension	
$C_{12,14}AE_{15}$ (Alfonic)	Inoculated medium	98% in 2 days	Surface tension	
C_nAE_5	Inoculated medium	98% in 2 days	Surface tension	
C_nAE_{15}	Inoculated medium	96% in 3 days	Surface tension	
C_nAE_5	Inoculated medium	98% in 2 days	Surface tension	
C_nAE_{15}	Inoculated medium	97% in 3 days	Surface tension	
br-$C_{13}AE_5$	Inoculated medium	15% in 8 days	Surface tension	
br-$C_{13}AE_{15}$	Inoculated medium	30% in 8 days	Surface tension	
$C_{10,12}AE_8$ (Alfonic)	Inoculated medium	94% in 4 days	Bismuth iodide	Brüschweiler, 1974
		92% in 4 days	Surface tension	
$C_{16,18}AE_{10}$ (tallow)	Inoculated medium	99% in 30 days	Bismuth iodide	Fischer et al., 1974
		75% in 30 days	Oxygen consumption	
$C_{16-20}AE_{14}$	Inoculated medium	95% in 5 days	Bismuth iodide	Rudling and Solyom, 1974
br-$C_{13}AE_9$	Inoculated medium	31% in 10 days	Bismuth iodide	Wencker et al., 1974
$C_{11-14}AE_n$	Continuous-flow activated sludge	Expressed as rates/3 hours	Gas chromatography	Wickbold, 1974
$C_{11-13}AE_{13}$	Continuous-flow activated sludge	96% in 3 hours	Bismuth iodide	
$C_{10}AE_5$	Batch or semi-continuous activated sludge	96-98% in 13 days	Chemical oxygen demand	Zahn and Wellens, 1974

Appendix (cont.)

Chemical[a]	Test System/Protocol	Extent of Biodegradation[b]	Measurement Method	Reference
$C_{10-12}AE_8$	Inoculated medium	97% in 8 days	Bismuth iodide	Brüschweiler, 1975
		95% in 8 days	Surface tension	
$C_{10-12}AE_{11}$	Inoculated medium	93% in 4 days	Surface tension	
$C_{10-12}AE_{23}$	Inoculated medium	96% in 4 days	Surface tension	
$C_{16,18}AE_{10}$ (tallow)	Inoculated medium	82% in 8 days	Surface tension	
$C_{16,18}AE_{23}$ (tallow)	Inoculated medium	95% in 8 days	Surface tension	
$C_{16}AE_{10}$	Inoculated medium	99% in 14 days	Bismuth iodide	Fischer et al, 1975
		100% in 30 days	Bismuth iodide	
		71% in 30 days	Oxygen consumption	
	Continuous-flow activated sludge	98% in 3 hours	Bismuth iodide	
		67% in 3 hours	Chemical oxygen demand	
		56% in 3 hours	Organic carbon	
$C_{16}AE_{31}$	Inoculated medium	84% in 14 days	Bismuth iodide	
		99% in 30 days	Bismuth iodide	
		59% in 30 days	Oxygen consumption	
	Continuous-flow activated sludge	98% in 3 hours	Bismuth iodide	
		33% in 3 hours	Chemical oxygen demand	
$C_{12}AE_9$	River-water dieaway	100% in 3 days	Cobalt thiocyanate	Kurata and Koshida, 1975
		98% in 30 days	Chemical oxygen demand	
		97% in 5 days	Foaming	
$C_{12-15}AE_9$ (60% linear)	River-water dieaway (polluted water)	92% in 30 days	Cobalt thiocyanate	
		86% in 25 days	Chemical oxygen demand	
		99% in 20 days	Foaming	
	River-water dieaway (cleaner water)	69% in 30 days	Cobalt thiocyanate	
		100% in 3 days	Chemical oxygen demand	
sec-$C_{12-14}AE_9$	River-water dieaway (polluted water)	100% in 3 days	Foaming	
		95% in 30 days	Cobalt thiocyanate	
		83% in 30 days	Chemical oxygen demand	
	River-water dieaway (cleaner water)	97% in 25 days	Foaming	
		78% in 30 days	Chemical oxygen demand	

Appendix (cont.)

Chemical[a]	Test System/Protocol	Extent of Biodegradation[b]	Measurement Method	Reference
$C_{12}AE_9$	Inoculated medium	98% in 3 days 100% in 31 days	Surface tension Organic carbon	Sekiguchi et al., 1975
$C_{12-14}AE_5$	Batch or semi-continuous activated sludge Continuous-flow activated sludge	82-92% in 14 days 0-70% in 3 hours	Chemical oxygen demand Organic carbon Chemical oxygen demand Organic carbon	Zahn and Huber, 1975
$C_{13-15}AE_n$	Inoculated medium	100% in 12 days 95% in 39 days 99% in 6 hours 76% in 6 hours	Bismuth iodide Organic carbon Bismuth iodide Organic carbon	Brown, 1976
$C_{13-15}AE_7$	Continuous-flow activated sludge Continuous-flow activated sludge	99% in 3 hours 64% in 3 hours 96% in 6 hours 82% in 6 hours	Bismuth iodide Organic carbon Bismuth iodide Organic carbon	
$C_{13-15}AE_{11}$	Continuous-flow activated sludge	99% in 3 hours 60% in 3 hours	Bismuth iodide Organic carbon	
$C_{13-15}AE_{19}$	Continuous-flow activated sludge	97% in 3 hours 40% in 3 hours	Bismuth iodide Organic carbon	
$C_{13-15}AE_{25}$	Continuous-flow activated sludge	95% in 3 hours 33% in 3 hours	Bismuth iodide Organic carbon	
C_nAE_m	Soil	~100% in 30 days	—	Citernesi et al., 1976
sec-$C_{12-14}AE_9$	Continuous-flow activated sludge	100% in 1.5 days 93-98% in 1.5 days 80-90% in 1.5 days	Cobalt thiocyanate Chemical oxygen demand Organic carbon	Kurata and Koshida, 1976
$C_{12}AE_9$ sec-$C_{13}AE_{10}$ (C-7 branch)	Inoculated medium Inoculated medium	97% in 20 days 98% in 45 days	Organic carbon Organic carbon	Kuwamura and Takahaski, 1976

Appendix (cont.)

Chemical[a]	Test System/ Protocol	Extent of Biodegradation[b]	Measurement Method	Reference
$C_{12}AE_3$	Inoculated medium	93-96% in 42 days	Organic carbon	Laboureur et al., 1976
$C_{13}AE_{80}$	Inoculated medium	58-96% in 28 days	Organic carbon	
$C_{17}AE_{11}$	Inoculated medium	89-96% in 42 days	Organic carbon	
$C_{14}AE_3$	Inoculated medium	90-94% in 14 days	Organic carbon	
$C_{14}AE_{11}$	Inoculated medium	86-93% in 42 days	Organic carbon	
$C_{14}AE_{23}$	Inoculated medium	82-92% in 28 days	Organic carbon	
br-$C_{10}AE_3$	Inoculated medium	0-49% in 42 days	Organic carbon	
br-$C_{10}AE_{11}$	Inoculated medium	3-81% in 42 days	Organic carbon	
br-$C_{10}AE_{25}$	Inoculated medium	12-70% in 42 days	Organic carbon	
br-$C_{10}AE_{50}$	Inoculated medium	5-41% in 42 days	Organic carbon	
br-$C_{13}AE_2$	Inoculated medium	80-89% in 42 days	Organic carbon	
br-$C_{13}AE_3$	Inoculated medium	21-25% in 42 days	Organic carbon	
br-$C_{13}AE_{11}$	Inoculated medium	5-23% in 42 days	Organic carbon	
br-$C_{13}AE_{25}$	Inoculated medium	1-68% in 42 days	Organic carbon	
$C_{13-14}AE_{10}$ (45% linear)	Pond water	96% in 33 days	Cobalt thiocyanate	Mann and Schöberl, 1976
$C_{12-16}AE_3$	Continuous-flow activated sludge	72-87% in 3 hours	Phosphotungstate	Miksch, 1976; 1980
$C_{14-15}AE_7$ (Dobanol 24-7)	River-water dieaway	95% in 14 days 98% in 24 days	Surface tension Bismuth iodide	Reiff, 1976
$C_{14-15}AE_{11}$ (Dobanol 45-11)	River-water dieaway	96% in 24 days 99% in 24 days	Surface tension Bismuth iodide	
sec-$C_{11-15}AE_9$	River-water dieaway	96-97% in 11 days 94-96% in 11 days	Cobalt thiocyanate Foaming, Surface tension	Ruiz Cruz and Dobarganes Garcia, 1976
$C_{11-14}AE_7$ (45% linear)	Seawater Pond water	85% in 16 days 92% in 23 days	Bismuth iodide	Schöberl and Mann, 1976
$C_{11-14}AE_9$ (45% linear)	Seawater Pond water	95% in 30 days 92% in 36 days	Bismuth iodide	
$C_{11-14}AE_{11}$ (45% linear)	Seawater pond water	95% in 30 days 86% in 36 days	Bismuth iodide	

Appendix (cont.)

Chemical[a]	Test System/ Protocol	Extent of Biodegradation[b]	Measurement Method	Reference
$C_{12-14}AE_8$	Continuous-flow activated sludge	97% in 3 hours	Bismuth iodide	Stache, 1976
$C_{13-14}AE_9$ (45% linear)	Inoculated medium	60% in 18 days	Organic carbon	
$C_{16-18}AE_5$	Batch or semi-continuous activated sludge	98% in 3 days	Chemical oxygen demand	Stühler and Wellens, 1976
$C_{16-18}AE_{11}$	Batch or semi-continuous activated sludge	95% in 6 days	Chemical oxygen demand	
$C_{16-18}AE_{15}$	Batch or semi-continuous activated sludge	95% in 9 days	Chemical oxygen demand	
$C_{16-18}AE_{36}$	Batch or semi-continuous activated sludge	93% in 14 days	Chemical oxygen demand	
$C_{12-15}AE_9$ (Dobanol 25-9)	Inoculated medium	100% in 15 days	Bismuth iodide	Tobin et al., 1976a; 1976b
		70% in 22 days	Carbon dioxide formation	
	Continuous-flow activated sludge	95% in 3 hours	Bismuth iodide	
	Lake water *in situ*	26% in 3 hours	Gas chromatography	
	Inoculated medium	90% in 15 days	Bismuth iodide	
		88% in 7 days	Foaming	
$C_{13-18}AE_8$	Inoculated medium	98% in 7 days	Bismuth iodide	
		95% in 7 days	Foaming	
$C_{12-14}AE_9$	Inoculated medium	98% in 7 days	Bismuth iodide	
		97% in 7 days	Foaming	
$C_{16}AE_{16}$	Inoculated medium	100% in 7 days	Bismuth iodide	
		98% in 7 days	Foaming	
$C_{16}AE_{20}$	Inoculated medium	98% in 7 days	Bismuth iodide	
		89% in 7 days	Foaming	
$C_{13-18}AE_{22}$	Inoculated medium	91% in 7 days	Bismuth iodide	
$C_{12-15}AE_9$ (Dobanol 25-9)	Inoculated medium	90% in 7 days	Bismuth iodide	Anthony and Tobin, 1977

Appendix (cont.)

Chemical[a]	Test System/Protocol	Extent of Biodegradation[b]	Measurement Method	Reference
$C_{12}AE_{12}$	River-water dieaway	99-100% in 7 days	Cobalt thiocyanate	Dobarganes Garcia and Ruiz Cruz, 1977
sec-$C_{11-15}AE_9$	River-water dieaway	94% in 9 days	Cobalt thiocyanate	
	River-water dieaway	96% in 7 days	Cobalt thiocyanate	
$C_{12}AE_5$	Inoculated medium	53% in 10 days	Oxygen consumption	Inoue et al., 1977
$C_{12}AE_9$	Inoculated medium	38% in 10 days	Oxygen consumption	
$C_{13}AE_8$	Inoculated medium	41% in 10 days	Oxygen consumption	
$C_{14}AE_6$	Inoculated medium	55% in 10 days	Oxygen consumption	
sec-$C_{13}AE_6$	Inoculated medium	45% in 10 days	Oxygen consumption	
sec-$C_{13}AE_7$	Inoculated medium	44% in 10 days	Oxygen consumption	
sec-$C_{13}AE_9$	Inoculated medium	43% in 10 days	Oxygen consumption	
sec-$C_{13}AE_{12}$	Inoculated medium	42% in 10 days	Oxygen consumption	
$C_{16}AE_{10}$ (spermyl)	Inoculated medium	91-97% in 2 days	Bismuth iodide	Janicke and Hilge, 1977
		25-50% in 2 days	Chemical oxygen demand	
$C_{12-15}AE_9$	River-water dieaway	99% in 4 days	Cobalt thiocyanate	Kurata et al., 1977
sec-$C_{12-14}AE_9$	River-water dieaway	99% in 4 days	Cobalt thiocyanate	
$C_{16-18}AE_{50}$	Soil	95-100%	Bismuth iodide	Rizet et al., 1977
$C_{12-14}AE_9$	River-water dieaway	94% in 4 days	Cobalt thiocyanate	Ruiz Cruz and Dobarganes Garcia, 1977
$C_{12}AE_6$	River-water dieaway	91% in 4 days	Cobalt thiocyanate	
$C_{12}AE_8$	River-water dieaway	90% in 3 days	Cobalt thiocyanate	
$C_{18}AE_8$	River-water dieaway	93% in 7 days	Cobalt thiocyanate	
C_nAE_6	River-water dieaway	99% in 6 days	Cobalt thiocyanate	
C_nAE_9	River-water dieaway	99% in 6 days	Cobalt thiocyanate	
C_nAE_{12}	River-water dieaway	99% in 6 days	Cobalt thiocyanate	
C_nAE_{23}	River-water dieaway	70% in 10 days	Cobalt thiocyanate	
$C_{12-15}AE_9$ (80% linear)	River-water dieaway	90% in 6 days	Cobalt thiocyanate	
sec-$C_{11-15}AE_9$	River-water dieaway	88% in 6 days	Cobalt thiocyanate	
br-$C_{13}AE_9$	River-water dieaway	83% in 12 days	Cobalt thiocyanate	

Appendix (cont.)

Chemical[a]	Test System/ Protocol	Extent of Biodegradation[b]	Measurement Method	Reference
br-$C_{13}AE_9$	Shake flask	48% in 8 days	Bismuth iodide	Ruiz Cruz and Dobarganes Garcia, 1977 (cont.)
	Batch or semi-continuous activated sludge	68% in 1 day	Bismuth iodide	
	Inoculated medium	52% in 19 days	Bismuth iodide	
	Continuous-flow activated sludge	72% in 3 hours	Bismuth iodide	
$C_{9-11}AE_6$ (80% linear)	Inoculated medium	90% in 29 days	CO_2 formation	Kravetz et al., 1978
		91% in 29 days	Organic carbon	
		78–86% in 29 days	Biological oxygen demand	
$C_{12-15}AE_7$ (75% linear)	Inoculated medium	89% in 29 days	CO_2 formation	
		92% in 29 days	Organic carbon	
		81–84% in 29 days	Biological oxygen demand	
$C_{12-15}AE_9$ (75% linear)	Inoculated medium	79% in 29 days	CO_2 formation	
		95% in 29 days	Organic carbon	
		71–81% in 29 days	Biological oxygen demand	
$C_{12-15}AE_9$ (45% linear)	Inoculated medium	74% in 29 days	CO_2 formation	
		87% in 29 days	Organic carbon	
		68–82% in 29 days	Biological oxygen demand	
$C_{12-15}AE_{12}$ (75% linear)	Inoculated medium	84% in 29 days	CO_2 formation	
		92% in 29 days	Organic carbon	
		81–87% in 29 days	Biological oxygen demand	
$C_{14}AE_9$ (75% linear)	Inoculated medium	76% in 29 days	CO_2 formation	
		96% in 29 days	Organic carbon	
		70–88% in 29 days	Biological oxygen demand	
sec-$C_{11-15}AE_9$ (Tergitol 15-S-9)	Inoculated medium	69% in 29 days	CO_2 formation	
		77% in 29 days	Organic carbon	
		65–77% in 29 days	Biological oxygen demand	

Appendix (cont.)

Chemical[a]	Test System/ Protocol	Extent of Biodegradation[b]	Measurement Method	Reference
$C_{12}AE_6$ (lauryl)	River-water dieaway	100% in 6 days	Cobalt thiocyanate	Ruiz Cruz and Dobarganes Garcia, 1978
	Shake flask	99% in 8 days	Bismuth iodide	
	Activated sludge	99% in 1 day	Bismuth iodide	
	Inoculated medium	98% in 19 days	Bismuth iodide	
	Continuous-flow activated sludge	98% in 3 hours	Bismuth iodide	
$C_{16,18}AE_{10}$ (tallow)	River-water dieaway	97% in 5 days	Cobalt thiocyanate	
	Shake flask	98% in 8 days	Bismuth iodide	
	Activated sludge	97% in 1 day	Bismuth iodide	
	Inoculated medium	96% in 19 days	Bismuth iodide	
	Continuous-flow activated sludge	96% in 3 hours	Bismuth iodide	
n-pri-$C_{12-14}AE_9$	River-water dieaway	98% in 7 days	Cobalt thiocyanate	
	Shake flask	98% in 8 days	Bismuth iodide	
	Activated sludge	100% in 1 day	Bismuth iodide	
	Inoculated medium	99% in 19 days	Bismuth iodide	
	Continuous-flow activated sludge	98% in 3 hours	Bismuth iodide	
$C_{12-14}AE_9$ (80% linear)	River-water dieaway	96% in 9 days	Cobalt thiocyanate	
	Shake flask	97% in 8 days	Bismuth iodide	
	Activated sludge	99% in 1 day	Bismuth iodide	
	Inoculated medium	95% in 19 days	Bismuth iodide	
	Continuous-flow activated sludge	96% in 3 hours	Bismuth iodide	
sec-$C_{11-15}AE_9$	River-water dieaway	98% in 10 days	Cobalt thiocyanate	
	Shake flask	95% in 8 days	Bismuth iodide	
	Activated sludge	98% in 1 day	Bismuth iodide	
	Inoculated medium	97% in 19 days	Bismuth iodide	
	Continuous-flow activated sludge	97% in 3 hours	Bismuth iodide	
$C_{14-15}AE_7$ (Dobanol 45-7)	Activated sludge	99% in 3 days	Thin-layer chromatography	Cook, 1979
		99% in 14 days	Organic carbon	

Appendix (cont.)

Chemical[a]	Test System/Protocol	Extent of Biodegradation[b]	Measurement Method	Reference
$C_{12-15}AE_9$ (75% linear)	Inoculated medium	85% in 28 days	CO_2 formation	Kravetz et al., 1979
$C_{12-15}AE_9$ (45% linear)	Inoculated medium	85% in 28 days	CO_2 formation	Scharer et al., 1979
$C_{12-15}AE_7$ (Neodol 25-7)	Inoculated medium	100% in 6 days 96% in 20 days 81% in 35 days	Cobalt thiocyanate Organic carbon CO_2 formation	
$C_{12-15}AE_{18}$	Inoculated medium	95% in 6 days 87% in 35 days 85% in 35 days	Cobalt thiocyanate Organic carbon CO_2 formation	
$C_{12-15}AE_{30}$	Inoculated medium	100% in 7 days 100% in 27 days 91% in 35 days	Cobalt thiocyanate Organic carbon CO_2 formation	
$C_{12-15}AE_{100}$	Inoculated medium	91% in 14 days 21% in 40 days 21% in 40 days	Cobalt thiocyanate Organic carbon CO_2 formation	
sec-$C_{11-15}AE_9$	Inoculated medium	79% in 28 days	CO_2 formation	
$C_{11-12}AE_{6.5}$	Inoculated medium	99% in 24 days 100% in 24 days 100% in 1 day	CO_2 formation Organic carbon Organic carbon	Larson, 1979
$C_{13-14}AE_7$	Activated sludge Inoculated medium Activated sludge	95-100% in 29 days 100% in 29 days 100% in 1 day	CO_2 formation Organic carbon Organic carbon	
C_8AE_{31}	Inoculated medium	77% in 12 days	$^{14}CO_2$ formation	Lötzsch et al., 1979
$C_{12}AE_{8.5}$	Inoculated medium	98% in 5 days 50% in 15 days 35% in 12 days	Cobalt thiocyanate Oxygen uptake Organic carbon	Miura et al., 1979

Appendix (cont.)

Chemical[a]	Test System/ Protocol	Extent of Biodegradation[b]	Measurement Method	Reference
$C_{12,14}AE_n$ (Alfonic 1412-70)	Continuous-flow activated sludge	100% in 3 hours 78% in 3 hours	Bismuth iodide Chemical oxygen demand	Moreno Danvila, 1979
sec-$C_{11,15}AE_9$ (Tergitol 15-S-9)	Continuous-flow activated sludge	95% in 3 hours 70% in 3 hours	Bismuth iodide Chemical oxygen demand	
br-$C_{13}AE_5$ br-$C_{13}AE_{10}$	Activated sludge Activated sludge	37% in 14 days 63% in 14 days	Chemical oxygen demand Chemical oxygen demand	Zahn and Wellens, 1980
$C_{13-14}AE_{12}$ (40% linear)	—	95% 60%	Bismuth iodide Oxygen uptake	Schöberl and Bock, 1980
$C_{12}AE_4$	Inoculated medium	75% in 30 days	Oxygen uptake	Fischer, 1981
$C_{12}AE_6$	Inoculated medium	80% in 30 days	Oxygen uptake	
$C_{12-14}AE_6$	Inoculated medium	96% in 5 days	Bismuth iodide	
$C_{12-18}AE_6$	inoculated medium	94% in 5 days	Bismuth iodide	
$C_{12-18}AE_7$	Inoculated medium	98% in 19 days 97% in 28 days 86% in 30 days	Bismuth iodide Organic carbon Oxygen uptake	
$C_{12}AE_8$	Inoculated medium	99% in 5 days 78–83% in 30 days	Bismuth iodide Oxygen uptake	
$C_{14}AE_8$	Inoculated medium	92% in 30 days	Oxygen uptake	
$C_{12-18}AE_8$	Inoculated medium	85% in 30 days	Oxygen uptake	
$C_{18}AE_8$	Inoculated medium	60% in 30 days	Oxygen uptake	
$C_{12}AE_{10}$	Inoculated medium	85% in 30 days	Oxygen uptake	
$C_{12}AE_{12}$	Inoculated medium	78% in 30 days	Oxygen uptake	
$C_{12}AE_{18}$	Inoculated medium	57% in 30 days	Oxygen uptake	
$C_{12}AE_{20}$	Inoculated medium	50% in 30 days	Oxygen uptake	
$C_{16,18}-AE_5$ (cetyl, oleyl)	Inoculated medium	99% in 19 days 93% in 30 days 94% in 28 days	Bismuth iodide Oxygen uptake Organic carbon	

Appendix (cont.)

Chemical[a]	Test System/Protocol	Extent of Biodegradation[b]	Measurement Method	Reference
$C_{16,18}$-AE_{10} (cetyl, oleyl)	Inoculated medium	94% in 19 days 77% in 30 days 89% in 28 days	Bismuth iodide Oxygen uptake Organic carbon	Fischer, 1981 (cont.)
$C_{16,18}AE_{14}$ (tallow)	Inoculated medium	99% in 19 days 86% in 30 days 94% in 28 days	Bismuth iodide Oxygen uptake Organic carbon	
$C_{13-15}AE_7$ (65% linear)	Inoculated medium	94% in 19 days	Bismuth iodide	
$C_{13-15}AE_9$ (65% linear)	Inoculated medium	78% in 30 days 87% in 19 days	Oxygen uptake Bismuth iodide	
sec-$C_{11-15}AE_8$ sec-$C_{11-15}AE_9$ sec-$C_{14-15}AE_8$	Inoculated medium Inoculated medium Inoculated medium	52% in 30 days 47% in 30 days 47% in 30 days	Oxygen uptake Oxygen uptake Oxygen uptake	
$C_{11}AE_9$ $C_{11.5}AE_7$	River-water dieaway River-water dieaway	88% in 14 days 82% in 20 days	CO_2 formation Oxygen uptake	Larson and Perry, 1981
C_{12}*AE_9 *$C_{12}AE_9$	River-water dieaway (shake flask)	88% in 18 days 90% in 18 days (1-100 µg/L)	$^{14}CO_2$ formation $^{14}CO_2$ formation	Larson and Games, 1981
C_{16}*AE_3 *$C_{16}AE_3$	River-water dieaway (shake flask)	86% in 18 days 89% in 18 days (1-100 µg/L)	$^{14}CO_2$ formation $^{14}CO_2$ formation	
$C_{17}AE_{22}$	Inoculated medium	96% in 30 days (20 mg/L)	CO_2 formation	
$C_{13-14}AE_{12}$ (40% linear)	Continuous-flow activated sludge	94-97% 40-60%	Bismuth iodide Organic carbon	Schöberl et al., 1981
$C_{12-18}AE_{11}$ $C_{12-15}AE_9$ (Dobanol 25-9)	Inoculated medium Inoculated medium	100% in 7 days 100% in 7 days	Cobalt thiocyanate Cobalt thiocyanate	Tabak and Bunch, 1981

Appendix (cont.)

Chemical[a]	Test System/Protocol	Extent of Biodegradation[b]	Measurement Method	Reference
n-pri-C_nAE_{10}	Dieaway (OECD)	99%	Bismuth iodide	Birch, 1982; 1984
n-pri-C_nAE_{20}	Dieaway (OECD)	99%	Bismuth iodide	
n-pri-C_nAE_{30}	Dieaway (OECD)	98%	Bismuth iodide	
n-pri-C_nAE_{40}	Dieaway (OECD)	98%	Bismuth iodide	
n-pri-C_nAE_{50}	Dieaway (OECD)	98%	Bismuth iodide	
$C_{13-15}AE_7$ (50% linear)	Activated sludge	97-100% in 6 hours	Bismuth iodide	
C_nAE_9 (75% linear)	Natural sewage	99% in 3 hours	Bismuth iodide	
	Synthetic sewage	94% in 3 hours	Bismuth iodide	
C_nAE_{10}	Inoculated medium	75% in 28 days	CO_2 formation	
C_nAE_{20}	Inoculated medium	68% in 28 days	CO_2 formation	
C_nAE_{30}	Inoculated medium	52% in 28 days	CO_2 formation	
C_nAE_{40}	Inoculated medium	34% in 28 days	CO_2 formation	
C_nAE_{50}	Inoculated medium	21% in 28 days	CO_2 formation	
sec-$C_{11-15}AE_9$	Activated sludge	88-98% in 6 hours	Bismuth iodide	
sec-C_nAE_{10}	River-water dieaway	96%	Bismuth iodide	
sec-C_nAE_{20}	River-water dieaway	64%	Bismuth iodide	
sec-C_nAE_{30}	River-water dieaway	59%	Bismuth iodide	
sec-C_nAE_{40}	River-water dieaway	65%	Bismuth iodide	
$C_{12-15}AE_9$ (80% linear)	Continuous-flow activated sludge	98-100% in 8 hours >95% in 8 hours	Cobalt thiocyanate Foaming, surface tension	Kravetz et al., 1982
$C_{12}*AE_9$	River-water dieaway	97% in 14 days 90% in 14 days	Organic ^{14}carbon $^{14}CO_2$ formation	Larson and Wentler, 1982
$C_{16}*AE_3$	River-water dieaway	96% in 14 days 90% in 14 days	Organic ^{14}carbon $^{14}CO_2$ formation	
$C_{12.5}AE_{6.5}$	Inoculated medium	90% in 24 days	CO_2 formation	
$C_{14.5}AE_7$	Inoculated medium	90% in 24 days	CO_2 formation	

Appendix (cont.)

Chemical[a]	Test System/ Protocol	Extent of Biodegradation[b]	Measurement Method	Reference
$C_{18}*AE_5$	Inoculated medium	97% in 12 days 74% in 12 days	Organic ^{14}carbon $^{14}CO_2$ formation	Neufahrt et al., 1982
$C_{18}*AE_{10}$	Inoculated medium	97% in 12 days 60% in 12 days	Organic ^{14}carbon $^{14}CO_2$ formation	
$C_{18}*AE_{17}$	Inoculated medium	98% in 12 days 68% in 12 days	Organic ^{14}carbon $^{14}CO_2$ formation	
$*C_{18}AE_{31}$	Inoculated medium	97% in 12 days 77% in 12 days	Organic ^{14}carbon $^{14}CO_2$ formation	
$*C_{16}AE_3$ $C_{12}*AE_9$	Estuary water Estuary water	87% in 18 days 77% in 18 days	$^{14}CO_2$ formation $^{14}CO_2$ formation	Vashon and Schwab, 1982
$C_{16-20}AE_{15}$	Continuous-flow activated sludge	98% in 10 hours 100% in 10 hours	Polarography Bismuth iodide	Kozarac et al., 1983
$C_{18}AE_7$	Continuous-flow activated sludge	99% in 3 hours 40-60% in 3 hours	Bismuth iodide $^{14}CO_2$ formation	Steber and Wierich, 1983
$C_{16-18}AE_{10}$	Continuous-flow activated sludge	98% in 3 hours 62% in 3 hours	Bismuth iodide Organic carbon	Berth et al., 1984
$C_{13-14}AE_{12}$	Continuous-flow activated sludge	96% in 3 hours 59% in 3 hours	Bismuth iodide Organic carbon	
$sec\text{-}C_{11-13}AE_9$	Continuous-flow activated sludge	86% in 3 hours 36% in 3 hours	Bismuth iodide Organic carbon	
C_nAE_3	Continuous-flow activated sludge	90-97% in 6 hours	Bismuth iodide	Birch, 1984
C_nAE_7	Continuous-flow activated sludge	100% in 6 hours	Bismuth iodide	
C_nAE_{11}	Continuous-flow activated sludge	100% in 6 hours	Bismuth iodide	
C_nAE_{15}	Continuous-flow activated sludge	99% in 6 hours	Bismuth iodide	
C_nAE_{20}	Continuous-flow activated sludge	99% in 6 hours	Bismuth iodide	

Appendix (cont.)

Chemical[a]	Test System/ Protocol	Extent of Biodegradation[b]	Measurement Method	Reference
$C_{12}AE_8$	Inoculated medium	70% in 28 days 100% in 28 days	Oxygen uptake Organic carbon	Gerike, 1984
$C_{16,18}AE_{10}$ (tallow)	Continuous-flow activated sludge	62% in 3 hours	Organic carbon	Gerike and Jasiak, 1984
$C_{12-14}AE_{30}$	Continuous-flow activated sludge	59% in 3 hours	Organic carbon	
$C_{12-15}AE_9$ (80% linear)	Continuous-flow activated sludge biotreater units	93-98% in 9 hours 98% at 25°C 97% at 12°C 93% at 8°C	Cobalt thiocyanate	Kravetz et al., 1984
*$C_{12-15}AE_9$ (^{3}H label)	Activated sludge biotreater units	79% at 25°C 76% at 8°C	^3H_2O	
C_{12-15}*AE_9 (^{14}C label)	Activated sludge biotreater units	58% at 25°C 50-58% at 12°C 10% at 8°C	$^{14}CO_2$	
*C_{12-15}*AE_7 (Neodol 25-7)	Hanford fine sandy loam-laboratory columns	50% in 2 days 90% in 2 weeks	$^{14}CO_2$, ^3H_2O	Howells et al., 1984
$C_{12}AE_9$ $C_{16}AE_9$	Groundwater Groundwater	60% in 50 days 60% in 50 days	$^{14}CO_2$ formation $^{14}CO_2$ formation	Larson, 1984
*$C_{16}AE_7$ C_{18}*AE_7	Batch anaerobic digester Batch anaerobic digester	87% in 28 days 96% in 28 days 84% in 28 days 93% in 28 days	$^{14}CH_4 + {}^{14}CO_2$ Gas + metabolites $^{14}CH_4 + {}^{14}CO_2$ Gas + metabolites	Steber and Wierich, 1984; Steber and Wierich, 1987
$C_{14,15}AE_7$ (Dobanol 45-7) $C_{14,15}AE_{11}$ (Dobanol 45-11)	Trickling filter Trickling filter	96-98% removal 96-98% removal	Bismuth iodide Bismuth iodide	Turner et al., 1985
$C_{12}AE_7$	River-water dieaway	98% in 5-10 days	Cobalt thiocyanate, HPLC	Yoshimura, 1986

Appendix (cont.)

Chemical[a]	Test System/ Protocol	Extent of Biodegradation[b]	Measurement Method	Reference
$C_{12}AE_{23}$	Inoculated medium	80% in 37 days	CH_4 production	Wagener and Schink, 1987
	Anaerobic reactor	90%	CH_4 production	
$C_{10-12}AE_{7.5}$	Inoculated medium	70% in 37 days	CH_4 production	
$C_{12}AE_{8.9}$	Laundry pond sediment	50% in 2.8-8.6 days	$^{14}CO_2$ formation	Federle and Pastwa, 1988
	Control sediment	50% in 3.5-137 days	$^{14}CO_2$ formation	
$C_{13}AE_7$	Laundry pond sediment	37% in 35 days	$^{14}CO_2$ formation	
	Control sediment	35% in 35 days	$^{14}CO_2$ formation	
$C_{12}AE_7$	Activated sludge	100% in 5-10 days	Bismuth iodide	Janicke, 1988
$C_{17}AE_{23}$	Activated sludge	100% in 5-10 days	Bismuth iodide	
$C_{13}AE_9$ (linear, 2-methyl branched)	Semi-continuous activated sludge	>99% in 32 days	Foam height	Naylor et al., 1988
$C_{13}AE_{4.5}PO_2EO_{4.5}$[c]	Semi-continuous activated sludge	>90% in 32 days	Foam height	
$C_{13}AE_{5.2}PO_{3.3}EO_{5.4}$[c]	Semi-continuous activated sludge	>90% in 32 days	Foam height	
$C_{13}AE_{4.9}PO_5EO_{6.9}$[c]	Semi-continuous activated sludge	<80% in 32 days	Foam height	
$C_{13}AE_{11}$ (3-alkyl branched; both branches linear)	Semi-continuous activated sludge	99% in 32 days	Foam height	
$C_{13}AE_5PO_2EO_{5.8}$[c]	Semi-continuous activated sludge	80-90% in 32 days	Foam height	
$C_{14}AE_9$ (2-alkyl branched + methyl branching)	Semi-continuous activated sludge	>90% in 32 days	Foam height	
$C_{14}AE_3(EO_{9.3}PO_{0.2}$ mixed)[c]	Semi-continuous activated sludge	>90% in 32 days	Foam height	
$C_{14}AE_3(EO_{4.3}PO_{0.4}$ mixed)[c]	Semi-continuous activated sludge	80-90% in 32 days	Foam height	

Appendix (cont.)

Chemical[a]	Test System/ Protocol	Extent of Biodegradation[b]	Measurement Method	Reference
$C_{14}(EO_{10.6}PO_{0.7}$ mixed)[c]	Semi-continuous activated sludge	<80% in 32 days	Foam height	Naylor et al., 1988 (cont.)
n-pri-$C_{12-13}AE_{6.5}$ (Neodol 23-6.5)	Semi-continuous activated sludge	100% (days 20-85)[d] 97% (days 8-85)[d]	Foam height Dissolved organic carbon	
$C_{10-12}AE_{3.6}PO_{1.5}EO_{3.6}$[c] (Surfonic JL-80-X)	Semi-continuous activated sludge	96% (days 20-85)[d] 96% (days 8-85)[d]	Foam height Dissolved organic carbon	
$C_{10-12}AE_{3.7}PO_{2.8}EO_{3.7}$[c]	Semi-continuous activated sludge	77% (days 20-85)[d] 95% (days 43-85)[d]	Foam height Dissolved organic carbon	
$C_{10-12}AE_{3.8}PO_{4.2}EO_{5.6}$[c]	Semi-continuous activated sludge	53% (days 57-85)[d] 52% (days 8-85)[d]	Foam height Dissolved organic carbon	
$C_{12-14}AE_9PO_9$[c] (Surfonic LF-17)	Semi-continuous activated sludge	71% (days 41-85)[d] 40% (days 71-85)[d]	Foam height Dissolved organic carbon	
$C_{12-15}AE_7$ (Neodol 25-7)	Continuous-flow activated sludge Inoculated medium	100% >95% 70-75% in 14 days	Cobalt thiocyanate Biological oxygen demand CO_2 formation	Salanitro et al., 1988
n-$C_{15}AE_7$	Plant-associated microbiota	>35% in 35 days	$^{14}CO_2$ formation	Federle and Schwab, 1989
br-$C_{12}AE_7$ (DDA-7)	Standard Method 507 Closed bottle test	28% in 5 days (2 mg/L) 98% in 28 days (2 mg/L)	Oxygen consumption Oxygen consumption	Markarian et al., 1989
br-$C_{13}AE_7$ (TDA-7)	Shake flask (Gledhill) Standard Method 507 Closed bottle test	74% in 28 days (10 mg/L) 11% in 5 days (2 mg/L) 79% in 28 days (2 mg/L)	CO_2 formation Oxygen consumption Oxygen consumption	
n-$C_{13-15}AE_7$ (L-1315-7)	Shake flask (Gledhill) Standard Method 507 Closed bottle test	75% in 28 days (10 mg/L) 51% in 5 days (2 mg/L) 94% in 28 days (2 mg/L)	CO_2 formation Oxygen consumption Oxygen consumption	
$C_{21}AE_{11}$	Shake flask (Gledhill) Closed bottle test	54% in 29 days (15 mg/L) 10% in 28 days (2 mg/L)	CO_2 formation Oxygen consumption	
$C_{12-15}AE_7$ (Neodol 25-7)	Standard Method 507 Closed bottle test	46,62% in 5 days (2 mg/L) 91% in 28 days (2 mg/L)	Oxygen consumption Oxygen consumption	
n-$C_{13}AE_7/C_{13}AE_7$ (70/30 blend)	Shake flask (Gledhill) Closed bottle test	52, 62% in 29 days 86, 95% in 28 days (2 mg/L)	Oxygen consumption CO_2 formation	
$C_{12-14}AE_7$	Closed bottle test	101% in 28 days (2 mg/L)	Oxygen consumption	

Appendix (cont.)

Chemical[a]	Test System/Protocol	Extent of Biodegradation[b]	Measurement Method	Reference
$EO_3PO_{30}EO_3$ (Pluronic L-61)[c]	Shake flask (SCAS inoculum)	74.5, 74.2% in 28 days (10, 20 mg/L)	TCO_2 formation	BASF Corporation, 1990
$EO_8PO_{30}EO_8$ (Pluronic L-62)[c]	Shake flask (SCAS inoculum)	66.9, 59.5% in 32 days (10, 20 mg/L)	TCO_2 formation	
$EO_{13}PO_{30}EO_{13}$ (Pluronic L-64)[c]	Shake flask (SCAS inoculum)	81.2, 70.1% in 28 days (10, 20 mg/L)	TCO_2 formation	
$EO_{75}PO_{30}EO_{75}$ (Pluronic F-68)[c]	Shake flask (SCAS inoculum)	2.9, 1.5% in 28 days (10, 20 mg/L)	TCO_2 formation	
$EO_{31}PO_{54}EO_{31}$ (Pluronic P-104)[c]	Shake flask (SCAS inoculum)	0.7, 1.1% in 28 days (10, 20 mg/L)	TCO_2 formation	
$PO_{22}EO_{11}PO_{22}$ (Pluronic 25R2)[c]	Shake flask (SCAS inoculum)	65.1, 47.9% in 32 days (10, 20 mg/L)	TCO_2 formation	
$C_{12}*AE_{8.9}$	Soils	25-69% in 20 days	$^{14}CO_2$ formation	Knaebel et al, 1990
n-$C_{12-15}AE_7$	Closed bottle	92% in 30 days	Biological oxygen demand	Kravetz et al., 1991
n-$C_{12-15}AE_9$	Closed bottle	88% in 30 days	Biological oxygen demand	
	Closed biotreater units	64% in 28 days (10 mg/L)	CO_2 formation	
	Continuous-flow activated sludge	95-99% over 240 days	Cobalt thiocyanate	
n-$C_{14,15}AE_7$	Closed bottle	83% in 30 days	Biological oxygen demand	
br-$C_{12}AE_7$	Closed bottle	41% in 30 days	Biological oxygen demand	
br-$C_{13}AE_7$	Closed bottle	44% in 30 days	Biological oxygen demand	
	Closed biotreater units	48% in 28 days (10 mg/L)	CO_2 formation	
	Continuous-flow activated sludge	62-99% over 240 days	Cobalt thiocyanate	

Appendix (cont.)

Chemical[a]	Test System/Protocol	Extent of Biodegradation[b]	Measurement Method	Reference
$C_{8-10}AE_a$ (Tergitol 80-L-97)	Batch jars	88, 90% in 29 days (20, 40 mg/L)	CO_2 formation	Procter & Gamble, 1985
	Semi-continuous activated sludge	101% in 7 days	Soluble organic carbon	
n-$C_{12-13}AE_{6.5}$	Batch jars	52-60% (3-6 mg/L)	CO_2 formation	
$C_{12}AE_a$ (Perosperse 200 ML)	Batch jars	67, 74% (10, 20 mg/L)	CO_2 formation	
$C_{14-15}AE_7$ (Neodol 45-7)	Batch jars	71, 73% in 26 days (10, 20 mg/L)	CO_2 formation	
$C_{16-18}AE_{18}$	Semi-continuous activated sludge	100% in 7 days	Soluble organic carbon	
	Batch jars	103, 88% in 25 days (10, 20 mg/L)	CO_2 formation	
$C_{16-18}AE_{22}$	Semi-continuous activated sludge	58.5% over 7 days	Soluble organic carbon	
	Batch jars	114, 96% in 30 days (10, 20 mg/L)	CO_2 formation	
$C_{16-18}AE_{30}$	Semi-continuous activated sludge	98% in 7 days	Soluble organic carbon	
	Batch jars	74, 87% in 25 days (10 and 20 mg/L)	CO_2 production	
$C_{18}AE_a$ (Brij 93)	Semi-continuous activated sludge	91-98% in 7 days	Soluble organic carbon	
$C_{18}AE_{8.25}$	Semi-continuous activated sludge	>100% in 7 days	Soluble organic carbon	
	Batch jars	78-88% in 25 days	CO_2 formation	
	River-water dieaway	100% in 6 days (1 mg/L)	Cobalt thiocyanate	
$C_{13-15}AE_7$ (Lutensol AO7)	Semi-continuous activated sludge	2.4% in 7 days	Soluble organic carbon	
Pluronic L-35[c] ($EO_{11}PO_{16}EO_{11}$)	Batch jars	40, 37% in 25 days (10, 20 mg/L)	CO_2 formation	

Appendix (cont.)

Chemical[a]	Test System/ Protocol	Extent of Biodegradation[b]	Measurement Method	Reference
Pluronic L-35[c] ($EO_{11}PO_{16}EO_{11}$)	Semi-continuous activated sludge Batch jars	63% in 18 days (20 mg/L) 75, 87% in 29 days (10, 20 mg/L)	Soluble organic carbon CO_2 formation	Procter & Gamble, 1991 Procter & Gamble, 1985
Pluronic L-61[c] ($EO_3PO_{30}EO_3$)	Semi-continuous activated sludge Batch jars	1.3% in 7 days (20 mg/L) 6, 7% in 25 days (10, 20 mg/L)	Soluble organic carbon CO_2 formation	Procter & Gamble, 1991 Procter & Gamble, 1985
Pluronic L-62[c] ($EO_8PO_{30}EO_8$)	Semi-continuous activated sludge Batch jars	78% in 18 days (20 mg/L) 76, 69% in 29 days (10, 20 mg/L)	Soluble organic carbon CO_2 formation	Procter & Gamble, 1991 Procter & Gamble, 1985
Pluronic P-85[c] ($EO_{27}PO_{39}EO_{27}$)	Semi-continuous activated sludge Batch jars	4.7% in 18 days (20 mg/L) -1, 2% in 25 days (10, 20 mg/L)	Soluble organic carbon CO_2 formation	Procter & Gamble, 1991 Procter & Gamble, 1985
Pluronic 25R2[c]	Semi-continuous activated sludge Batch jars	5-15% in 22 days (20 mg/L) 59, 60% in 28 days (10, 20 mg/L)	Soluble organic carbon CO_2 formation	Procter & Gamble, 1991 Procter & Gamble, 1985
Plurafac RA 20[f] (EO/PO ratio 55/45)	Semi-continuous activated sludge Batch jars	43-70% in 7 days (20 mg/L) 8-45% in 26 days (10, 20 mg/L)	Soluble organic carbon CO_2 formation	Procter & Gamble, 1991 Procter & Gamble, 1985
C_nAE_7 C_nAE_{11} C_nAE_{15} C_nAE_{20}	Activated sludge units[e] Activated sludge units[e] Activated sludge units[e] Activated sludge units[e]	97, 100% (5, 25 mg/L) 97, 100% (5, 25 mg/L) 96, 99% (5, 25 mg/L) 93, 97% (5, 25 mg/L)	Bismuth iodide Bismuth iodide Bismuth iodide Bismuth iodide	Birch, 1991a

Appendix (cont.)

Chemical[a]	Test System/ Protocol	Extent of Biodegradation[b]	Measurement Method	Reference
$C_{14-15}AE_{10}$	Modified Sturm test	78% in 28 days	CO_2 formation	Birch, 1991b
$C_{14-15}AE_{20}$	Modified Sturm test	65% in 28 days	CO_2 formation	
$C_{14-15}AE_{30}$	Modified Sturm test	38% in 28 days	CO_2 formation	
$C_{14-15}AE_{40}$	Modified Sturm test	32% in 28 days	CO_2 formation	
$C_{14-15}AE_{50}$	Modified Sturm test	28% in 28 days	CO_2 formation	
n-pri-$C_{10-12}AE_5$	Microcosms containing sterilized soil, groundwater	>98% in 17-36 days (180-1000 mg/L)	HPLC peak area	Ang and Abdul 1992

[a]In the absence of specific information, the alcohol chain is probably linear.
[b]Numbers in parentheses are initial concentrations.
[c]Contains propylene oxide.
[d]85-day test.
[e]Retention time 6 days; temperature 15°C.
$C_{18}^=$ designates a double bond.
[e]Radiolabeled moiety of surfactant.

REFERENCES

Albanese, P. and R. Capuci. 1974. Biodegradation of nonionic surfactants. Note 2: Measuring biodegradation. Riv. Ital. Sost. Grasse. 51:70-81.

Allred, R.C. and R.L. Huddleston. 1967. Microbial oxidation of surface active agents. Southwest Water Works J. 49:26-28.

Ang, C.C. and A.S. Abdul. 1992. A laboratory study of the biodegradation of an alcohol ethoxylate surfactant by native soil microbes. J. Hydrology 138:191-209.

Anthony, D.H.J. and R.S. Tobin. 1977. Immiscible solvent extraction scheme for biodegradation testing of polyethoxylated nonionic surfactants. Anal. Chem. 49:398-401.

Arpino, A. 1969. Detergent products and the evaluation of their biodegradability. Riv. Ital. Sost. Grasse. 46:347-355.

Arpino, A., E. Fedeli, G.B. Borsari, F. Buosi and E.P. Fuochi. 1974. Evaluation of synthetic primary fatty alcohol derivatives in relation to chemical structure. II. Applicability of Wickbold method to the evaluation of the biodegradability of ethoxylate nonionic surfactants. Riv. Ital. Sost. Grasse. 51:253-265.

Baleux, B. and P. Caumette. 1974. Biodegradation of nonionic surfactants. Study of a new experimental method and screening test. Rev. Inst. Pasteur Lyon. 7:278-297.

Barbaro, R.D. and J.V. Hunter. 1965. Effect of clay minerals on surfactant biodegradability. Purdue Conf. 20:189-196.

BASF Corporation. 1990. CO_2 production tests performed for BASF Corporation by Roy F. Weston, Inc. Unpublished data.

Berth, P. P. Gerike, P. Gode, and J. Steber. 1984. Ecological evaluation of important surfactants. Proceedings of the Eighth International Congress on Surface-Active Substances, 1:227-236. (Cited in Swisher, 1987).

Birch, R.R. 1982. The biodegradability of alcohol ethoxylates. XIII Jornadas Com. Español Deterg. 33-48.

Birch, R.R. 1984. Biodegradation of nonionic surfactants. J. Am. Oil. Chem. Soc. 61:340-343.

Birch, R.R. 1991a. Prediction of the fate of detergent chemicals during sewage treatment. J. Chem. Tech. Biotechnol. 50:411-422.

Birch, R.R. 1991b. Recent developments in the biodegradability testing of nonionic surfactants. Riv. Ital. Sost. Grasse 68:433-437.

Blankenship, F.A. and V.M. Piccolini. 1963. Biodegradation of nonionics. Soap Chem. Specialties. 39(12):75-78,181.

Booman, K.A., D.E. Daugherty, J. Duprè and A.T. Hagler. 1965. Degradation studies on branched-chain EO surfactants. Soap Chem. Specialties. 41(1):60-63, 116, 118-119.

Booman, K.A., J. Duprè and E.S. Lashen. 1967. Biodegradable surfactants in the textile industry. Am. Dyestuff Reptr. 56(3):P82-P88.

Borsari, G.B., G. Buosi and E.P. Fouchi. 1974. Evaluations of synthetic primary fatty alcohol derivatives in relation to their chemical structure. I. Biodegradability and fish toxicity of surfactants. Riv. Ital. Sost Grasse. 51:193-207.

Borstlap, C. and C. Kortland. 1967. Biodegradability of nonionic surfactants under aerobic conditions. Fette Siefen Anstrichmittel 69:736-738.

Brown, D. 1976. The assessment of biodegradability: A consideration of possible criteria for surface-active substances. Proceedings of the Seventh International Congress on Surface-Active Substances, Moscow, 4:44-57.

Brüschweiler, H. 1974. Remarks on primary biodegradation on nonionic surfactants. Communication to OECD expert group on biodegradability of nonionic surfactants, January 16, 1974.

Brüschweiler, H. 1975. Properties and biodegradation characteristics of surfactants. Chimia 29:31-42.

Bunch, R.L. and C.W. Chambers. 1967. A biodegradability test for organic compounds. J. Water Poll. Control Fed. 39:181-187.

Citernesi, U., R. Materassi, F. Favilli. 1976. Interaction of synthetic detergents with anaerobic microorganisms in the soil. Nuovi Ann. Ig. Microbiol. 27:39-49.

Conway, R.A., C.A. Vath and C.E. Renn. 1965. New detergent nonionics-biodegradable. Water Works Wastes Eng. 2:28-31.

Conway, R.A. and G.T. Waggy. 1966. Biodegradation testing of typical surfactants in industrial usage. Am. Dyestuff Reptr. 55:P607-P614.

Cook, K.A. 1979. Degradation of the nonionic surfactant Dobanol 45-7 by activated sludge. Water Res. 13:259-266.

Cuta, J. and J. Hanusová. 1964. Hygienic problems of detergents. IV. Biochemical oxidation of anionic and nonionic surfactants. Cesk. Hyg. 9:507-516.

Dobarganes Garcia, M.C. and J. Ruiz Cruz. 1977. XI. Influence of experimental variables in the biodegradation of nonionic surfactants in river water. Grasas y Aceitas. 28:161-172.

Federle, T.W. and G.M. Pastwa. 1988. Biodegradation of surfactants in saturated subsurface sediments: A field study. Groundwater 26:761-770.

Federle, T.W. and B.S. Schwab. 1989. Mineralization of surfactants by microbiota of aquatic plants. Appl. Environ. Microbiol. 55:2092-2094.

Fischer, W.K. 1971. Testing and evaluation of biodegradability of nonionic surfactants. II. Testing nonionic surfactants for biodegradability in the closed bottle test. Tenside 8:177-182.

Fischer, W.K. 1972. Correlation between chemical constitution and biodegradability for nonionic surfactants. Proceedings of the Sixth International Congress on Surface-Active Substances, Zurich, Switzerland, September 11-15, 1972. Carl Hanser Verlag, Munich, 3:735-752.

Fischer, W.K. 1981. Important aspects of the ecological evaluation of fatty alcohols and their derivatives. In Fettalkohole-Rohstoffe, Verfahren und Verwendung, Henkel, Düsseldorf pp. 193-230; English edition pp. 181-222.

Fischer, W.K., P. Gerike and R. Schmid. 1974. Method combination for sequential testing and evaluation of biodegradability of synthetic substances, e.g., organic chelants, through generally applicable gross parameters (BOD, CO_2, COD, TOC). Z. Wass.-Abwass.-Forsch. 4:99-118.

Fischer, W.K., P. Gerike and W. Holtmann. 1975. Biodegradation determinations via unspecific analyses (COD, DOC) in coupled units of the OECD confirmatory test. II. Results. Water Res. 9:1131-1137.

Frazee, C.D., Q.W. Osburn and R.O. Crisler. 1964. Application of infrared spectroscopy to surfactant degradation studies. J. Am. Oil Chem. Soc. 41:808-812.

Garrison, L.J. and R.D. Matson. 1964. A comparison by Warburg respirometry and die-away studies of the degradability of select nonionic surfactants. J. Am. Oil Chem. Soc. 41:799-804.

Gebril, B.A. and H.M. Naim. 1969a. Biodegradable nonionic surfactants from chlorinated Egyptian kerosine. Indian J. Technol. 7:365-369.

Gebril, B.A. and H.M. Naim. 1969b. Nonionic surfactants from chlorinated kerosene. Indian J. Technol. 7:370-372.

Gerike, P. 1984. The biodegradability testing of poorly water soluble compounds. Chemosphere 13:169-190.

Gerike, P. and R. Schmid. 1973. Determination of nonionic surfactants with the Wickbold method in biodegradation research and in river waters. Tenside 10:186-189.

Gerike, P. and W. Jasiak. 1984. Surfactants in the recalcitrant metabolite test. Proceedings of the Eighth International Congress on Surface-Active Substances, 1:195-208.

Han, K.W. 1967. Determination of biodegradability of nonionic surfactants by sulfation and methylene blue extraction. Tenside 4:43-45.

Hartmann, L., P. Wilderer and W. Staub. 1967. Reaction-kinetic investigations into the biodegradation of modern surfactants through trickle filter organisms. Tenside 4:138-143.

Heinz, H.J. and W.K. Fischer. 1967. II. Current test methods and suggestions for a combined international standard method. Fette Siefen Anstrichmittel 69:188-196.

Howells, W.G., L. Kravetz, D. Loring, C.D. Piper, B.W. Poovaiah, E.C. Seim, V.P. Rasmussen, N. Terry, and L.J. Waldron. 1984. The use of nonionic surfactants for promoting the penetration of water into agricultural soils. Proceedings of the World Surfactants Congress, Munich, West Germany, 10 pp.

Huddleston R.L. 1966. Biodegradable detergents for the textile industry. Am. Dyestuff Reporter 55:P52-P54.

Huddleston, R.L. and R.C. Allred. 1964a. Evaluation of detergents by using activated sludge. J. Am. Oil Chem. Soc. 41:732-735.

Huddleston, R.L. and R.C. Allred. 1964b. Effect of structure on biodegradation of nonionic surfactants. Proceedings of the Fourth International Congress on Surface-Active Substances, 3:871-882.

Huddleston, R.L. and R.C. Allred. 1965. Biodegradability of ethoxylated alkyl phenol surfactants. J. Am. Oil Chem. Soc. 42:983-986.

Hunter, J.V. and H. Heukelekian. 1964. Determination of biodegradability using Warburg respirometric techniques. Purdue Conf. 19:616-627.

Huyser, H.W. 1960. Relation between the structure of detergents and their biodegradation. Proceedings of the Third International Congress on Surface-Active Substances, 3:295-301.

Inoue, Z., J. Fukuyama and A. Honda. 1977. Toxicity and biodegradation of surfactants. Mizu Shori Gijutsu 18:119-132.

Janicke, W. 1988. Biologic degradation of nonionic surfactants: Biologic degradation of surfactants of nonaromatic ethoxylates in the confirmatory test. Communication 10. On the behavior of synthetic organic compounds for wastewater treatment. Tenside 25:345-355.

Janicke, W. and G. Hilge. 1977. Determination of the elimination-degree of water-endangering substances. (General water-elimination-test). Z. Wass.-Abwass.-Forsch. 10:4-9.

Jenkins, S.H., N. Harkness, A. Lennon and K. James. 1967. The biological oxidation of synthetic detergents in recirculating filters. Water Res. 1:51-53.

Knaebel, D.B., T.W. Federle and J.R. Vestal. 1990. Mineralization of linear alkylbenzene sulfonate (LAS) and linear alcohol ethoxylate (LAE) in 11 contrasting soils. Environ. Toxicol. Chem. 9:981-988.

Knaggs, E.A. 1964. Alkylolamides in soft detergents. Soap Chem. Spec. 40(12):79-82.

Kozarac, Z.D., D. Hrsak, B. Cosovic and J. Vrzina. 1983. Electroanalytical determination of the biodegradation of nonionic surfactants. Env. Sci. Technol. 17:268-272.

Kravetz, L., H. Chung, J.C. Rapean, K.F. Guin and W.T. Shebs. 1978. Ultimate biodegradability of detergent range alcohol ethoxylates. Presented at the 69th Annual Meeting of the American Oil Chemists' Society, St. Louis, May 1978. Shell Chemical Company Technical Bulletin SC:321-81.

Kravetz, L., H. Chung, K.F. Guin and W.T. Shebs. 1979. Ultimate biodegradation of alcohol ethoxylates. Surfactant concentration and polyoxyethylene chain effects. Presented at the 70th Annual Meeting of the American Oil Chemists' Society, San Francisco, CA. Shell Chemical Company Technical Bulletin SC:442-80.

Kravetz, L., H. Chung, K.F. Guin, W.T. Shebs and L.S. Smith. 1982. Ultimate biodegradation of an alcohol ethoxylate and a nonylphenol ethoxylate under realistic conditions. Household Pers. Prod. Ind. 19(3;4):46-52, 72:62-70. Soap Cosm. Chem. Spec. 58:34-42, 102B.

Kravetz, L., H. Chung, K.F. Guin, W.T. Shebs and L.S. Smith. 1984. Primary and ultimate biodegradation of an alcohol ethoxylate and an alkylphenol ethoxylate under average winter conditions in the USA. Tenside 21:1-6.

Kravetz, L., J.P. Salanitro, P.B. Dorn and K.F. Guin. 1991. Influence of hydrophobe type and extent of branching on environmental response factors of nonionic surfactants. J. Am. Oil Chem. Soc. 68:610-618.

Kurata, N. and K. Koshida. 1975. Biodegradability of nonionic surfactants in river die away test using Tama river water. Yukagaku 24:879-881.

Kurata, N. and K. Koshida. 1976. Biodegradability of sec-alcohol ethoxylate: Experimental results in continuous-flow activated sludge tests. Yukagaku 25:499-500.

Kurata, N., K. Koshida and T. Fujii. 1977. Biodegradation of surfactants in river water and their toxicity to fish. Yukagaku 26:115-118.

Kuwamura, T. and H. Takahaski. 1976. Structural effects of hydrophobe on the surfactant properties of polyglycol monoalkyl ethers. Proceedings of the Seventh International Congress on Surface-Active Substances, Moscow, 1:33-43.

Laboratory of the Government Chemist. 1966. Report of the Government Chemist 1966, London.

Laboureur, P., M. Bechet and J. Emeraud. 1976. Ultimate biodegradability of ethoxylated and propoxylated alcohols. Proceedings of the Seventh International Congress on Surface-Active Substances, Moscow, 4:139-148.

Larson, R.J. 1979. Evaluation of biodegradation potential of xenobiotic organic chemicals. Appl. Env. Microbiol. 38:1153-1161.

Larson, R.J. 1984. Biodegradation of detergent chemicals in groundwater/subsurface systems. Household Pers. Prod. Ind. 21:55-58.

Larson, R.J. and L.M. Games. 1981. Biodegradation of linear alcohol ethoxylates in natural waters. Env. Sci. Technol. 15:1488-1493.

Larson, R.J. and R.L. Perry. 1981. Use of the electrolytic respirometer to measure biodegradation in natural waters. Water Res. 15:697-702.

Larson, R.J. and G.E. Wentler. 1982. Biodegradation of detergent materials in natural systems at realistic concentrations. Soap Cosm. Chem. Spec. 58:33-42,127.

Lashen, E.S., F. A. Blankenship, K.A. Booman and J. Dupre. 1966. Biodegradation studies on a p-t-octylphenoxypolyethoxyethanol. J. Am. Oil Chem. Soc. 43:371-376.

Lötzsch, K., A. Neufahrt and G. Täuber. 1979. Comparative tests on the biodegradation of secondary alkane sulfonates using C-labeled preparations. Tenside 16:150-155.

Mann, A.H., and V.W. Reid. 1971. Biodegradation of synthetic detergents evaluation by community trials. II. Alcohol and alkylphenol ethoxylates. J. Am. Oil. Chem. Soc. 48:588-594.

Mann, H., and P. Schöberl. 1976. Biodegradation of a nonionic surfactant in estuary water. Arch. Fischerei Wiss. 26:177-180.

Markarian, R.K., K.W. Pontasch, D.R. Peterson and A.I. Hughes. 1989. Review and analysis of environmental data on Exxon surfactants and related compounds. Unpublished Technical Report, Exxon Biomedical Sciences, Inc.

Mausner, M. J.H. Benedict, K.A. Booman, T.E. Brenner, et al. 1969. The status of biodegradability testing of nonionic surfactants. J. Am. Oil Chem. Soc. 46:432-444.

Miksch, K. 1976. Application of dehydrogenase activity determination on activated sludge in studies of surfactant biodegradation. Proceedings of the Seventh International Congress on Surface-Active Substances, Moscow, 4:305-309.

Miksch, K. 1980. Use of dehydrogenase activity in studies of surfactant biodegradation. Tenside 17:124-125.

Miura, K., K. Yamanaka, T. Sangai, K. Yoshimura and N. Hayashi. 1979. Application of the biological oxygen consumption measurement technique to the biodegradation test of surfactants. Yukagaku 28:351-355.

Moller, U.J. 1972. Estimation of biological effects of nonionic emulsifiers in split-water phase from metal working emulsions. Erdöl & Kohle 25:451-456.

Moreno Danvila, A. 1979. Toxicity of surfactants during biodegradation. Behavior of *Daphnia magna* in activated sludge effluents. X Jornadas Com. Espanol Deterg. pp. 59-80.

Myerly, R.C., J.M. Rector, E.C. Steinle, C.A. Vath and H.T. Zika. 1964. Secondary alcohol ethoxylates as degradable detergent materials. Soap Chem. Spec. 40(5):78-82.

Naylor, C.G., F.J. Castaldi and B.J. Hayes. 1988. Biodegradation of nonionic surfactants containing propylene oxide. J. Am. Oil Chem. Soc. 65:1669-1676.

Neufahrt, A., K. Lötzsch and D. Gantz. 1982. Biodegradability of C-labelled ethoxylated fatty alcohols. Tenside 19:264-268.

Nooi, J.R., M.C. Testa and S. Willemse. 1970. Biodegradation mechanisms of fatty alcohol nonionics. Experiments with some C-labelled stearyl alcohol/EO condensates. Tenside 7:61-65.

Patterson, S.J., K.B.E. Tucker and C.C. Scott. 1966. Nonionic detergents and related substances in British waters. WPR Conf. #3 2:103-116.

Patterson, S.J., C.C. Scott and K.B.E. Tucker. 1967. Nonionic detergent degradation. I. TLC and foaming properties of alcohol polyethoxylates. J. Am. Oil Chem. Soc. 44:407-412.

Pitter, P. 1964. Synthetic surfactants in wastewater. VIII. Relation between molecular structure and the susceptibility of anionic surfactants to biochemical oxidation (Summary of results). Sb VSChT 8(2):13-39. (Cited in Swisher, 1987.)

Pitter, P. 1968a. Relation between degradability and chemical structure of nonionic polyethylene oxide compounds. Proceedings of the Fifth International Congress on Surface-Active Substances 1:115-123.

Pitter, P. 1968b. Biodegradability of surface-active alkyl polyethylene glycol ethers. Coll Czech. Chem. Comm. 33:4083-4088.

Pitter, P. 1968c. Synthetic surfactants in wastewaters. XIII. Evaluation of surfactant biodegradability by the COD technique and ultraviolet spectra. Sb VSChT F14:7-17.

Pitter, P. and J. Trauc. 1963. Synthetic surfactants in wastewaters. IV. Biodegradation of nonionic agents in laboratory models of aeration tanks. Sb. VSChT 7(1):201-216.

Procter & Gamble. 1985. Unpublished data.

Procter & Gamble. 1991. Unpublished data.

Reiff, B. 1976. The effect of biodegradation of three nonionic surfactants on their toxicity to rainbow trout. Proceedings of the Seventh International Congress on Surface-Active Substances, Moscow, 4:163-176.

Rizet, M., J. Mallevialle and J.C. Cournarie. 1977. Pilot plant investigation of the evolution of various pollutants during artificial recharge of an aquifer by a basin. Prog. Water Technol. 9:203-215.

Rudling, L. 1972. Biodegradability of Nonionic Surfactants: A Progress Report. Swedish Water and Air Pollution Research Laboratory (IVL), Report No. B134.

Rudling, L. and P. Solyom. 1974. The investigation of branched NPE's. Water Res. 8:115-119.

Ruiz Cruz, J. and M.C. Dobarganes Garcia. 1976. Pollution of natural waters by synthetic detergents. X. Biodegradation of nonionic surfactants in river water. Grasas Aceites 27:309-322.

Ruiz Cruz, J. and M.C. Dobarganes Garcia. 1977. Pollution of natural waters by synthetic detergents. XII. Relation between structure and biodegradation of nonionic surfactants in river water. Grasas Aceites 28:325-331.

Ruiz Cruz, J. and M.C. Dobarganes Garcia. 1978. Pollution of natural waters by synthetic detergents. XIII. Biodegradation of nonionic surfactants in river water and determination of their biodegradability by different test methods. Grasas Aceites 29:1-8.

Ruschenberg, E. 1963. Structure elements of detergents and their influence on biodegradation. Vom Wasser 30:232-248.

Salanitro, J.P., G.C. Langston, P.B. Dorn and L. Kravetz. 1988. Activated sludge treatment of ethoxylate surfactants at high industrial use concentrations. Water Sci. Technol. 20:126-130.

Scharer, D.H., L. Kravetz and J.B. Carr. 1979. Biodegradation of nonionic surfactants. Paper presented at the TAPPI Environmental Conference, April, 1979, Houston, TX. J. Tech. Asso. Pulp Paper Ind. 62:8p.

Schöberl, P. and K.J. Bock. 1980. Surfactant biodegradation and its metabolites. Tenside 17:262-266.

Schöberl, P., and H. Mann. 1976. Temperature influence on the biodegradation of nonionic surfactants in sea-and freshwater. Arch. Fischereiwiss 27:149-158.

Schöberl, P., E. Kunkel and K. Espeter. 1981. Comparative investigations on the microbial metabolism of a nonylphenol and an oxoalcohol ethoxylate. Tenside 18:64-72.

Sekiguchi, H., K. Miura, K. Oba and A. Mori. 1975. Biodegradation of α-olefin sulfonates and other surfactants. Yukagaku 24:145-148.

Stache, H. 1976. Properties of cycloaliphatic alcohol derivatives. Proceedings of the Seventh International Congress on Surface-Active Substances, Moscow, 1:378-391.

Stead, J.B., A.T. Pugh, I.I. Kaduji and R.A. Morland. 1972. A comparison of biodegradability of some alkylphenol ethoxylates using three methods of detection. Proceedings of the Sixth International Congress on Surface-Active Substances, Zurich, Switzerland, September 11-15, 1972. Carl Hauser Verlag, Munich, 3:721-734.

Steber, J. and P. Wierich. 1983. The environmental fate of detergent range fatty alcohol ethoxylates: Biodegradation studies with a C-labelled model surfactant. Tenside 20:183-188.

Steber, J. and P. Wierich. 1984. Anaerobic biodegradation of C-labeled fatty alcohol ethoxylates. Proceedings of the Eighth International Congress on Surface-Active Substances, 1:176-187.

Steber, J. and P. Wierich. 1987. The anaerobic degradation of detergent range fatty alcohol ethoxylates. Studies with ^{14}C-labelled model surfactants. Water Res. 21:661-667.

Steinle, E.C. R.C. Myerly and C.A. Vath. 1964. Surfactants containing ethylene oxide: Relationship of structure to biodegradability. J. Am. Oil Chem. Soc. 41:804-807.

Stiff, M.J., R.C. Rootham and G.E. Culley. 1973. The effect of temperature on the removal of nonionic surfactants during small scale activated sludge sewage treatment. I. Comparison of alcohol ethoxylates with a branched chain alkylphenol ethoxylate. Water Res. 7:1003-1010.

Stühler, H. and H. Wellens. 1976. Biodegradation of nonionic surfactants (polyglyxol ethers). Proceedings of the Seventh International Congress on Surface-Active Substances, Moscow, 4:106-115. (Cited in Swisher, 1987).

Sturm, R.N. 1973. Biodegradability of nonionic surfactants: screening test for predicting rate and ultimate biodegradation. J. Am. Oil Chem. Soc. 50:159-167.

Tabak, H.H. and R.L. Bunch. 1981. Measurement of nonionic surfactants in aqueous environments. Proceedings of the Purdue Industrial Waste Conference, 36:888-907.

Tobin, R.S., F.I. Onuska, B.G. Brownlee, D.H.J. Anthony and M.E. Comba. 1976a. The application of an ether cleavage technique to a study of the biodegradation of a linear alcohol ethoxylate nonionic surfactant. Water Res. 10:529-535.

Tobin, R.S., F.I. Onuska, D.H.J. Anthony and M.E. Comba. 1976b. Nonionic surfactants: Conventional biodegradation test methods do not detect persistent polyglycol products. Ambio 5:30-31.

Treccani, V., G. Braggi, E. Galli, G. Pensotti and V. Andreoni. 1973. The determination of biodegradability of surfactants: Elective culture test. Riv. Ital. Sost. Grasse. 50:418-422.

Truesdale, G.A., G.V. Stennett and G.E. Eden. 1968. Assessment of biodegradability of synthetic detergents: A comparison of methods. Proceedings of the Fifth International Congress on Surface-Active Substances, 1:91-101.

Turner, A.H., F.S. Abram, V.M. Brown and H.A. Painter. 1985. The biodegradability of two primary alcohol ethoxylate nonionic surfactants under practical conditions, and the toxicity of the biodegradation products to rainbow trout. Water Res. 19:45-51.

Vashon, R.D. and B.S. Schwab. 1982. Mineralization of linear alcohol ethoxylates and linear alcohol ethoxy sulfates at trace concentrations in estuarine water. Env. Sci. Technol. 16:433-436.

Vath, C.A. 1964. A sanitary engineer's approach to biodegradation of nonionics. Soap Chem. Spec. 40(2):56-58.

Wagener, S. and B. Schink. 1987. Anaerobic degradation of nonionic and anionic surfactants in enrichment cultures and fixed-bed reactors. Water Res. 21:615-622.

Weil, J.K. and A.J. Stirton. 1964. Biodegradation of some tallow-based surfactants in river water. J. Am. Oil Chem. Soc. 41:355-358.

Wencker, D.E. E. Allenbach and P. Laugel. 1974. Some observations on the biodegradation of surfactants. Bull. Soc. Pharm. Strasbourg 17:135-145.

Wickbold, R. 1966. Analysis for nonionic surfactants in water and wastewater. Vom Wasser 33:229-241.

Wickbold, R. 1974. Analytical comments on surfactant biodegradation. Tenside 11:137-144.

Yoshimura, K. 1986. Biodegradation and fish toxicity of nonionic surfactants. J. Am. Oil Chem. Soc. 63:1590-1596.

Zahn, R. and W. Huber. 1975. Ring test on biodegradability of products. Tenside 12:266-270.

Zahn, R. and H. Wellens. 1974. A simple method for testing the biodegradability of products and effluents. Chem. Ztg. 98:228-232.

Zahn, R. and H. Wellens. 1980. Determining biodegradability in a static test: Further experiments and new possibilities. Z. Wass.-Abwass-Forsch. 13:1-7.

Zika, H.T. 1971. The use of biodegradable linear alcohol surfactants in textile wet processing. J. Am. Oil Chem. Soc. 48:273-278.

PART II

ALKYLPHENOL ETHOXYLATES

SYNOPSIS

Alkylphenol ethoxylates (APE) compose one of two major classes of nonionic surfactants; the other major class is alcohol ethoxylates (AE). In 1988, approximately 450 million pounds of APE surfactants were sold in the U.S. The most important markets are industrial, including uses in plastics and elastomers, textiles, agricultural chemicals, and paper products (55%); institutional cleaning products (30%); and household cleaning and personal care products (15%). Of the APE, branched nonylphenol ethoxylates (NPE) with 9-10 ethylene oxide (EO) units are most commonly used in cleaning products and, as such, are most likely to enter the environment.

Alkylphenols are manufactured by the reaction of branched olefins with phenol in the presence of an acid catalyst. Commercial production of APE involves the base-catalyzed reaction of alkylphenols with ethylene oxide.

In the past, the determination of primary biodegradation of nonionic surfactants in the environment was accomplished indirectly, using chemically non-specific techniques. The two most widely used methods include the detection of cobalt thiocyanate active substances (CTAS) and the detection of bismuth iodide active substances (BIAS). The physical properties, foaming potential and surface tension, have also been used.

A concern over the possible persistence of some APE and their metabolites in the environment has led, over the last 12 years, to the development of sensitive analytical techniques for the extraction, separation, detection, and measurement of specific APE surfactants and their degradation products. These instrumental techniques include high performance liquid chromatography (HPLC) and gas chromatography (GC) separation techniques coupled with various detectors, as well as mass spectroscopy using chemical or electron impact ionization. These techniques, used alone or in combination, focus on the separation, detection, and measurement of the EO chain distribution of APE; identification of the alkyl chain; and identification and measurement of the lower molecular weight homologues and degradation products of APE, specifically alkylphenols and alkylphenol mono- and diethoxylates (APE_1 and APE_2) as well as the alkylphenol carboxylic acids (APEC). In the U.S., the degradation intermediates of most interest are nonylphenol (NP) and nonylphenol mono- (NPE_1) and diethoxylate (NPE_2).

National criteria for concentrations of APE in natural waters of the U.S. do not presently exist. The Alkylphenol and Ethoxylates Program Panel of the Chemical Manufacturers Association has undertaken a comprehensive environmental monitoring survey of NP and NPE in U.S. rivers in order to establish current concentrations in the water column and sediments of natural waterways within a 90% confidence limit. Recent data, based on a survey of 30 U.S. rivers considered likely to contain NPE, showed average concentrations of NP, NPE_1, NPE_2, and NPE_{3-17} of 0.12 µg/L, 0.09 µg/L, 0.10 µg/L, and 2.0 µg/L, respectively. Nonylphenol and NPE concentrations in 60 and 75% of the river water samples, respectively, were below detection limits. The highest concentrations detected in water samples were 0.64 µg/L for NP, 0.60 µg/L for NPE_1, 1.2 µg/L for NPE_2, and 14.9 µg/L for total NPE_{3-17}. Average concentrations of NP and NPE_1 in sediments of the same rivers were 161.9 µg/kg and 18.1 µg/kg dry weight, respectively. Sediment concentrations ranged from non-detectable to 2960 µg/kg for NP and from non-detectable to 175 µg/kg for NPE_1. Calculated sediment interstitial water concentrations of NP were similar to concentrations measured in the water column.

The primary biodegradation of APE surfactants has been observed in a variety of laboratory test systems and field studies. In laboratory studies, primary biodegradation, largely a measure of polyethoxy chain oxidation, may be >90% complete within a few hours to 28 days, depending on the starting material, type of test system, and source of bacteria or type of sludge. The mechanism of sequential ethoxy unit cleavage proceeds through NPE_2, NPE_1, and NP, less soluble components than the higher molecular weight NPE. Significant levels of NP have not been reported in laboratory systems. Several studies cite evidence of phenol ring degradation, as determined by disappearance of its UV absorption band. Laboratory studies relating rate of biodegradation and chemical structure demonstrated that commercially-produced branched-alkyl chain APE biodegrade more slowly than linear chain APE and p-substituted APE biodegrade faster than o-substituted APE.

At wastewater treatment plants, the higher molecular weight APE components (NPE_{3-20}) are effectively removed (often >95%) under normal secondary treatment conditions. The original ethoxylate oligomer distribution is slightly to significantly skewed in effluent, suggesting that the more lipophilic biodegradation intermediates NP, NPE_1, and NPE_2 are degraded more slowly than the higher molecular weight NPE. These species tend to adsorb to sewage sludge where they are carried into the sludge digestion and disposal process. The few field studies reported on APE biodegradation under winter conditions show variable results, ranging from 20% to 94% APE biodegradation in the winter. Nonylphenol, NPE, and the nonylphenol carboxylic acids have been measured in secondary effluents. Nonylphenol has also been measured in sewage sludge. Concentrations of NP measured in digested sludges at a limited number of U.S. plants ranged from 10 to 134 mg/kg dry weight. In studies of European wastewater treatment plants, concentrations of NP in anaerobically stabilized sludge were higher than in aerobically stabilized sludge, averaging 1000 mg/kg dry weight compared to 400 mg/kg dry weight, respectively. Reaeration studies show that NP and the NPE contained in sewage sludge degrade when the sludge is applied to soil.

Isolation of primary biodegradation products from both laboratory and field studies indicates a single pathway of metabolism proceeding via initial oxidation of the EO units. Some evidence for oxidation metabolism of the alkyl chain has been reported, as has evidence of opening of the aromatic ring. In both aerobic and anaerobic systems, sequential cleavage of the ether bonds of the EO units is rapid and results in formation of NPE_2, NPE_1, their respective carboxylic acids, and NP. All of these metabolites continue to biodegrade to CO_2 and water under aerobic conditions. Anaerobic biodegradation to methane and CO_2 is probably extremely slow.

In laboratory toxicity assays, the acute toxicity (96-hr LC_{50}) of APE to sensitive species of fish such as the bluegill sunfish (*Lepomis macrochirus*) ranged from 1.3 mg/L for C_9APE_4 to >1000 mg/L for C_9APE_{30}. Toxicity values of commercially available APE fell between 4 and

14 mg/L for most species of fish. Nonylphenol and lower ethoxylates such as NPE_1 are more toxic than the higher ethoxylates. The toxicity of NP ranged from 0.13 mg/L for juvenile Atlantic salmon (*Salmo salar*) to 3.0 mg/L for cod (*Gadus morhua*). The NPE_1C and NPE_2C were intermediate in toxicity, with LC_{50} values of 8.9 and 9.6 mg/L for the killifish (*Oryzias latipse*). For the crustacean, *Daphnia magna*, LC_{50} values of APE ranged from 2.9 mg/L for C_9APE_9 to 10,000 mg/L for C_9APE_{30}. For other fresh- and saltwater invertebrates, values for commercially available products ranged from 1.2 to >100 mg/L. The 48-hr LC_{50} of NPE_1 + NPE_2, tested on *Ceriodaphnia dubia*, was 1.0 mg/L; the 96-hr LC_{50} values for NP ranged from 0.043 mg/L for the marine shrimp, *Mysidopsis bahia*, to 3.0 mg/L for the bivalve, *Mytilus edulis*. Data on the toxicity of APE to algae varied widely. Concentrations of APE of ≤10 mg/L did not inhibit the growth of most species of algae, although in one test the freshwater green alga, *Selenastrum capricornutum*, was sensitive to an octylphenol ethoxylate, OPE_{10}, at 0.21 mg/L (96-hr EC_{50} for growth). Nonylphenol was generally toxic to algae at concentrations of ≥0.5 mg/L; the lowest 96-hr EC_{50} for growth was 0.027 mg/L for the marine diatom, *Skelatonema costatum*. Some of the studies indicated that toxicity to aquatic species was greater in laboratory waters than in water from natural sources. For all of the above species, toxicity generally decreased with increasing EO chain length.

Reported sublethal effects of C_9APE_{10} to fish (*Gadus morhua*) included erratic swimming, loss of equilibrium, decreased breathing rate, and eventual immobility at concentrations of 5 and 10 mg/L. Sublethal effects for several marine molluscs and crustaceans included decreased byssal thread formation, impaired locomotion and reduced burrowing activity, heart rate, and syphon activity at 0.5-20 mg/L. In chronic tests, no effects were noted in the behavior of *Gadus morhua* at 1 mg/L C_9APE_{10}, fathead minnows (*Pimephales promelas*) at 1 mg/L br-C_9APE_9, and *Daphnia magna* at 10 mg/L br-C_9APE_9. Toxicity values for NP were much lower with maximum acceptable toxicant concentrations (MATC) of 5.1 µg/L for 28-day growth of *Mysidopsis bahia* and 10 µg/L for 33-day survival of *Pimephales promelas*. Although the log K_{ow} value of >3 for the lower molecular weight APE and NP would indicate potential bioaccumulation, both uptake and excretion of the APE is rapid. Bioconcentration factors (BCF) varied widely among laboratory studies; however, when exposure conditions were held constant, BCF of NP ranged from 100 in shrimp (*Crangon crangon*) to 3430 in mussels (*Mytilus edulis*). The BCF for mussels exposed under field conditions was 340.

The toxicity and irritation potential of APE relative to human safety have been studied using several species of laboratory animals. In acute exposures, APE are non- to slightly toxic in rats, mice, guinea pigs, and rabbits by oral administration. In the rat, LD_{50} values ranged from 1410 mg/kg for C_9APE_9 to >28,000 mg/kg for C_8APE_{40}. A sex difference in toxicity was apparent in one study, with female rats being more than twice as sensitive to C_8APE_1 as males. For APE with <20 EO units, no correlation between structure and toxicity was apparent. Orally, octylphenol (OP) and NP did not appear to be more toxic than the APE. The dermal LD_{50} in rabbits ranged from >2000 mg/kg for C_9APE_5 to >10,000 mg/kg for C_9APE_{40}.

In subchronic and chronic studies with rats, the oral administration of APE with 4 to 30 EO units at doses up to 1000 mg/kg/day resulted in no toxicological effects. However, in a

90-day study with dogs, feeding of 1000 mg/kg/day of APE in the range of APE_{15} - APE_{20} resulted in cardiotoxicity as evidenced by focal myocardial necrosis. Cardiotoxicity was not observed in dogs with APE of <15 and >20 EO units.

Results of chronic (two-year) feeding studies in which APE was administered to rats and dogs were negative for carcinogenicity. In these studies, C_9APE_4 was administered to rats and dogs at doses up to 1000 mg/kg/day, C_9APE_9 was administered to rats and dogs at doses up to 200 and 88 mg/kg/day, respectively, and C_8APE_{40} was administered to rats at doses up to 700 mg/kg/day.

Alkylphenol ethoxylates were not mutagenic, either with or without metabolic activation, in the standard Ames (*Salmonella typhimurium*) test system. They also tested negative in most other genotoxicity assays. Although inflammatory effects were noted in the reproductive tracts of some female rats and rabbits administered high doses of APE, oral or intravaginal administration did not result in teratogenicity. Single vaginal applications of ≥ 25 mg/kg of C_9APE_9 to rats during pregnancy resulted in embryo- and fetotoxic effects in some studies, but not in others.

Orally and intravaginally administered APE are rapidly absorbed, metabolized, and excreted by mammalian species. Excretion rates of radiolabeled compounds indicated that intestinal absorption decreases with increasing EO chain length. Within one to seven days, ~90% of the radiolabel, present on either the EO chain or the phenol ring, was eliminated in the urine, feces, and as CO_2 in expired air. The latter route was observed only when the label was present on the EO chain. Urinary metabolites as well as expired CO_2 indicated extensive metabolism of the EO chain, but little metabolism of the NP except conjugation with glucuronic acid.

Undiluted APE were generally moderately to severely irritating to both the skin and eyes of test rabbits. Irritation to the eye was a function of concentration, with concentrations of 10 to 25% being generally non- to minimally irritating. Undiluted NP was severely irritating to the skin and eyes of rabbits and may cause permanent eye damage.

Available data from human studies indicated that APE applied as neat compounds are not dermal irritants or sensitizers. Their use in contraceptive formulations has not produced congenital effects.

NOMENCLATURE AND ABBREVIATIONS

Throughout this report the abbreviation APE has been used to designate alkylphenol ethoxylates. Alkylphenol ethoxylates are composed of an alkyl chain, usually branched (br-), attached to a phenol ring (hydrophobe moiety) which is combined, via an ether linkage, with one or more ethylene oxide or (poly)oxyethylene units (hydrophile moiety). The average number or range of carbon atoms in the alkyl chain (C) and the average number or average range of ethoxylate units (APE or EO) are designated by subscripts (for example, C_8APE_9, C_9APE_{9-10}, or EO_{5-6}). The most commonly used and commercially important APE are the nonylphenol ethoxylates (C_9APE or NPE). Important degradation products are nonylphenol (NP) and its mono-ethoxylate (NPE_1) and di-ethoxylate (NPE_2) adducts. The terms C_9APE and NPE are used interchangeably and reflect use in the cited papers.

If provided, surfactant tradenames for alkylphenol ethoxylates, enclosed in parentheses, follow the chemical names. In-text references to the products are not capitalized and do not carry the registered trademark symbol. Product tradenames for U.S. and European-manufactured products discussed in the text and the manufacturers/suppliers follow:[1]

Product	Company
AGRIMUL	Henkel Corporation
AKTAFLO	Barold Drilling Fluids, Inc.
ALKASURF, ANTAROX	Rhone-Poulenc
ARDRIL	Aquaness Chemicals
ARMUL	Witco Corporation
ATLAS	Atlas Refinery, Inc.
BASOPON	BASF Corporation
BLOPAL	Rhone-Poulenc
CALOXYLATE	Pilot Chemicals Company
CAPCURE EMULSIFIER	Henkel Corporation
CARSONON®	Lonza, Inc.
CEDEPAL	Stepan Canada, Inc.
CENEGEN, CENEKOL®	Crompton & Knowles
CHEMAX	Chemax Inc.
COREXIT	Exxon Chemical Company
DESONIC®	Witco Corporation
DIAZOPON®	Rhone-Poulenc

---

[1]*McCutcheon's Emulsifiers and Detergents.* 1992. North American Ed., MC Publishing Company, McCutcheon Division, 175 Rock Pond, Glen Rock, New Jersey.

Product	Company
ECCOTERGE	Eastern Color & Chemical Company
EMULSIFIER 632	Ethox Chemicals Inc.
EMULSOGEN	Hoechst Celanese Corporation
GRADONIC	Graden Chemical
HETOXIDE	Heterene Chemical
HOSTAPAL	Hoechst Celanese Corporation
HYONIC	Henkel Corporation
IBERPAL, IBERSCOUR, INERWET	A. Harrison & Co., Inc.
ICONOL	BASF Corporation
IGEPAL®	Rhone-Poulenc
MACOL®	PPG/Mazur
MAKON, NEUTRONYX	Stepan Company
NONIONIC	Hodag Chemical Company
NORFOX	Norman, Fox & Company
NUTROL	Clough Chemical, Inc.
POLYTERGENT®	Olin Chemical Corporation
POLYSTEP	Stepan Company
REXOL	Hart Chemical Ltd.
SANDOXYLATE	Sandoz Chemicals Corporation
SIPONIC	Rhone-Poulenc
STANDAPON	Henkel Corporation
STEROX®	Monsanto Industrial Chemicals Company
SURFLO	Exxon Chemical Company
SURFONIC®	Texaco Chemical Company
SYN FAC	Milliken Chemical
SYNPERONIC, SYNTHRAPOL	ICI Americas Inc.
T-DET®	Harcros Chemicals Inc.
TERGITOL®	Union Carbide Corporation
TEX-WET	Intex Chemical, Inc.
TRITON®	Union Carbide Chemicals and Plastics Company, Inc.
TRYCOL	Henkel Corporation
VALDET	Valchem
WITBREAK™	Witco Corporation

NOMENCLATURE AND ABBREVIATIONS

The following abbreviations have been used throughout the text:

AE	Alcohol ethoxylates
APE	Alkylphenol ethoxylates
APE$_n$	Ethylene oxide moiety of n units or moles
APEC	Alkylphenoxy (ethoxy) acetic acids
APHA	American Public Health Association
ASTM	American Society for Testing and Materials
BAS	Batch activated sludge
BCF	Bioconcentration factor
BIAS	Bismuth iodide active substances
BOD	Biochemical oxygen demand
br-	Branched
CAS	Chemical Abstract Service
CFAS	Continuous-flow activated sludge
CI	Chemical ionization
C_n	Alkyl group with n carbons
CMA	Chemical Manufacturers Association
CMC	Critical micelle concentration
COD	Chemical oxygen demand
CTAS	Cobalt thiocyanate active substances
DOC	Dissolved organic carbon
EC$_{50}$	Concentration effective to 50% of organisms
ECD	Electron capture detector
EI	Electron impact
ELS	Evaporative light scattering
EO	Ethylene oxide
EPCRA	Emergency Planning and Community Right to Know Act
FAB	Fast atom bombardment
FD	Field desorption
FID	Flame ionization detector
g	Gram
GC	Gas chromatography
HPLC	High performance liquid chromatography
HRGC	High resolution gas chromatography
IR	Infrared
L	Liter
LC$_{50}$	Concentration lethal to 50% of organisms
LD$_{50}$	Dose lethal to 50% of organisms
LOEC	Lowest-observed-effect concentration
log K_{ow}	Octanol/water partition coefficient (also log P_{ow})
m-	*meta*
MATC	Maximum acceptable toxicant concentration
mg	Milligram
mL	Milliliter

MS	Mass spectroscopy
n-	Normal, linear
ng	Nanogram
NICI	Negative ion chemical ionization
NOEC	No-observed-effect concentration
NP	Nonylphenol
NPE	Nonylphenol ethoxylates (also C_9APE)
NPE_1	Nonylphenol monoethoxylate
NPE_2	Nonylphenol diethoxylate
NPEC	Nonylphenoxy (ethoxy) acetic acids (Nonylphenol carboxylic acids)
o-	ortho
OECD	Organization for Economic Cooperation and Development
OP	Octylphenol
OPE	Octylphenol ethoxylates (also C_8APE)
p-	para (same as 4-)
P_a	Adsorption partition coefficient
PEG	Polyethylene glycols
PICI	Positive ion chemical ionization
PPAS	Potassium picrate active substances
ppm	Parts per million (=mg/L)
pri-	Primary
sec-	Secondary
SCAS	Semi-continuous activated sludge
SDA	Soap and Detergent Association
TLC	Thin-layer chromatography
TOC	Total organic carbon
t-, tert-	Tertiary
tp-	Tripropylene derived (highly branched)
μg	Microgram
UV	Ultraviolet

I. INTRODUCTION

A. CHEMISTRY AND MANUFACTURE

Alkylphenol ethoxylates are a class of nonionic surfactants composed of a branched alkyl chain attached to a phenol ring and combined, via an ether linkage, with ethylene oxide. The alkylphenol moiety is manufactured by reacting branched olefins with phenol in the presence of an acid catalyst (McKetta, 1977; Reed, 1978).

The predominant positional isomer of monoalkylphenols is the *para* isomer:

para-alkylphenol
$\geq 90\%$

ortho-alkylphenol
$\leq 10\%$

The R group consists of an alkyl group which can occur in a multitude of highly branched isomers for all alkylphenols except when R is derived from diisobutylene. In that case a single isomer results.

By far the most important alkylphenol is nonylphenol in which case the R group is C_9H_{19}; the nonene used in its manufacture is propylene trimer. Also significant commercially are octylphenol, dodecylphenol and dinonylphenol. Industrial nonylphenol (4-nonylphenol, branched) is assigned the CAS Registry No. 84852-15-3.

Commercial APE are manufactured by the base-catalyzed ethoxylation of alkylphenols (Cahn and Lynn, 1978). The reaction with ethylene oxide is rapid and consumes all of the phenol. Nonylphenol ethoxylate (NPE) surfactants are depicted as follows:

$$C_9H_{19}-\text{C}_6\text{H}_4-O-[CH_2-CH_2-O]_{n-1}-CH_2-CH_2-OH$$

where n = number of moles of ethylene oxide per mole of nonylphenol and ranges from 1 to 100. Water solubility is increased by alkyl branching and is directly proportional to n. Nonylphenol ethoxylates are water soluble for values of $n \geq 7$. The NPE most commonly used in cleaning products has n = 9-10.

Other nomenclature for NPE includes:
- nonylphenoxy[poly(ethyleneoxy)]ethanol
- nonoxynol
- α-(*p*-nonylphenyl)-*w*-hydroxypoly(oxyethylene)

According to suppliers' Material Safety Data Sheets (MSDS, 1991), a conventional commercial grade of NPE may typically contain 1.00 to 3.99% weight "glycol ethers" as defined under EPCRA Section 313, i.e., 1, 2, and 3-mole ethoxylate adducts. In addition, trace levels of certain regulated chemicals such as ethylene oxide may be present at concentrations below 10 ppm.

B. USES

Alkylphenol ethoxylate surfactants have been used in cleaning products and industrial processing for more than forty years. Total U.S. sales of APE exceeded 450 million pounds in 1988 (USITC, 1989).

A brief overview of the U.S. markets for APE follows (CMA, 1988):

> Industrial - 55% of total
> Four industries make up 90% of this category, in the following order:
>> Plastics and elastomers
>> Textiles
>> Agricultural chemicals
>> Paper

The principal industrial end use for APE is in plastics and elastomers. APE dominate the nonionic market for emulsion polymerization. Mainly 20-40 mole EO APE are used in acrylic and some vinyl acetate polymerization processes. The APE act as a stabilizer for the final latex.

APE serve a variety of functions in the textile industry aimed at adapting the general processes of cleaning, spinning, weaving, and finishing to the full range of fiber types. They are excellent wetting agents with good handling and rinsing characteristics.

Agricultural chemicals provide a large market for APE. APE are used as emulsifiers in the production of liquid pesticides. They are also used as enhancers and wetting agents to improve adhesion of the toxicant to the target organism.

APE have a diversity of uses in the paper industry. They are used as dispersants in the pulping process, for paper deinking, and dissolving pulp.

> Institutional - 30% of total

This category consists entirely of cleaning products. Three basic types contribute over 90% of the total, in the following order:
>> Metal and commercial vehicle cleaners
>> Commercial laundry products
>> Hard surface cleaners

INTRODUCTION

Household and personal care - 15% of total
Over 80% of this category consists of heavy duty powders, liquid laundry detergents, and hard surface cleaners.

APE markets are thus dominated by industrial processing and institutional cleaners. Sales to the household markets, while significant, are very small compared to those of the other major class of nonionic surfactants, alcohol ethoxylates (AE).

C. PRODUCT CAS IDENTIFICATION

The names, CAS (Chemical Abstract Service) numbers, and number of EO units of commercially available APE surfactants are listed in Table 1-1 (TSCA, 1979). Also listed are some octylphenols, nonylphenols, and carboxylated (acetic acid) biodegradation products of APE surfactants.

Table 1-1. Nomenclature and CAS Numbers of Alkylphenols, Alkylphenol Ethoxylates, and Alkylphenol Carboxylates

Chemical	CAS Number	EO Units
4-Nonylphenoxy ethanol	104-35-8	1
p-Nonylphenol; 4-nonylphenol	104-40-5	0
p-Dodecylphenol, 4-docecylphenol	104-43-8	0
o-Nonylphenol; 2-nonylphenol*	136-83-4	0
p-tert-Octylphenol; 4-tert-octylphenol	140-66-9	0
p-(1,1,3,3-Tetramethylbutylphenol)	140-66-9	0
o-Octylphenol; 2-octylphenol	949-13-3	0
Octylphenol monoethylene oxide	1322-97-0	1
p-Octylphenol; 4-octylphenol	1806-26-4	0
p-Octylphenol diethoxylate	2315-61-9	2
p-Octylphenol ethoxylate	2315-62-0	3
p-Octylphenol ethoxylate	2315-63-1	4
p-Octylphenol ethoxylate	2315-63-2	5
p-Octylphenol monoethoxylate	2315-67-5	1
2-(4-(1,1,3,3-tetramethylbutyl)phenoxy) ethanol	2315-67-5	1
p-Nonylphenoxy acetic acid	3115-49-9	—
o-tert-Octylphenol	3884-95-5	0
4-Nonylphenol polyethylene glycol ether	7311-27-5	4
p-tert-Octylphenoxypolyethoxyethanol	9002-93-1	>1
Polyethylene glycol mono(p-(1,1,3,3-tetramethylbutyl)phenyl)ether	9002-93-1	>1
Isooctylphenyl polyethylene glycol ether	9004-87-9	>1
Dodecylphenol polyethylene glycol ether	9014-92-0	>1

Table 1-1. (cont.)

Chemical	CAS Number	EO Units
Nonylphenol polyethylene glycol ether	9016-45-9	>1
Nonylphenol polyethylene oxide	9016-45-9	>1
Nonylphenol polyglycol ether	9016-45-9	>1
Nonylphenol polyethylene glycol ether	9016-45-9	>1
Nonylphenoxypoly(oxyethylene)ethanol	9016-45-9	>1
Octylphenoxypoly(ethoxyethanol)	9036-19-5	>1
4-Nonylphenol diethylene glycol ether	20427-84-3	2
4-Nonylphenoxy ethoxy ethanol	20427-84-3	2
2-(2,4-nonylphenoxy)ethoxy ethanol	20427-84-3	2
n-Nonylphenol (mixed isomers)	25154-52-3	0
p-Nonylphenol polyethylene glycol ether	26027-38-3	>1
α-(p-Nonylphenyl)-w-hydroxypoly(oxyethylene)	26027-38-3	>1
Mono(p-dodecylphenyl)polyethylene glycol ether	26401-47-8	>1
Nonylphenol octa(oxyethylene) ethanol	26571-11-9	9
Decylphenol	27157-66-0	0
Nonylphenoxydiglycol	27176-93-8	2
Nonylphenol hepta(oxyethylene) ethanol	27177-05-5	8
Nonylphenol octaethoxylate	27177-05-5	8
Nonylphenol octaglycol ether	27177-05-5	8
Nonylphenol decaethylene glycol ether	27177-08-8	10
Nonylphenolnona(oxyethylene) ethanol	27177-08-8	10
Octylphenols	27193-28-8	0
Dodecylphenol	27193-86-8	0
Nonylphenoxyglycol	27986-36-3	1
(Nonylphenoxy)ethanol	27986-36-3	1
Nonylphenol monoethoxylate	27986-36-3	1
(C_9) Branched alkylphenol ethoxylate	68412-54-4	>1
(C_8) Branched alkylphenol ethoxylate	68987-90-6	>1
Nonylphenols (br-C_9 alkylphenols) (industrial nonylphenol)**	84852-15-3	0
4-Nonylphenol, branched	84852-15-3	0
4-Nonylphenoxyethoxy acetic acid	106807-78-7	—
4-Nonylphenoxydiethoxy acetic acid	108149-59-3	—
p-Octylphenoxyethoxy acetic acid	108241-00-5	—
p-Octylphenoxydiethoxy acetic acid	121057-06-5	—
p-Octylphenoltriethoxy acetic acid	121057-07-6	—

*2-Nonylphenol is no longer manufactured.
**Commercially-manufactured product comprised of primarily *para*-branched C_9 alkylphenols

REFERENCES

Cahn, A. and J.L. Lynn, Jr. 1978. Surfactants and Detersive Systems. In: *Kirk-Othmer Encyclopedia of Chemical Technology*, Vol. 22. J. Wiley & Sons, Inc., New York, p. 363-365.

Chemical Manufacturers Association (CMA). 1988. CMA Panel on Alkylphenol and Ethoxylates.

McKetta, J.J., Ed. 1977. *Encyclopedia of Chemical Processing and Design*. Vol. 2. Marcel Dekker, Inc., p. 399.

MSDS (Material Safety Data Sheets). 1991. Suppliers material safety data sheets on NPE.

Reed, H.W.B. 1978. Alkylphenols. In: *Kirk-Othmer Encyclopedia of Chemical Technology*, Vol. 2. John Wiley & Sons, Inc., New York, p. 72.

TSCA (Toxic Substances Control Act) Chemical Substances Inventory. 1979. U.S. Environmental Protection Agency, Washington, DC.

USITC (U.S. International Trade Commission). 1989. Report on Synthetic Organic Chemicals, Publication No. 2219.

II. ENVIRONMENTAL LEVELS

A. ANALYTICAL METHODS

Before the development of chemical-specific methods for the measurement of APE in biodegradation studies and environmental samples, primary biodegradation was demonstrated by methods that measured the loss of a chemical or physical property of the parent molecule. These standard nonspecific techniques such as measurement of foaming potential, changes in surface tension, and detection of cobalt thiocyanate, bismuth iodide, and potassium picrate active substances do not differentiate between AE and APE surfactants and do not detect the lower EO oligomers of AE or APE.

The concern over the possible persistence of some APE and their metabolites in the environment has led, over the last 12 years, to the development of sensitive and specific techniques for the determination of APE and their metabolites in environmental media. Chemical and chemical-moiety specific methods for APE as well as their metabolites include thin-layer, gas, and high performance liquid chromatography coupled with various detection methods. Nonylphenols can also be identified by their fragmentation patterns with mass spectrometry. These analytical techniques focus on the separation, detection, and measurement of the EO distribution of APE; identification of the alkyl chain; and identification and measurement of the lower molecular weight homologs of APE, specifically nonylphenol (NP) and nonylphenol mono- and diethoxylates (NPE_1 and NPE_2) as well as the nonylphenol carboxylated (acetic acid) metabolites (NPE_1C and NPE_2C). Most of these techniques have been reviewed by Hummel (1962), Rosen and Goldsmith (1972), Longman (1976), Arthur D. Little (1977), Kuo and Mottola (1980), Goyer et al. (1981), Llenado and Jamieson (1981), Garti et al. (1983), Cross (1987), and Swisher (1987).

A monitoring survey of U.S. rivers for NP and NPE was performed by the Radian Corporation under contract to the Alkylphenol and Ethoxylates Program Panel (APE Program Panel) of the Chemical Manufacturers Association (CMA) (Radian Corporation, 1990). Two extraction/chromatography methods developed by Texaco Chemical Corporation (Kubeck and Naylor, 1990) were validated according to EPA guidelines. Detection limits were determined for NP, NPE_1, NPE_2, and NPE_{3-17} in river water and sediment.

1. Physical Methods

The two principal physical procedures used to monitor the primary degradation of nonionic surfactants in the laboratory and in environmental samples are measurement of foaming and surface tension. The usefulness of the foam method is limited since foaming is complex, transient, and not a linear function of surfactant concentration. For APE, the polyoxyethylene chain may be considerably shortened by microbial attack, yet the intermediate products may produce a considerable amount of foam. Furthermore, other substances present in environmental samples such as protein, partial degradation products, and other surfactants make quantification impossible. Measurement of nonionics with the foam method is further confounded by the fact that the nonionics are generally low in foaming potential compared to anionic surfactants (Swisher, 1987).

The surface tension of water is lowered significantly at a concentration of a few ppm surfactant. It decreases further with increasing concentration of surfactant up to a critical micelle concentration (CMC), but the relationship is not linear. Above the CMC, there is little change in surface tension with increasing surfactant concentration. As surfactants degrade, surface tension tends to rise. As with foaming properties, surface tension lacks specificity and is subject to interferences from other substances (Swisher, 1987).

The accuracy and repeatability of these two physical methods were reviewed by the Soap and Detergent Association (SDA) (Mausner et al., 1969). Greater consistency among and within testing laboratories was achieved through the use of a foam loss procedure.

2. Chemical Methods

The two most widely used analytical tests to determine the primary biodegradability of nonionic surfactants are detection of cobalt thiocyanate and bismuth iodide active substances. Both methods are based on the formation of metal complexes with the oxygen atoms in the polyoxyethylene chain. The most serious limitation of these methods is their inability to differentiate between different surfactants containing the polyoxyethylene chain. Thus, when these methods are employed for environmental samples, the results apply to all nonionics present. These two methods as well as some less frequently used chemical methods are described here. Also described are cleanup steps necessary for removal of interfering substances from environmental samples.

Cobalt Thiocyanate Active Substances (CTAS). Cobalt thiocyanate active substances such as nonionics form a blue-colored complex with the ammonium cobaltothiocyanate reagent. These complexes are formed with polyethers higher than EO_5. The complex can be extracted into an organic solvent and the nonionic content determined spectrophotometrically by comparing the absorbance of the complex with standard curves for the individual ethoxylates (Crabb and Persinger, 1964; Swisher, 1987). Quantities as low as 1 mg/L (1 ppm) can be detected. Sensitivity of the method is dependent on the number of moles of EO and on the extraction efficiency of the complex. The procedure is most sensitive for surfactants having 6 to 25 EO groups per mole of hydrophobe. Thus, the mildly persistent, lower molecular weight metabolites of APE are not detected. Additionally, the molar absorbance is not linear below 6 EO units or for higher molecular weight complexes. The absorption coefficients of the latter may be 10-15 times greater than lower molecular weight complexes. Without the use of a sublation preconcentration technique, the method is subject to interferences from strongly acidic or basic solutions, cationic surfactants, biodegradation intermediates such as polyethylene glycols (PEG), and naturally occurring substances (Arthur D. Little, 1977; Swisher, 1987).

In 1977 the SDA (Boyer et al., 1977) developed an improved CTAS procedure for determination of nonionic surfactants in biodegradation and environmental studies. Their procedure involves an improved sublation step with dichloromethane as the solvent and an ion-exchange step prior to reaction with cobalt thiocyanate. The method is applicable to both linear and branched APE with carbon chain lengths of C_8-C_{18}. The limit of detection is 0.1 mg/L. The ion-exchange step is essential, since without it CTAS values for environmental samples more than doubled, due to anionic surfactant interferences. A round robin series of tests among six laboratories using both the CTAS and bismuth iodide active substances (BIAS) methods on the same environmental samples indicated good intralaboratory reproducibility, but significant interlaboratory variation. The SDA procedure has been adopted by the American Public Health Association (APHA, 1989) as the reference method for nonionic surfactants.

Bismuth Iodide Active Substances (BIAS). In the BIAS method, Dragendorff reagent, a preparation containing barium tetraiodobismuthate, forms a precipitate with polyoxyethylene compounds by interaction with the ethoxylate oxygen atoms. Interference from PEG, the major degradation products of AE surfactants, can be eliminated by an initial butanone extraction step (Bürger, 1963; Swisher, 1987).

Wickbold (1966, 1971, 1972, 1973) modified the BIAS method into the now widely accepted procedure for determination of nonionic surfactants in environmental samples. The procedure, as adopted by the Organization for Economic Cooperation and Development (OECD, 1976), involves these steps: removal of particulates by filtration, sublation into ethyl acetate using a stream of nitrogen bubbles presaturated with ethyl acetate vapor, removal of cationics with cation exchange resin, precipitation of nonionics by barium bismuth iodide reagent, and measurement of bismuth content of the precipitate by potentiometric titration with pyrrolidine dithiocarbamate solution. This method has a detection limit of approximately 0.01 mg/L. In Europe, the BIAS procedure of Wickbold has been officially adopted for determination of nonionic surfactants in biodegradation test studies (EEC, 1982). Although the BIAS method is widely used in laboratory biodegradation tests, it generally overestimates the presence of nonionics in environmental samples.

For environmental samples, the BIAS method is preferable over the CTAS method in that it provides more accurate results for samples that require an arbitrary calibration factor. The BIAS precipitates show a fairly constant ratio of one barium atom to about 9.8 ether oxygens whereas the CTAS response depends on both the number of ether oxygens and the extractability of the cobalt complex. Thus, less variation is expected with the BIAS method (Swisher, 1987).

Invernizzi and Gafà (1973) compared five different analytical methods. They concluded that surface tension and the Wickbold sublation/BIAS method were the best followed by foaming potential and the Patterson TLC method. CTAS was the least preferred method. In river-water tests on nonionics, however, Ruiz Cruz and Dobarganes Garcia (1976) found that the results from foam, surface tension, and CTAS were generally in agreement.

Potassium Picrate Active Substances (PPAS). Favretto and co-workers (Favretto and Tunis, 1976; Favretto et al., 1980, 1983) developed and improved a paired ion extraction method for nonionic surfactants in water and wastewaters. In the improved PPAS method, the sample is purified and/or concentrated and extracted into ethylene dichloride; the ethoxylate chain is then complexed with aqueous K^+ and paired with picrate anion; the ion pair is extracted into an organic solvent such as methylene chloride and the picrate content of the organic phase is determined by its spectrophotometric absorbance. The lower limit of detection is 0.1 mg/L. This method can be used for ethoxylate chains ranging from 7-28 units. Of several potentially interfering substances tested, only cationic surfactants gave positive results at low concentrations (Favretto et al., 1982).

Other Methods. In addition to the above methods, polyoxyethylene or polyoxypropylene units form water-insoluble complexes with mercuric and cadmium chlorides and with the calcium and barium salts of phosphosilicic, phosphomolybdic, and silico- and phosphotungstic acids. The complexes are analyzed by gravimetric or spectrometric methods. These methods are not commonly used since they are time consuming and subject to interferences (Cross, 1987; Swisher, 1987). Han (1967) developed a method that involved extraction of polyethoxylates into chloroform and sodium chloride, sulfation of the polyethoxylates, and determination by methylene blue activity which is routinely used to measure anionic surfactants.

Thin-layer chromatography (TLC) is a useful physicochemical separation technique. Patterson et al. (1964) was one of the first investigators to apply TLC to the analysis of nonionic surfactants by developing a method for the analysis of APE of 8-10 ethylene oxide units. A 250-ml river water or wastewater sample is extracted four times with chloroform using magnesium sulfate as the salting-out agent. The combined extracts are washed sequentially with acidic and basic solutions and then evaporated to dryness. The residue, dissolved in chloroform, is applied to silica-gel plates and developed. A modified Dragendorff reagent is used for detection and the resultant spots are compared to standard solutions. Concentrations as low as 0.5 μg/sample can be detected.

In the past, TLC has been useful for the determination of the extent of primary biodegradation. With the development of rapid and sensitive HPLC and GC techniques for the analysis of multiple environmental samples, TLC is seldom used.

3. Instrumental Analyses

Several types of separation techniques interfaced with ultraviolet (UV), infrared (IR), or mass spectroscopy (MS) detection have been used to identify APE and their degradation intermediates in aqueous environmental samples and in sewage sludge as indicated in Table 2-1. Generally a combination of methods, preceded by specific extraction or concentration techniques, is required for complete identification of APE in environmental samples. These methods provide greater specificity and sensitivity than physical and chemical methods.

ENVIRONMENTAL LEVELS

Table 2-1. Analytical Methods for APE Surfactants and Metabolites

Chemical/Moiety	Enrichment/Cleanup	Separation/Detection	Detection Limit[a]	Reference
APE: Alkyl chain EO distribution	—	Reversed-phase HPLC/UV Field desorption MS	50 µg/L	Otsuki and Shiraishi, 1979
APE_{1-20}: EO distribution	—	Normal-phase HPLC	—	Allen and Linder, 1981a; 1981b
Nonylphenols, C_5APE_1, C_5APE_2, C_5APE_3	Steam distillation/ solvent extraction	GC/EI-MS	10 µg/L	Giger et al, 1981; Schaffner et al., 1982; Stephanou and Giger, 1982
APE	—	FD-MS, FAB-MS, CI-MS	—	Schneider et al., 1983
Alkylphenols, NPE_1, NPE_2, NPE_1C, NPE_2C	—	HPLC/HRGC/MS	1 µg/L	Giger et al, 1984
APE: EO distribution	Solvent extraction	Normal-phase HPLC/fluorescence	0.2 mg/L	Kudoh et al., 1984
APE: EO distribution Alkyl chain	—	GC/CI-MS	—	Stephanou, 1984
APE_{1-18}: EO distribution Alkyl chain	Gaseous stripping	Normal-phase HPLC/UV Reversed-phase HPLC/UV	1 µg/L	Ahel and Giger, 1985b
Alkylphenols, Alkylphenol mono- and diethoxylates	Steam distillation/ solvent extraction	Normal-phase HPLC/UV	0.5 µg/L	Ahel and Giger, 1985a
Alkylphenols, APE, APE_1C, APE_2C	Solvent extraction	GC/MS of trimethylsilyl derivatives	—	Stephanou, 1986
APE: EO distribution	Gaseous stripping; ion-exchange column	Normal-phase HPLC/fluorescence	0.2 ng	Holt et al., 1986
APE_{1-18}: Nonylphenols	Soxhlet extraction	Normal-phase HPLC/UV Reversed-phase HPLC/fluorescence	65-95 ng	Marcomini and Giger, 1987

Table 2-1. (cont.)

Chemical/Moiety	Enrichment/Cleanup	Separation/Detection	Detection Limit[a]	Reference
Nonylphenol, NPE_1, NPE_2	Octadecylsilica cartridges	Normal-phase HPLC/fluorescence	4 µg/L (total)	Marcomini et al., 1987
NPE_1C, NPE_2C Nonylphenol isomers	Solvent extraction; gaseous stripping	Normal-phase HPLC/UV HRGC HRGC/MS	1 µg/L	Ahel et al., 1987
Nonylphenol NPE_1 NPE_2 NPE_{3-17}	Steam distillation	Normal-phase HPLC/fluorescence	0.11 µg/L 0.07 µg/L 0.06 µg/L 1.6 µg/L	Kubeck and Naylor, 1990; Radian Corporation, 1990
Nonylphenol isomers	Ion-exchange plus octadecylsilica column	GC/MS		
Nonylphenol, NPE_{1-3}	Batch extraction	GC/EI-MS, GC/CI-MS, GC/ECD	0.1 µg/L 0.2-1 µg/L	Wahlberg et al., 1990

[a]Individual oligomers.
A dash (—) indicates no data.

Prior to analysis of environmental samples, several widely used extraction and clean-up steps have been employed. These include solvent extraction, gaseous stripping (commonly called solvent sublation) and ion exchange (Crisp, 1987; Swisher, 1987). In the solvent extraction preconcentration technique, the sample is extracted from water with a suitable solvent such as methylene chloride, concentrated by evaporation, and subjected to analysis. Because of the evaporation step, solvent extraction cannot be used for volatile compounds.

Gaseous stripping involves bubbling an inert gas, such as nitrogen, through a column of water to strip volatile organics from the water into the gas phase. The gas phase passes into an overlying organic solvent or into a cold or lipophilic trap. The extract in the solvent may be further purified by passage through ion-exchange columns. Using this method, Wickbold (1971; 1972) achieved almost complete removal of nonionic surfactants from river water and test samples within minutes.

In the ion-exchange column technique, large samples of water are passed through a column packed with lipophilic material such as charcoal, XAD resin, or Tenax. Specific anion- and cation-exchange resins remove anionic and cationic surfactants which may interfere with many analysis methods. The nonionic surfactants are removed from the column by a suitable solvent. Jones and Nickless (1978a; 1978b) evaluated Amberlite XAD-4 resin as an extraction medium. They achieved 80-100% adsorption/desorption efficiencies although exhaustive purification of the resin was necessary to remove contaminants prior to use.

Waters et al. (1986) developed a method for the cleanup of environmental samples prior to BIAS analysis, that can be used prior to other analytical techniques. Unfiltered environmental samples were subjected to four 10-minute gaseous stripping steps followed by a cation/anion ion-exchange step.

Nonylphenols and lower molecular weight NPE can be isolated from environmental samples by exhaustive steam distillation/solvent extraction (Giger et al., 1981). The system utilizes a closed-loop apparatus that is based on the work of earlier investigators. Water or sludge samples were mixed with sodium chloride and refluxed for several hours with cyclohexane. The samples were then dried with sodium sulfate and directly analyzed by HPLC or GC. No additional cleanup steps were needed. Schaffner et al. (1984) used the same method with the addition of a silica column clean-up step for the analysis of NP in sewage sludge. Marcomini et al. (1991) also extracted NP and NPE_1 from dried sludge after addition of NaOH (20% w/w) using Soxhlet extraction. Methanol was the extraction solvent. Recently Menges et al. (1992) proposed a method for the extraction of NPE from wastewater that uses centrifugal partition chromatography. Ethyl acetate was used as the extractant; the addition of salt to the aqueous phase increased the extraction efficiency.

High Performance Liquid Chromatography (HPLC). HPLC is a useful technique for separating and identifying the ethoxy components of complex mixtures of nonionic surfactants and their polyethylene glycol metabolites as well as the alkyl and ethoxy chain lengths of individual APE. Using normal-phase HPLC, APE can be separated according to their EO chain lengths on a silica gel matrix using organic solvent gradient or isocratic elutions. The eluents used are usually combinations of water with a polar organic solvent such as methanol, ethanol, or acetonitrile. The separated components are detected by UV absorption or fluorescence and quantitated with external standards. Ultraviolet absorption has commonly been used, but the more sensitive fluorescence allows measurement of individual components in environmental samples down to one-tenth to one-hundredth microgram (μg), depending on the individual chemical. Because these compounds contain a benzene ring, derivatization is not necessary for detection. Garti et al. (1983) have reviewed various types of columns, eluents, detectors, and retention times for EO adducts under specified conditions.

Reversed-phase HPLC can be used to separate APE according to their alkyl chains. In the reversed-phase mode, the APE are loaded onto a C_{18} column with water as the solvent. The APE are removed isocratically with an organic solvent such as methanol or with gradient elution with water/methanol. The compounds are separated in the order octylphenol, nonylphenol, and dodecylphenyl ethoxylates (Otsuki and Shiraishi, 1979; Ahel and Giger, 1985a). These authors also discuss enrichment/cleanup steps appropriate to the method and chemical being monitored.

In 1975, Nakamura and Matsumoto demonstrated the feasibility of using adsorption HPLC for the separation of ethoxylate units of commercially available AE. The acetate derivatives were detected by adapting a flame ionization (FI) detector to the HPLC instrument. Retention time on the column was directly proportional to the number of moles of ethylene oxide present. This method has been applied to APE. Allen and Linder (1981a; 1981b) resolved APE with ethoxy chains of 3 to 15 units into well defined peaks using UV absorption as the detection method. Analysis by GC was necessary to determine the free alcohol and alkyl distributions.

Kudoh et al. (1984) and Holt et al. (1986) were the first to use fluorescence detection, a more sensitive method than UV absorption, following the separation of APE by normal-phase HPLC. The limit of detection in the first study was 0.2 μg/g. The minimum level of detection of individual homologues of APE in the second study was 0.2 ng. Holt used the method on both a commercial product and sewage influent and effluent and detected APE with from 1 to 19 EO units. Prior to analyses, the extraction and clean-up method of Waters et al. (1986) together with chromatography on a Zorbax NH_2 column were used.

Otsuki and Shiraishi (1979) used field desorption mass spectrometry (FD-MS) following reversed-phase liquid chromatography (for alkyl chain structure) to identify and determine the number of ethylene oxide units in commercial products. Using reversed-phase HPLC, the recovery of APE from spiked samples was 96% at 1 mg/L and 71% at 50 μg/L. They constructed tables for identification of the APE by FD-MS of their standard samples.

High performance liquid chromatography and FD-MS were used to separate and identify non-volatile organics in river and drinking water (Watts et al., 1984). Organics were adsorbed on XAD-2 resin or freeze dried and extracted, separated by HPLC, and examined, as whole samples or HPLC fractions by FD-MS. The FD-MS method is based on the assumption that different compounds desorb at different emitter heating currents. Although only approximate quantitative information could be obtained, NPE with 6 to 15 EO units were identified in both drinking and river water.

Researchers at the Swiss Federal Institute for Water Resources and Water Pollution Control developed analytical methods for the analysis of specific APE and their known metabolites. Ahel and Giger (1985a; 1985b) and Ahel et al. (1987) developed HPLC methods for the determination of alkylphenol polyethoxylates ($C_{8,9}APE_3$ - $C_{8,9}APE_{20}$), alkylphenol mono- and diethoxylates ($C_{8,9}APE_1$ and $C_{8,9}APE_2$), alkylphenols ($C_{8,9}APE_0$ such as OP and NP), and alkylphenoxy carboxylic acids (APEC) in environmental samples.

In these studies, alkylphenol polyethoxylates were extracted from wastewater by gaseous stripping into ethyl acetate. The compounds were separated by normal-phase HPLC using bonded-phase aminosilica columns and detected by UV absorption. The estimated limit of detection for individual oligomers was 1 μg/L with an 87% recovery for total APE. The alkyl substituents - OPE and NPE - eluted separately as individual peaks from octylsilica columns (reversed-phase HPLC). The o- and p-NP isomers co-eluted, but preliminary results indicated that isomers with different branched side chains may appear as separate peaks (Ahel and Giger, 1985b).

For alkylphenols and alkylphenol mono- and diethoxylates, enrichment from water and solid samples was by an exhaustive combined steam distillation/cyclohexane extraction. No additional cleanup was necessary. Separation was by normal-phase HPLC and detection was by UV absorption. Reversed-phase HPLC resolved the alkyl chains. The limit of detection for each component was 0.5 μg/L of water with recoveries of >80% (Ahel and Giger, 1985a). Compared to gaseous stripping used by Ahel and Giger (1985b) for total APE, steam distillation is less likely to lose the volatile lower homologs. The authors suggest that both methods be used together for the analysis of environmental samples.

In a similar manner, Ahel et al. (1987) developed a routine method for the quantitative determination of alkylphenoxy mono- and dicarboxylic acids in sewage and environmental samples. Good precision and reproducibility were attained with two extraction methods: (1) double extraction with chloroform in a separatory funnel (recovery 95%) or (2) gaseous stripping into ethyl acetate (recovery 65%). A cleanup step employing a silica column was used. Quantitative determination was by normal-phase HPLC/UV absorption or high resolution (HR) GC of methylated derivatives. Identification of branched nonyl isomers was by HRGC/MS. Both HPLC and HRGC methods achieved a sensitivity of 1 μg/L.

Marcomini and Giger (1987) and Marcomini et al. (1987) reported procedures for the simultaneous determination of linear alkylbenzenesulfonates (LAS), NP, and NPE by reversed-phase HPLC. Following Soxhlet extraction or separation by steam-distillation, separation was on an octylsilica column with water/acetonitrile for gradient elution. In the second paper, an enrichment procedure involving percolation of water samples through octadecylsilica cartridges as a simpler, alternative concentration technique was evaluated and found comparable to other extraction methods. Quantification of NP and associated ethoxylates required additional information from normal-phase HPLC. Detection limits with UV fluorescence were 95 ng for NP and 65 ng for the ethoxylates.

Kubeck and Naylor (1990) refined earlier HPLC methods so that NPE could be rapidly and quantitatively measured in environmental samples. Their method involved isolation of the NPE from water samples by a dual column extraction procedure: the first column, a mixed-bed ion-exchange resin, removed ionic species and the second column, octadecylsilica, adsorbed the NPE. The samples needed to be protected from air and dissolved oxygen. Nonylphenol and low NPE oligomers were extracted from water samples by steam distillation. Sample extracts from both procedures were assayed by HPLC with a fluorescence detector. Recovery of spiked samples was good: 96% for laboratory-water samples and 84% for river-water samples. The procedure was applied to river water and treated wastewater samples.

The extraction and analytical methods developed by Kubeck and Naylor (1990) for NP and its ethoxylates in environmental samples as well as sample preservation methods were validated by the Radian Corporation (1990). Using this validated technique, they measured concentrations in 30 U.S. rivers. The minimum detection limits of NP in water and sediment were 0.11 μg/L and 2.93 μg/kg, respectively. The minimum detection limits for NPE_1 were 0.07 μg/L in water and 2.26 μg/kg in sediment. For NPE_2 and the higher ethoxylates the values for water were 0.06 μg/L and 1.6 μg/L, respectively; limits for sediment were not determined.

Recently Bear (1988) investigated the use of evaporative light scattering (ELS) as a "universal" detector of both anionic and nonionic surfactants following HPLC separation. Results of analyses of EO distribution of C_9APE_{11} by both ELS and UV absorption showed similar distribution profiles and comparable sensitivities; however, the UV detector gave a higher response for the lower molecular weight components. The detection limits for ELS were in the low nmole range.

Gas chromatography (GC). GC/MS has become one of the most useful methods for the identification of volatile contaminants in environmental samples and was used in many of the studies discussed above to identify the alkyl chain structure. For example, Radian Corporation (1990) used GC/MS to confirm the presence of isomers of both OP and NP in the Mohawk River. Ethoxylates larger than APE_9 could not be gas chromatographed due to their lack of volatility.

Various ionization techniques have been combined with MS following separation by GC for analysis of APE. Flame ionization is a common detector technique for GC. For APE with a low degree of ethoxylation, GC/MS using electron impact (EI) and chemical ionization (CI) methods of ionizing are commonly used. One of the first groups to use GC for the separation of APE was Nadeau et al. (1964) who used this method to separate NPE according to its EO distribution. Recorded peak heights were based on thermal conductivities of the homologs. The procedure was quantitative for EO_{1-5}.

Austern et al. (1975) devised a method for the detection of low levels of NP in wastewater. The method consisted of extraction with freon, concentration of the extract in a Kuderna-Danish apparatus, and separation/detection by GC with an FI detector. Recovery of spiked raw and treated wastewater samples was 99.9%, with a minimum detection level of 2.2 ng.

In their monitoring study of organic contaminants in river systems, Hites et al. (1979) used a variety of GC and HPLC techniques to identify several hundred organic compounds. Preliminary analyses were carried out on a GC equipped with FI and electron capture (EC) detectors. Low resolution mass spectra were obtained with a GC/MS system with a dual EI/CI source. High resolution mass spectra were obtained with photographic plate detection. Where high concentrations of some components in the samples precluded identification of minor components, HPLC separations followed by mass spectrometry were used.

Giger et al. (1981) used glass capillary gas chromatography/mass spectrometry (GC/MS) to identify NP and NPE with one, two, or three EO units in effluents from activated sludge plants. Sewage samples needed no further cleanup following continuous reflux steam distillation with cyclohexane as the organic solvent. Gas chromatography separations were performed on glass capillary columns and identification was by mass spectrometry, comparing chromatograph peaks to those of standard samples. A companion paper (Stephanou and Giger, 1982) reports additional information on sewage effluents. The detection limit was 10 μg each for NP and NPE_1 and NPE_2. Aliphatic hydrocarbons and carboxylic acids were found to interfere with determinations. As reported earlier, Giger et al. (1987) used GC/MS as well as HPLC to determine NPEC in sewage effluents.

Based on comparisons of GC/MS chromatograms of a standard sample of 4-NP, Schaffner et al. (1984) identified and quantified 4-NP in sewage sludge. Dry samples of sludge weighing 1 g were diluted with 1 liter of water, mixed with 30 g NaCl, and refluxed for 3 hours in a closed-loop steam distillation/solvent extraction apparatus. The clean-up step was performed on a silica column. Using this system, reproducibility was $\pm 4\%$ on a 1.18 g of NP/kg dry matter sample with recoveries of spiked samples between 93-105%.

Giger et al. (1984) advocated the application of HRGC subsequent to a preparative separation either by normal-phase or reversed-phase HPLC for the complete analysis of environmental samples. They identified alkylphenols, nonylphenol mono- and diethoxylates, and nonylphenol mono- and diethoxy carboxylic acids in sewage sludge. Detection limits were not given.

Stephanou (1986) simultaneously determined both the acid and neutral degradation products of APE. The procedure involved liquid-liquid extraction with methylene chloride, evaporation, and GC/MS analysis of the trimethylsilyl derivatives. Detection limits were not given.

Stephanou (1984) used GC separation combined with CI induced MS (CI-MS) as an alternative to the GC/EI-MS technique for the analysis of commercial products and wastewater effluent. Although the GC/CI-MS method proved more reliable for higher molecular weight NPE, operational problems made the use of EI more practical.

Recently, Wahlberg et al. (1990) developed a method for the determination of NP and NPE containing 1 to 3 ethoxy units in water, sludge, and biota. After isolation, the substances are converted into their corresponding pentafluorobenzoyl derivatives and quantified by GC/EC or GC/MS.

Other Instrumental Methods. Many methods described in the literature rely on instrumentation without the HPLC or GC separation techniques described above. Only a few of these methods have been applied to environmental samples and thus will be only briefly mentioned here.

Molecular absorption spectroscopy, including IR and UV absorption, can be used to identify nonionic surfactants containing a benzene ring. However, because of interfering spectra from other organic compounds in environmental samples, this type of analysis is more aptly applied to biodegradation studies of pure compounds in the laboratory.

Frazee et al. (1964) first applied IR spectroscopy to surfactant biodegradation studies in the laboratory. Osburn and Benedict (1966) followed biodegradation of an APE in the laboratory by both UV and IR spectroscopy, but Jones and Nickless (1978b) found the use of IR spectra limited for environmental samples because of similar spectra from APE and secondary AE.

Mass spectrometry with newer ionization techniques are increasingly being tested. Nonylphenols and APE can be identified by their fragmentation patterns with MS (Giger et al., 1981). Schneider et al. (1983) compared results of analyses by mass spectrometry using three ionization techniques: (1) field desorption (FD), (2) fast atom bombardment (FAB), and (3) desorption chemical ionization (DCI). Only APE with short ethoxylate chains could by measured with FD and FAB instruments, with FD being the more sensitive technique. Compounds with higher molecular weights, typically used in detergent formulations, could be measured using the DCI technique. DCI spectra did not provide information on the alkylphenol group, but this group can be identified by EI spectra. The sensitivity for the FAB technique depended on the matrix used to dissolve the compound. Without a glycerol matrix, good spectra could be obtained at $>5 \times 10^{-4}$ g; with a glycerol matrix, quasimolecular ions were detectable at $\geq 10^{-7}$ g on the target. The authors suggest combining these ionization techniques with collisionally activated decomposition (CAD) in a tandem mass spectrometer for a direct mixture analysis.

Rivera et al. (1989) used FAB/MS alone to successfully identify polyglycols and anionic, cationic and nonionic surfactants in river and drinking water. They used HPLC separation coupled with MS (HPLC/MS) in the plasmaspray ionization mode to tentatively identify propoxylated nonylphenols.

The combination of FD and CAD in a tandem mass spectrometer was used for the characterization of mixtures of surfactants without prior separation (Weber et al., 1982). Cationic, anionic and nonionic surfactants desorb at distinct emitter heating currents. Although the interpretation of the FD/CAD spectra for nonionics were not as straight forward as for cationic and anionic surfactants, it allowed identification of the branched alcohol, ethoxylate, and benzene ring moieties. Schneider et al. (1984) applied this approach to the analysis of surfactants in surface water. However, the CAD spectra could not be used to identify nonionic surfactants.

B. STANDARDS AND REGULATIONS

1. Water Quality Standards

There are presently no national criteria for regulation of nonionic surfactants or their degradation products in waters of the United States.

The U.S. Environmental Protection Agency (U.S. EPA) has signed a Consent Order with the APE Program Panel of the CMA whereby the APE Program Panel has agreed to perform chemical fate and environmental effects tests on 4-NP (CAS No. 84852-15-3) (Fed. Reg. 55:5991-5994, February 21, 1990). This group has conducted a voluntary environmental monitoring survey on NP in U.S. rivers (Radian Corporation, 1990).

2. Air Emission Standards

The U.S. EPA has proposed standards of performance for volatile organic compound emissions from the synthetic organic chemical manufacturing industry (SOCMI) reactor processes. The proposed standards implement section 111 of the Clean Air Act. Facilities that produce NP (CAS No. 25154-52-3) and NPE (CAS No. 9016-45-9) would be affected by the proposed standards (Fed. Reg. 55:26953-26980, June 29, 1990).

C. NONIONIC SURFACTANTS IN NATURAL WATER BODIES

1. Water Column.

In earlier studies, documented by Arthur D. Little (1977) and Goyer et al. (1981), analytical methods did not distinguish between intact APE, AE, other nonionic surfactants, and any degradation products; thus, it was not possible to ascertain which type of surfactant contributed to levels of nonionics in waterways. With more recent analytical techniques, specific APE and their degradation products can be identified and quantified. The older studies are reviewed below and are followed by more recent quantitative monitoring results. Studies that list concentrations in natural waterways are listed in Table 2-2.

Between 1965 and 1974, nonionic surfactants in the Calder and Aire Rivers located in the Yorkshire wool-treatment region of England ranged from 0.2-1.0 mg/L as measured by the Patterson TLC method while concentrations in the Lippe and Rhine Rivers of Germany ranged from 0.014-0.110 mg/L BIAS as measured by the Wickbold method. Nonionics were not monitored in U.S. rivers during this time (Arthur D. Little, 1977). A recent study documented that the load of nonionics in the Rhine River decreased by 67% between 1974 and 1987 with an average concentration of 0.01 mg/L in 1987 (BIAS/DIN method) (Gerike et al., 1989). The limit of detection in this study was 0.005 mg/L.

Table 2-2. Concentrations of Alkylphenol Ethoxylates and Alkylphenols in Rivers

Chemical	Site	Concentration	Measurement Method	Reference
Nonylphenol	Savannah River tributary	1000 μg/L	Not reported	Gustafson, 1970
4-Nonylphenol	Savannah River	2 μg/L	Gas chromatography/ flame ionization	Garrison and Hill, 1972
4-Nonylphenol	Tennessee River	325 μg/L	Gas chromatography/MS	Goodley and Gordon, 1976
Total nonionics	Rivers, Great Britain	200-1000 μg/L	Thin-layer chromatography	Arthur D. Little, 1977
Total nonionics	Lippe, Rhine Rivers, Germany	14-110 μg/L[a]	BIAS	Arthur D. Little, 1977
Nonylphenols Octylphenols	Delaware River	0.04-2 μg/L 0.02-2 μg/L	Gas chromatography/MS	Sheldon and Hites, 1978 Hites et al., 1979
Nonionics	River Avon, Great Britain	8 μg/L	Thin-layer chromatography	Jones and Nickless, 1978b
4-Nonylphenol NPE_1 NPE_2	Glatt River, Switzerland	1.8 μg/L 12.7 μg/L 15.7 μg/L	HPLC/UV	Ahel et al., 1984
Nonylphenol NPE_1 NPE_2	Glatt River, Switzerland	<0.5[b]-1.5 μg/L <0.5[b]-18 μg/L <0.5[b]-16 μg/L	HPLC/UV	Ahel and Giger, 1985a
NPE_1C, NPE_2C	Glatt River, Switzerland	2-116 μg/L	HPLC/UV HRGC-MS	Ahel et al., 1987

Table 2-2. (cont.)

Chemical	Site	Concentration	Measurement Method	Reference
Nonylphenol NPE_1 NPE_2 NPE_{3-20} NPE_1C NPE_2C	Glatt River, Switzerland	$<0.3^b$-7.9 µg/L $<0.3^b$-20 µg/L $<0.3^b$-21 µg/L $<0.1^b$-7.1 µg/L $<0.1^b$-29 µg/L 2.0-59 µg/L	HPLC/UV	Ahel et al., 1991
Nonylphenol NPE_1 NPE_2	Five Swiss rivers	$<0.1^b$-55.4 µg/L $<0.1^b$-32.2 µg/L $<0.1^b$-37.2 µg/L	HPLC/UV	Ahel et al., 1991
Total nonionics	Rhine River, Germany	10 µg/L^a	BIAS	Gerike et al., 1989
Nonylphenol NPE_1 NPE_2 Nonylphenol NPE_1 NPE_2	Alabama River, Alabama Saginaw River, Michigan	$<0.2^b$ µg/L $<0.2^b$ µg/L $<0.2^b$ µg/L 0.2-1.0 µg/L $<0.2^b$-0.2 µg/L $<0.2^b$-3.0 µg/L	HPLC/fluorescence	Radian Corporation, 1989
NPE_{0-18}	Colorado River, Texas	1.1, 1.9 µg/L	HPLC/fluorescence	Kubeck and Naylor, 1990
Nonylphenol NPE_1 NPE_2 NPE_{3-17}	30 U.S. rivers	$<0.11^b$-0.64 µg/L $<0.07^b$-0.6 µg/L $<0.06^b$-1.2 µg/L $<1.6^b$-14.9 µg/L	HPLC/fluorescence	Radian Corporation, 1990; Naylor et al., 1992
NPE_n	Israel streams	1.6-2.6 mg/L^c	CTAS	Zoller et al., 1990

^{a}µg BIAS/L.
^{b}Limit of detection.
^{c}mg CTAS/L.

Ahel and coworkers (1984; 1985a; 1987; 1991) have measured APE and their metabolites in the Glatt River, Switzerland. In Switzerland, APE detergents are used in both household and industrial cleaning products. Concentrations of NP, NPE_1, and NPE_2 ranged from <0.3-1.8 µg/L, <0.3-20 µg/L, and <0.3-21 µg/L, respectively. Concentrations were ten times higher in treated wastewaters than in river water. Only minor amounts of octylphenol (C_8APE_n) and decylphenol ($C_{10}APE_n$) derivatives were present. Total alkylphenol mono- and dicarboxylic acids in the river ranged from 2-116 µg/L. Nonylphenol, NPE_1, and NPE_2 were measured in five other Swiss rivers.

The studies listed in Table 2-2 show that concentrations of NP in U.S. river waters range from below detection limits to a high of 1 mg/L downstream of a wool scouring plant on a tributary of the Savannah River (Gustafson, 1970).

In their study of organic compounds in the Delaware River, Sheldon and Hites (1978; 1979a; 1979b; Hites et al., 1979) identified the presence of NP, OP (including 4-[1,1,3,3-tetramethylbutyl]phenol), and t-octylphenol mono-, di-, and triethoxylates. The latter three degradation products were found in Philadelphia drinking water at concentrations of 0.03, 0.06, and 0.002 µg/L, respectively. There were no detectable levels of OP or NP downstream of a chemical manufacturing plant located on a small river, but two isomers of OP were present in one sample of sediment (Jungclaus et al., 1978; Hites et al., 1979).

Using a limited number of samples, Kubeck and Naylor (1990) assayed river and wastewater for NPE. The total NPE plus NP concentration in a single sample taken upstream of wastewater outfalls on the Colorado River, was 1.1 µg/L, with most of the sample composed of NPE_9. Total NPE concentrations in two samples taken downstream of wastewater treatment plants on the same river were both 1.9 µg/L, with most of the increase due to low oligomers, NPE_0-NPE_5.

In more recent studies, concentrations of total NPE (NP, NPE_1, NPE_2, and NPE_{3-17}) in 30 U.S. rivers ranged from below the limit of detection to 17.3 µg/L in the Grand Calumet River (Radian Corporation, 1990). As noted earlier, in 1987 the CMA formed an APE Program Panel which proposed to voluntarily conduct an environmental monitoring survey on NP and NPE. The survey was developed in conjunction with the U.S. EPA's Exposure Evaluation Division in order to meet the Agency's exposure data needs (Fed. Reg. 55:5991-5994). In the latter half of 1989, the APE Panel contracted with the Radian Corporation to monitor the concentrations of NP and NPE in a cross section of U.S. waterways (CMA, 1990).

During 1989, 30 U.S. rivers considered likely to contain NPE were monitored for concentrations of NP and NPE (Radian Corporation, 1990). The sites were chosen from the EPA River Reach File using a randomization procedure. Each site was just downstream from one or more outfalls of industrial or treated municipal wastewater. Thirty of the 5000 suitable sites was a number sufficient for giving nationwide statistical validity to the survey. Each stream was sampled at least three times along a transect perpendicular to the direction of flow. Rivers and sampling locations are listed in Table 2-3. Both river water and sediment samples were collected. During this study, analytical methods for NP and its ethoxylates, described in Section A (Kubeck and Naylor, 1990), were validated. Nonylphenol concentrations were greater than the detection limit in 30% of the water samples; the highest concentration was 0.64 μg/L (Table 2-2) and the average was 0.12 μg/L. NPE_1, NPE_2, and NPE_3-NPE_{17} were above the limit of detection in 33, 41, and 24% of the samples, respectively. Respective mean concentrations were 0.09, 0.10, and 2.0 μg/L.

Table 2-3. U.S. River Sampling Locations

River	Sampling Location
Mohawk	Utica, New York
Chattahoochee	West Point, Georgia
Chattooga	Trion, Georgia
Bernard Bayou	Gulfport, Mississippi
Red	Index, Arkansas
Grand Calumet	Gary, Indiana
Dragoon Creek	Deer Park, Washington
Brandywine Creek	Coatsville, Pennsylvania
Fish Creek, West Branch	Camden, New York
Great Egg Harbor	Berlin, New Jersey
Kennebec	Waterville, Maine
Pecos	Artesia, New Mexico
Palouse, South Fork	Colfax, Washington
Cuyahoga	Mantua, Ohio
Portneuf	Pocatello, Idaho
Perry Creek	Arlington, Georgia
Thames	Uncasville, Connecticut
Catawba	Morganton, North Carolina
Turkey Creek	Winnsboro, Louisiana
Delaware	Croydon, Pennsylvania
Shenandoah, North Fork	Mount Jackson, Virginia
Tallahaga Creek	Noxapater, Mississippi
South Anna	Ashland, Virginia
Potomac	Brunswick, Maryland
White	Sharon, Vermont
Youghiogheny	McKeesport, Pennsylvania
St Clair	Marysville, Michigan
Yellowstone	Gardiner, Montana
Machias	Machias, Maine
Muskegon	Freemont, Michigan

Using limited data on effluent concentrations at three publicly owned treatment works and the concentrations in the receiving waters presented in the 1989 Radian Corporation report, Labat-Anderson (1989) predicted the concentrations of NPE_{0-2} that would occur in the Alabama and Saginaw Rivers. Results of their model indicate that under all but the most severe drought conditions the concentrations of NP are expected to be <10 μg/L. The highest total concentrations of the mono- and diethoxylates of NP are expected to be comparable to the concentration of NP.

The fate (disappearance by dilution) of p-NP in receiving waters downstream of a carpet yarn mill was traced by Garrison and Hill (1972) (Table 2-4). The authors presumed that NP was produced in the anaerobic wastewater treatment ponds by biodegradation of a surfactant. Analysis was by GC/MS-FID.

Table 2-4. Concentration of 4-Nonylphenol in Textile Waste and Receiving Waters

Sample Location	Concentration (mg/L)[a]
Plant total untreated waste	0.05
Treatment pond effluent	4.0
Small creek (1 mi)[b]	3.0
Small river (1.5 mi)[b]	0.2
Small river (4.5 mi)[b]	0.03
Savannah River (6 mi)[b]	0.002

[a]Analysis by gas chromatography/flame ionization
[b]Distance below pond discharge.
Source: Garrison and Hill, 1972.

Using GC/MS and FAB/MS, Ventura et al. (1988) and Rivera et al. (1989) identified AE, APE, alkylphenol carboxylic acids and their brominated derivatives in the Llobregat River and tap water of Barcelona, Spain. Quantitative determinations were not made. It was shown by Reinhard and Ball (1985) that acidic and neutral metabolites of APE react during chlorination to produce brominated and chlorinated products.

Waldock and Thain (1991) followed the dispersion of 4-NP following dumping of sludge into the outer Thames Estuary, England. The dumped sludge contained 1000 µg/L of 4-NP. Concentrations of 4-NP in the water column were at or below the limit of detection (0.5 µg/L) within 30 min at a depth of 1 m and within one hour at a depth of 5 m. The authors indicated that the outer Thames Estuary is a highly dispersive region.

2. Sediments

Nonionic surfactants adsorb onto organic matter in soil and sediments. Using dried river sediments with a range of organic carbon content from 1.7 to 6.0%, Urano (1984) found a good correlation between organic carbon content and adsorption of several surfactants including NPE_{10}, as measured by CTAS. Using surface tension and spectrophotometric techniques, Liu et al. (1992) measured the adsorption of $C_9APE_{10.5}$ (Tergitol NP-10), $C_8APE_{9.5}$ (Triton X-100), and C_8APE_{12} (Igepal CA-720) to air-dried soil with a fractional organic carbon content of 0.0096. Adsorption data were found to fit a Freundlich isotherm.

In a laboratory study that simulated field conditions, Sundaram and Szeto (1981) determined that NP present in the water column is translocated to sediment. After four days NP was not detected in the water as analyzed by HPLC; however, about 50% of the added material was present in the sediment. Nonylphenol also adsorbs to sludge (See Chapter III, Section B). A survey study conducted in 1978 and several recent studies document the concentrations of OP, NP and NPE in river sediment; three of the studies compare water and sediment concentrations (Table 2-5). Analysis methods are listed in Table 2-1.

Two OP isomers were identified by GC and GC/MS in one sediment sample downstream of a chemical manufacturing plant on an unidentified U.S. river. The isomers were present in the wastewater from the plant (1-75 µg/L), but not in the downstream water. Nonylphenol was also present in the wastewater (50 µg/L) but was not identified in the river water or sediment (Jungclaus et al., 1978).

Marcomini and Giger (1987) measured dry weight concentrations of NP (900 µg/kg) and its mono- (800 µg/kg) and diethoxylates (700 µg/kg) in the surficial (0-3 cm) sediments of the Rhine River. For extraction and identification, they used Soxhlet extraction and combined reverse-phase and normal-phase HPLC/fluorescence techniques. Using the experimental data of Hellmann (1980) on Rhine River sediments, Swisher (1987) calculated the adsorption partition coefficient (P_a) of NPE_9 in sediments of the Rhine River. The P_a, defined as the ratio of the equilibrium concentrations in the adsorbed state and the dissolved state (or g of adsorbed substrate/g adsorbent)/(g of dissolved substrate/g solution), was 55,000-136,000. The equilibrium concentration of dissolved NPE_9 was 0.8-1 mg/L.

In the Radian Corporation (1989; 1990) studies, NP and NPE were analyzed in the water column and sediments at two selected river sites, the Alabama River at Montgomery, Alabama, and the Saginaw River at Saginaw, Michigan, and later at 30 U.S. rivers that received

Table 2-5. Concentrations of Nonylphenol and Nonylphenol Ethoxylates in Surface Waters and Sediments

Water Body/Analyte	Water			Sediment			Reference
	Number of Samples	Average (μg/L)	Range (μg/L)	Number of Samples	Average (μg/kg)	Range (μg/kg)	
Unidentified U.S. river							Jungclaus et al., 1978
Octyphenols	12	ND	—	13	—	ND-5000[a,b]	
Nonylphenol	12	ND	—	13	ND	—	
Rhine River, Germany							Marcomini and Giger, 1987
NP	—	—	—	—	900[c]	—	
NPE_1	—	—	—	—	800[c]	—	
NPE_2	—	—	—	—	700[c]	—	
Glatt River and tributaries, Switzerland							Ahel et al., 1991
NP	110	—	<0.2-45	—	—	190-13,100[c]	
NPE_1	110	—	<0.2-69	—	—	100-8850[c]	
NPE_2	110	—	<0.2-30	—	—	ND-2720[c]	
Besos River, Spain							Grifoll et al., 1990
NP	—	—	—	2	—	3000[c], <100[c,d]	
NPE_1	—	—	—	2	—	2400[c], <100[c,d]	
NPE_2	—	—	—	1	1200[c]	—	
Coastal waters, Spain							Valls et al., 1990
NPE_a	—	0.85[e]	—	—	6600[f]	—	
Marine lagoon, Venice, Italy NPE_0-NPE_{13}[g]	15	1.8	0.6-4.5	—	—	—	Marcomini et al., 1990
NP	—	—	—	20	—	5-42[c]	
NPE_1	—	—	—	20	—	9-82[c]	
NPE_2	—	—	—	20	—	3-20[c]	

Table 2-5. (cont.)

Water Body/Analyte	Water			Sediment			Reference
	Number of Samples	Average (μg/L)	Range (μg/L)	Number of Samples	Average (μg/kg)	Range (μg/kg)	
Alabama River, U.S.							Radian Corporation, 1989
NP	9	<0.2	—	3	—	<1.0-14[a]	
NPE$_1$	9	<0.2	—	3	—	<1.0-15[a]	
NPE$_2$	9	<0.2	—	3	—	<1.0-4[a]	
Saginaw River, U.S.							Radian Corporation, 1989
NP	7	—	0.2-1.0	3	—	32-249[a]	
NPE$_1$	7	—	<0.2-0.2	3	—	13-103[a]	
NPE$_2$	7	—	<0.2-0.3	3	—	<1.0-30[a]	
30 U.S. Rivers							Radian Corporation, 1990
NP	98	0.12	<0.11-0.64	81	161.9[c]	<2.9-2960[c]	
NPE$_1$	98	0.09	<0.07-0.60	81	18.1[c]	<2.3-175[c]	
NPE$_2$	101	0.10	<0.06-1.20	—	—	—	
NPE$_{3-17}$	101	2.00	<1.60-14.9	—	—	—	

[a]Wet weight.
[b]Single sample.
[c]Dry weight.
[d]Seacoastal sediment opposite river mouth.
[e]Primarily NPE$_2$.
[f]Primarily NP.
[g]NP + NPE$_1$ + NPE$_2$ = <10% of total NPE.
ND = not detected (GC analysis; detection limits not given).
A dash (—) indicates no data.

effluents. In the latter study, NP concentrations in the sediments ranged from not detectable in 28% of the samples to 2960 μg/kg. Limits of detection for NP and NPE_1 were 2.93 and 2.26 μg/kg dry weight, respectively. Calculation of interstitial water concentrations of NP from the measured sediment concentrations indicated that the interstitial concentrations follow essentially the same distribution as the water column concentrations of NP (Naylor et al., 1991).

Using HRGC-EI/PICI-MS, Grifoll et al. (1990) identified NP, NPE_1, and NPE_2 in sediments collected from the Besos River, Spain. Concentrations were 3.0, 2.4, and 1.2 μg/g dry weight, respectively. Concentrations in the marine sediment 500 m offshore from the river mouth were below the limit of detection (0.1 μg/g).

In their extensive study of the Glatt River Valley in Switzerland, Ahel et al. (1991) measured concentrations of NP in sediments of up to 13,100 μg/kg dry weight. Mud rich in organic matter contained considerably higher concentrations than sand collected at the same location.

3. Groundwater.

Several studies investigated the behavior of nonylphenolic compounds during the infiltration of riverwater into groundwater. No studies concerning groundwater contaminated in this manner in the U.S. were located. In Israel, where 67% of the nonionic surfactants used in laundry formulations are of the APE type and NPE surfactants comprise more than 10% of the synthetic detergents found in municipal wastewaters, receiving streams are highly contaminated (see Table 2-2) (Zoller et al., 1990; Zoller, 1992). The groundwaters in the vicinities of these streams are contaminated with nonionic surfactants at concentrations of 0.1-1.8 mg/L as determined by CTAS. There was an inverse relationship between distance of groundwater-fed wells from the stream and the concentration of nonionic detergent in the wells.

Ahel et al. (1991) also measured nonylphenolic compounds in groundwater in the vicinity of two Swiss rivers that receive sewage plant effluents. Ranges of concentrations of NPE in the Glatt River were 0.7-26 μg/L for NP, 2.0-20 μg/l for NPE_1 and 0.8-21 μg/L for NPE_2 while respective ranges in groundwater within 13 meters of the river bed were <1.0-3.1 μg/L for NP, <0.1-4.8 μg/L for NPE_1 and <0.1-1.6 μg/L for NPE_2. The NPE_1C and NPE_2C were present in the river at concentrations of 8.4-20.1 and 20.6-28.9 μg/L and in groundwater at concentrations of <0.1-13.1 and <0.1-32.2 μg/L, respectively. Concentrations in groundwater rapidly decreased with distance from the river. Seasonal variations were apparent in both river and groundwater, with higher concentrations in winter.

REFERENCES

Ahel, M. and W. Giger. 1985a. Determination of alkylphenols and alkylphenol mono- and diethoxylates in environmental samples by high-performance liquid chromatography. Anal. Chem. 57:1577-1583.

Ahel, M. and W. Giger. 1985b. Determination of nonionic surfactants of the alkylphenol polyethoxylate type by high-performance liquid chromatography. Anal. Chem. 57:2584-2590.

Ahel, M., T. Conrad and W. Giger. 1987. Persistent organic chemicals in sewage effluents. 3. Determination of nonylphenoxy carboxylic acids by high-resolution gas chromatography/mass spectrometry and high-performance liquid chromatography. Environ. Sci. Technol. 21:697-703.

Ahel, M., W. Giger, and C. Schaffner. 1991. Environmental occurrence and behaviour of alkylphenol polyethoxylates and their degradation products in rivers and groundwaters. In: Swedish EPA Seminar on Nonylphenol Ethoxylate/Nonylphenol held in Saltsjobaden, Sweden, February 6-8, pp. 105-151.

Ahel, M., W. Giger, E. Molnar-Kubica and C. Schaffner. 1984. Analysis of organic micropollutants in surface waters of the Glatt Valley, Switzerland. In: Analysis of Organic Micropollutants in Water, G. Angeletti and A. Bjorseth, eds. Reidel Publishing Company, Dordrecht, Holland.

Allen, M.C. and D.E. Linder. 1981a. Ethylene oxide oligomer distribution in nonionic surfactants via high performance liquid chromatography (HPLC). J. Am. Oil Chem. Soc. 58:950-957.

Allen, M.C. and D.E. Linder. 1981b. An improved HPLC method for the determination of ethylene oxide distribution in nonionic surfactants. Presented at the 72nd annual meeting of the American Oil Chemists' Society, New Orleans, May 19, 1981, Paper No. 160, p.23.

APHA (American Public Health Association). 1989. Standard Methods for the Examination of Water and Wastewater, 17th Ed., L.S. Clesceri, Eds. American Public Health Association, Washington, DC.

Arthur D. Little. 1977. Human safety and environmental aspects of major surfactants. A report to the Soap and Detergent Association by Arthur D. Little, Cambridge, MA.

Austern, B.M., R.A. Dobbs and J.M. Cohen. 1975. Gas chromatographic determination of selected organic compounds added to wastewater. Environ. Sci. Technol. 9:588-590.

Bear, G.R. 1988. Universal detection and quantitation of surfactants by high-performance liquid chromatography by means of the evaporative light-scattering detector. J. Chromatog. 459:91-107.

Boyer, S.L., K.F. Guin, R.M. Kelley, et al. 1977. Analytical method for nonionic surfactants in laboratory biodegradation and environmental studies. Environ. Sci. Technol. 11:1167-1171.

Bürger, K. 1963. Methods for the micro determination and trace detection of surfactants. III. Trace detection and determination of surface active poly EO compounds and polyethylene glycols. Z. Anal. Chem. 196:251-259.

CMA (Chemical Manufacturers Association). 1990. Alkylphenol and ethoxylates: CMA Panel Progress Report.

Crabb. N.T. and H.E. Persinger. 1964. The determination of poly EO nonionic surfactants in water at the parts per million level. J. Oil Chem. Soc. 41:752-755.

Crisp, P.T. 1987. Trace Analysis of Nonionic Surfactants. In: J. Cross (Ed.), *Nonionic Surfactants: Chemical Analysis*. Marcel Dekker, Inc., New York, pp. 77-116.

Cross, J. 1987. *Nonionic Surfactants: Chemical Analysis*. Surfactant Science Series, Vol. 19, Marcel Dekker, Inc., New York.

EEC (European Economic Community). 1982. Directive No. 4311/82, Annex, Chapter 3, Brussels, January, 1982.

Favretto, L. and F. Tunis. 1976. Determination of polyoxyalkylene ether nonionic surfactants in waters. Analyst 101:198-202.

Favretto, L., B. Stancher and F. Tunis. 1980. Determination of polyoxyethylene alkyl ether nonionic surfactants in waters at trace levels as PPAS. Analyst 105:833-840.

Favretto, L., B. Stancher and F. Tunis. 1982. Investigations on possible interferences in trace determination of polyoxyethylene non-ionic surfactants in waters as potassium picrate active substances. La Rivista Ital. Delle Sos. Grasse 59:23-27.

Favretto, L., B. Stancher and F. Tunis. 1983. An improved method for the spectrophotometric determination of polyoxyethylene nonionic surfactants in water as PPAS in presence of cationic surfactants. Intl. J. Env. Anal. Chem. 14:201-214.

Frazee, C.D., Q.W. Osburn and R.O. Crisler. 1964. Application of infrared spectroscopy to surfactant degradation studies. J. Am. Oil Chem. Soc. 41:808-812.

Garrison, A.W. and D.W. Hill. 1972. Organic pollutants from mill persist in downstream waters. Am. Dyestuff. Reporter 61:21-25.

Garti, N., V.R. Kaufman, A. Aserin. 1983. Analysis of nonionic surfactants by high performance liquid chromatography. Sep. Purif. Methods 12:49-116.

Gerike, P., K. Winkler, W. Schneider and W. Jacob. 1989. Detergent components in surface waters of the Federal Republic of Germany. In: Seminar on the Role of the Chemical Industry in Environmental Protection, Geneva, Switzerland, November 13-17.

Giger, W., E. Stephanou and C. Schaffner. 1981. Persistent organic chemicals in sewage effluents: 1. Identification of nonylphenols and nonylphenolethoxylates by glass capillary gas chromatography/mass spectrometry. Chemosphere 10:1253-1263.

Giger, W., M. Ahel and C. Schaffner. 1984. Determination of organic wastewater pollutants by the combined used of high-performance liquid chromatography and high-resolution gas chromatography. In: Analysis of Organic Micropollutants in Water, G. Angeletti and A. Bjorseth, eds. Reidel Publishing Company, Dordrecht, Holland.

Giger, W., P.H. Brunner, M. Ahel, J. McEvoy, A. Marcomini and C. Schaffner. 1987. Organische Waschmittelinhaltsstoffe und deren Abbauproductke in Abwasser and Klärschlamm. Gas Wasser Abwasser 67:111-122.

Goodley, P.C. and M. Gordon. 1976. Characterization of industrial organic compounds in water. Trans. Ky. Acad. Sci. 37:11-15.

Goyer, M.M. J.H. Perwak, A. Sivak and P.S. Thayer. 1981. Human safety and environmental aspects of major surfactants (Supplement). A report to the Soap and Detergent Association by Arthur D. Little, Cambridge, MA.

Grifoll, M., A.M. Solanas and J.M. Bayona. 1990. Characterization of genotoxic components in sediments by mass spectrometric techniques combined with *Salmonella*/microsome test. Arch. Environ. Contam. Toxicol. 19:175-184.

Gustafson, C.G. 1970. Personal communication. (Cited in Garrison and Hill, 1972).

Han, K.W. 1967. Determination of biodegradability of nonionic surfactants by sulfation and methylene blue extraction. Tenside 4:43-45.

Hellmann, H. 1980. Trace analysis for nonionic surfactants in sewage and other sludges. Z. Anal. Chem. 300:44-47.

Hites, R.A., G.A. Jungclaus, V. Lopez-Avila and L.S.Sheldon. 1979. Potentially toxic organic compounds in industrial wastewaters and river systems: two case studies. In: Monitoring toxic substances. D. Scheutzle (ed.), American Chemical Society, Washington, DC, pp. 63-90.

Holt, M.S., E.H. McKerrell, J. Perry and R.J. Watkinson. 1986. Determination of alkylphenol ethoxylates in environmental samples by high-performance liquid chromatography coupled to fluorescence detection. J. Chromatog. 362:419-424.

Hummel, D. 1962. *Identification and Analysis of Surface-Active Agents*. Interscience, New York.

Invernizzi, F. and S. Gafà. 1973. Biodegradation of nonionic surfactants. I. Study of some analytical methods for nonionic surfactants applied in biodegradation tests. Riv. Ital. Sost. Grasse 50:365-372.

Jones, P. and G. Nickless. 1978a. Characterization of non-ionic detergents of the polyethoxylated type from water systems. 1. Evaluation of Amberlite XAD-4 resin as an extractant for polyethoxylated material. J. Chromatogr. 156:87-97.

Jones, P. and G. Nickless. 1978b. Characterization of non-ionic detergents of the polyethoxylated type from water systems. II. Isolation and examination of polyethoxylated material before and after passage through a sewage plant. J. Chromatogr. 156:99-110.

Jungclaus, G.A., V. Lopez-Avila and R.A. Hites. 1978. Organic compounds in an industrial wastewater: a case study of their environmental impact. Environ. Sci. Technol. 12:88-96.

Kubeck, E. and C.G. Naylor. 1990. Trace analysis of alkylphenol ethoxylates. J. Am. Oil Chem. Soc. 67:400-405.

Kudoh, M., H. Ozawa, S. Fudano and K. Tsuji. 1984. Determination of trace amounts of alcohol and alkylphenol ethoxylates by high-performance liquid chromatography with fluorimetric detection. J. Chromatog. 287:337-344.

Kuo, M. and H.A. Mottola. 1980. CRC Critical Reviews in Analytical Chemistry 9:297.

Labat-Anderson, Inc. 1989. Simulation of environmental concentrations of nonylphenol and its mono- and diethoxylates at river sites in Michigan and Alabama. Report to the Chemical Manufacturers Association, Washington, DC.

Liu, Z., D.A. Edwards and R.G. Luthy. 1992. Sorption of non-ionic surfactants onto soil. Water Res. 26:1337-1345.

Llenado, R.A. and R.A. Jamieson. 1981. Surfactants. Anal. Chem. 53:174R-182R.

Longman, G.F. 1976. *The Analysis of Detergents and Detergent Products*. John Wiley & Sons, New York.

Marcomini, A. and W. Giger. 1987. Simultaneous determination of linear alkylbenzenesulfonates, alkylphenol polyethoxylates, and nonylphenol by high-performance liquid chromatography. Anal. Chem. 59:1709-1715.

Marcomini, A., S. Capri and W. Giger. 1987. Determination of linear alkylbenzenesulphonates, alkylphenol polyethoxylates and nonylphenol in waste water by high-performance liquid chromatography after enrichment on octadecylsilica. J. Chromatogr. 403:243-252.

Marcomini, A., B. Pavoni, A. Sfriso and A.A. Orio. 1990. Persistent metabolites of alkylphenol polyethoxylates in the marine environment. Marine Chem. 29:307-323.

Marcomini, A., F. Cecchi and A. Sfriso. 1991. Analytical extraction and environmental removal of alkylbenzene sulphonates, nonylphenol and nonylphenol monoethoxylate from dated sludge-only landfills. Environ. Toxicol. 12:1047-1054.

Mausner, M., J.H. Benedict, K.A. Booman, et al. 1969. The status of biodegradability testing on nonionic surfactants. J. Am. Oil Chem. Soc. 46:432-444.

Menges, R.A., T.S. Menges, G.L. Bertrand, D.W. Armstrong and L.A. Spino. 1992. Extraction of nonionic surfactants from waste water using centrifugal partition chromatography. J. Liquid Chromatog. 15:2909-2925.

Nadeau, H.G., D.M. Oaks, Jr., W.P. Nichols and L.P. Carr. 1964. Separation and analysis of nonylphenoxy-EO adducts by programmed temperature gas chromatography. Anal. Chem. 36:1914-1917.

Nakamura, K. and I. Matsumoto. 1975. Analysis of nonionic surfactants. I. Ethylene oxide adducts by HPLC. J. Chem. Soc. Japan 8:1342-1347.

Naylor, C.G., J.P. Mieure, W.J. Adams, J.A. Weeks, F.J. Castaldi, L.D. Ogle, and R.R. Romano. 1991. Alkylphenol Ethoxylates in the Environment. Presented at the 1991 Annual AOCS Meeting, Chicago.

Naylor, C.G., J.P. Mieure, I. Morici, and R.R. Romano. 1992. Alkylphenol ethoxylates in the environment. Presented at the Third CESIO International Surfactants Congress held in London, U.K., June 1992.

OECD (Organization for Economic Cooperation and Development). 1976. Environment Directorate. Proposed method for the determination of the biodegradability of surfactants used in synthetic detergents, OECD, Paris.

Osburn, Q.W. and J.H. Benedict. 1966. Polyethoxylated alkyl phenols: Relationship of structure to biodegradation mechanism.

Otsuki, A. and H. Shiraishi. 1979. Determination of poly(oxyethylene) alkylphenyl ether nonionic surfactants in water at trace levels by reversed phase adsorption liquid chromatography and field desorption mass spectrometry. Anal. Chem. 51:2329-2332.

Patterson, S.J., K.B.E. Tucker and C.C. Scott. 1964. Non-ionic detergents and related substances in British waters. J. Water Pollut. Control Fed. 38:350-351.

Radian Corporation. 1989. Environmental sampling and analysis for nonylphenol ethoxylate species in the Alabama and Saginaw Rivers. Report to the APE Panel, Chemical Manufacturers Association. Radian Corporation, Austin Texas.

Radian Corporation. 1990. Nonylphenol and nonylphenol ethoxylates in river water and bottom sediments. Final Report. Prepared for The Alkylphenol and Ethoxylates Panel of the Chemical Manufacturers Association by Radian Corporation, Austin, TX.

Reinhard, M. and H.A. Ball. 1985. Discharge of halogenated octylphenol polyethoxylate residues in a chlorinated secondary effluent. In: Water Chlorination, Vol. 5, R.L. Jolly et al. (ed.). Lewis, Chelsea, Michigan. pp. 1504-1514.

Rivera, J., J. Caixach, I. Espadaler, J. Romero, F. Ventura, J. Guardiola and J. Om. 1989. New mass spectrometric techniques in the analysis of organic micropollutants in water: Fast atom bombardment and HPLC/MS. Water Supply 7:97-103.

Rosen, M.J. and H.A. Goldsmith. 1972. *Systematic Analysis of Surface-Active Agents*, 2nd ed., P.J. Elving and I.M. Kolthoff, eds. Wiley-Interscience, New York.

Ruiz Cruz, J. and M.C. Dobarganes Garcia. 1976. Pollution of natural waters by synthetic detergents. X. Biodegradation of nonionic surfactants in river water. Grasas Aceites 27:309-322.

Schaffner, C., E. Stephanou and W. Giger. 1982. Determination of nonylphenols and nonylphenolethoxylates in secondary sewage effluents. Comm. Eur. Communities Rep.

Schaffner, C., P.H. Brunner and W. Giger. 1984. 4-Nonylphenol, a highly concentrated degradation product on nonionic surfactants in sewage sludge. In: Proceedings of the Third International Symposium on Processing and Use of Sewage Sludge. Reidel Publishing Company, Dordrecht, Holland.

Schneider, E., K. Levsen, P. Dähling and F.W. Röllgen. 1983. Analysis of surfactants by newer mass spectrometric techniques. Part I. Cationic and non-ionic surfactants. Fresenius Z. Anal. Chem. 316:277-285.

Schneider, E., K. Levsen, A.J.H. Boerboom, P. Kistemaker, S.A. McLuckey, and M. Przybylski. 1984. Identification of cationic and anionic surfactants in surface water by combined field desorption - collisionally activated decomposition mass spectrometry. Anal. Chem. 56:1987-1988.

Sheldon, L.S. and R.A. Hites. 1978. Organic compounds in the Delaware River. Environ. Sci. Technol. 12:1188-1194.

Sheldon, L.S. and R.A. Hites. 1979a. Environmental occurrence and mass spectral identification of ethylene glycol derivatives. Sci. Total Environ. 11:279-286.

Sheldon, L.S. and R.A. Hites. 1979b. Sources and movement of organic chemicals in the Delaware River. Environ. Sci. Technol. 13:279-286.

Stephanou, E. 1984. Identification of nonionic detergents by GC/CI-MS: 1. A complementary method or an attractive alternative to GC/EI-MS and other methods? Chemosphere 13:43-51.

Stephanou, E. 1986. Determination of acidic and neutral residues of alkylphenol polyethoxylate surfactants using GC/MS analysis of their TMS derivatives. Comm. Eur. Communities Rep.

Stephanou, E. and W. Giger. 1982. Persistent organic chemicals in sewage effluents. 2. Quantitative determinations of nonylphenols and nonylphenol ethoxylates by glass capillary gas chromatography. Environ. Sci. Technol. 16:800-805.

Sundaram, K.M.S. and S. Szeto. 1981. The dissipation of nonylphenol in stream and pond water under simulated field conditions. J. Environ. Science Health B16:767-776.

Swisher, R.D. 1987. *Surfactant Biodegradation*, 2nd. Ed. Surfactant Science Series, Vol. 18. Marcel Dekker, Inc., New York. pp. 47-146.

Urano, K., M. Saito and C. Murata. 1984. Adsorption of surfactants on sediments. Chemosphere 13:293-300.

Valls, M., J.M. Bayona and J. Albaiges. 1990. Broad spectrum analysis of ionic and non-ionic organic contaminants in urban wastewaters and coastal receiving aquatic systems. Intl. J. Environ. Anal. Chem. 39:329-348.

Ventura, F., A. Figueras, J. Caixach, I. Espadaler, J. Romero, J. Guardiola and J. Rivera. 1988. Characterization of polyethoxylated surfactants and their brominated derivatives formed at the water treatment plant of Barcelona by GC/MS and FAB mass spectrometry. Water Res. 22:1211-1217.

Wahlberg, C., L. Renberg and U. Wideqvist. 1990. Determination of nonylphenol and nonylphenol ethoxylates as their pentafluorobenzoates in water, sewage sludge and biota. Chemosphere 20:179-195.

Waldock, M.J. and J.E. Thain. 1991. Environmental concentrations of 4-nonylphenol following dumping of anaerobically digested sewage sludges: A preliminary study of occurrence and acute toxicity. Unpublished paper, Ministry of Agriculture, Fisheries and Food, Fisheries Laboratory, Burnham-on-Crouch, Essex, Britain.

Waters, J., T.J. Garrigan and A.M. Paulson. 1986. Investigations into the scope and limitations of the bismuth active substances procedure (Wickbold) for the determination of nonionic surfactants in environmental samples. Water Res. 20:247-253.

Watts, C.D., B. Crathorne, M. Fielding and C.P. Steel. 1984. Identification of non-volatile organics in water using field desorption mass spectrometry and high performance liquid chromatography. In: Analysis of Organic Micropollutants in Water, G. Angeletti and A. Bjorseth, eds. D. Reidel Publishing Company, Boston, pp. 120-131.

Weber, R., K. Levsen, G.J. Louter, A.J.H. Boerboom, and J. Havercamp. 1982. Direct mixture analysis of surfactants by combined field desorption/collisionally activated dissociation mass spectrometry with simultaneous ion detection. Anal. Chem. 54:1458-1466.

Wickbold, R. 1966. Analysis for nonionic surfactants in water and wastewater. Vom Wasser 33:229-241.

Wickbold. 1971. Enrichment and separation of surfactants from surface waters by transport at the gas/water interface. Tenside 8:61-63.

Wickbold, R. 1972. On the determination of nonionic surfactants in river- and wastewaters. Tenside 9:173-177.

Wickbold, R. 1973. Analytical determination of small amounts of nonionic surfactants. Tenside 10:179-182.

Zoller, U., E. Ashash, G. Ayali, S. Shafir and B. Azmon. 1990. Nonionic detergents as tracers of ground water pollution caused by municipal sewage. Environ. Intl. 16:301-306.

Zoller, U. 1992. Distribution and survival of nonionic surfactants in the surface, sea and groundwater of Israel. J. Environ. Sci. Health A27:1521-1533.

III. BIODEGRADATION

In the following review of APE biodegradation, the distinction between primary and ultimate biodegradation follows the definitions of Swisher (1987). Primary degradation describes the loss of a measurable chemical characteristic of a compound while ultimate biodegradation describes the mineralization of a compound to CO_2, H_2O, and inorganic substances. Under anaerobic conditions, CO_2 and methane are generated.

The biodegradation or removal of APE surfactants has been extensively studied. It is generally agreed that APE undergo primary degradation under a variety of field conditions and test systems, provided sufficient acclimation time has occurred. Biodegradation of NPE proceeds through the formation of water-insoluble intermediates (NP, NPE_1 and NPE_2) which may be more toxic to aquatic biota than the intact surfactants based on laboratory studies of the aquatic toxicity of model compounds. There is mounting evidence for the biodegradation of these metabolites in wastewater treatment plants. The efficiency of removal appears to be dependent on the operating conditions of the treatment plant. Biodegradation metabolites may also be removed from wastewater effluents or natural systems by adsorption to lipophilic material such as sludge and sediments.

This section summarizes available information on biodegradation or removal under both laboratory and field situations. Although branched OPE and NPE are currently the only alkylphenol ethoxylates of commercial importance, information on APE with linear alkyl chains is included for comparative purposes.

A. LABORATORY INVESTIGATIONS

1. Test Methods

In addition to the analytical methods outlined in Chapter II, the ultimate biodegradation of surfactants may be followed in the laboratory by a number of chemically non-specific methods. These methods include chemical oxygen demand (COD), biological oxygen demand (BOD; closed-bottle test), total organic carbon (TOC), and CO_2 formation. The application of these methods to environmental samples is limited because of their inability to determine organic carbon from a specific substrate. Standard laboratory methods for BOD, COD, and TOC are published by the American Public Health Association (Part 5000: Determination of Organic Constituents) (APHA, 1989).

Three widely used test systems for the measurement of biodegradability are the river-water dieaway, the shake-flask, and the activated-sludge tests (Schick, 1967; Swisher, 1987). The first test involves adding a specific quantity of surfactant to river water in a glass jar and allowing the solution to incubate at room temperature. Degradation is determined by measurement of the final substrate concentration using a suitable method. Although the natural microbial populations and solids of river water are usually low, this test imitates conditions in natural bodies of water. However, the natural variability of microbial populations and the range of pollution of natural waters makes this test inherently variable.

In the shake-flask test (ASTM, E1297), surfactant is added to a medium containing either adapted or unacclimated microbial cultures and inorganic supplements. In simple systems, samples are withdrawn and analyzed for primary biodegradation (ASTM, 1991a). In more complex systems, ultimate biodegradation via CO_2 evolution or BOD may be measured. The Sturm (1973) test, in which raw sewage is used as a source of microorganisms, CO_2-free air is bubbled through the test units, and the effluent gas is absorbed and titrated for CO_2, has become a standard method. The microbial populations in these units are dilute compared to sewage treatment conditions.

In the activated sludge test, a specific concentration of surfactant and sewage containing nutrients are continuously fed into an activated-sludge unit (CFAS units) and the overflow effluent is collected and analyzed for surfactant and/or metabolites by a suitable method. These units are designed to simulate municipal activated-sludge sewage treatment plants. In semi-continuous units (SCAS), which are operated on a 24-hr cycle, the nutrients and test surfactant are fed once a day. In the batch activated sludge system (BAS), settled activated sludge is added to sewage containing the test compound, air is bubbled through, and the dieaway of the test compound is monitored over periods lasting several hours to days. The source and age of the sludge vary among studies. The SDA shake flask and SCAS procedures are the two-stage "screening" and "confirmatory" tests in which branched sulfonates and linear alkylbenzene sulfonates are tested for 90% biodegradability (SDA, 1965). This two-step process has been adopted by the ASTM (D2667) (ASTM, 1991b).

The U.S. EPA outlines guidelines for the determination of the rate and extent of aerobic biodegradation that might occur when chemicals are released into aquatic environments (Fed. Reg. 40 CFR 796, 50:39277-39280, September 27, 1985). Their closed shake-flask system for CO_2 evolution is based on the method of Gledhill (1975) and is often referred to as the "Gledhill" test. Carbon dioxide values should be in the range of 80-100% of the theoretical CO_2. Flasks should also be monitored for dissolved organic carbon (DOC) removal. The TOC is analyzed in order to calculate the percent of theoretical yield of CO_2 and percent of DOC loss. The closed bottle test for measurement of oxygen uptake as measured by BOD is an earlier variation of this test.

Hughes et al. (1989) compared the biodegradation of C_9APE_{12} in three standard test systems. The ultimate biodegradation (percent conversion to CO_2) in 28-day tests ranged from approximately 30-65% and was in the order Sturm > Gledhill (sludge inoculum) > Gledhill (acclimated bacteria) > closed bottle. The Sturm test had the lowest variability among triplicate samples.

Primary degradation of the benzene ring of APE can be followed in aqueous solutions by the disappearance of its UV absorption band at 275 nm or its IR absorption at 6.1 and 11-15 μ (Schick, 1967; Swisher, 1987), but, because of interfering substances, these techniques are impractical for environmental samples. Nuclear magnetic resonance (NMR) can be used to characterize commercial formulations and to identify changes in structure during biodegradation. NMR spectra provide quantitative information on the number, kinds, and relative positions of various hydrogen-containing functional groups and allow identification of the length and branching of the hydrophobe, the o- or p-substitution of the ring substituents, and the average length of the EO units without the need for standard samples (Schick, 1967; Swisher, 1987). ^{13}Carbon NMR was employed by Kravetz et al. (1991) to identify the branching of the alkyl moiety of nonionic surfactants.

The use of radiolabeled compounds, as in the studies of Kravetz et al. (1982; 1984) provides a sensitive technique for both ultimate biodegradation and identification of biodegradation intermediates in laboratory studies. Recent studies also address routes other than biodegradation for dissipation or removal of APE from wastewaters. These routes include adsorption to sludge and volatilization.

2. Primary and Ultimate Biodegradation

A variety of test methods and microbial test systems have been employed in laboratory studies of biodegradation. Because of the large number of studies, they have been summarized in tabular form (Appendix). Test systems listed in Column 2 of the Appendix include uninoculated natural media such as river water and synthetic media inoculated with acclimated or unacclimated bacteria, as well as sludge-inoculated and continuous and semi-continuous model sewage systems. Because of the variety of test systems and conditions, the data are diverse and are not comparable from study to study. In addition, accurate measurement of biodegradation may be complicated by the adsorption of surfactants onto organic matter in some systems and the metabolic uptake of degradation products into the biomass present (Birch, 1984; Steber and Wierich, 1987; Swisher, 1987). Studies that illustrate the affect of various parameters on biodegradation rates are briefly described.

Lashen and Booman (1967) pointed out the importance of acclimation of bacteria to the surfactant in laboratory tests. The acclimation of microflora depends on the test method employed and the source of river water or sludge. Saeger et al. (1980) point out that under the same laboratory conditions, tests carried out at different times have variable results. In a comparison of three separate studies with C_9APE_9 (Sterox NJ), dissolved organic carbon removals in SCAS assays were 20-50%, 25-45%, and 58%.

Using a laboratory-scale activated-sludge plant, Birch (1991a; 1991b) demonstrated the importance of temperature and sewage retention time (SRT) during sewage plant treatment. Primary biodegradation of NPE_{10} was extensive at temperatures of 15° and 11°C, but at 7°C, high levels of nonionic were observed operating at SRTs of 2, 4, and 6 days. The results suggest a critical SRT of 6 days at low temperatures and are consistent with results of field studies carried out at different temperatures.

Although most studies show that primary biodegradation is often >90% complete under varying conditions, the available data indicate that in laboratory systems ultimate biodegradation of APE is often incomplete within the study time periods. Using systems with the highest biodegradation potential such as continuous-flow activated sludge units, ultimate biodegradation of br-NPE_{9-10} surfactants cited in recent studies in the Appendix ranged from as low as 14% as measured by CO_2 evolution in a single 28-day test (Kravetz et al., 1991) to 89% as measured by COD in simulated activated sludge plant units (Neufahrt et al., 1987). Initial concentrations were 10 and 12 mg/L. Most investigators reported results between 30 and 70%.

Studies on the ultimate biodegradation of APE - oxidation of the aromatic ring of APE as measured by disappearance of the UV absorption band at 275 nm and the recovery of radiolabeled metabolites have been reported. Lashen et al. (1966) found no loss of ^3H from the tritium-labeled ring of C_8APE_{10} in a simulated septic tank with a retention time of 67 hours. On the other hand, in a continuous-flow activated sludge system, Kravetz et al. (1982) accounted for approximately 97% of the tritium activity from ring-labeled C_9APE_9 as ^3H_2O (28.3%), soluble ^3H metabolites (36.7%), and ^3H (31.5%) in the accumulated biomass, indicating substantial aromatic transformation. The amount of ^3H_2O rose to ~42% by day 26. Carbon dioxide formation from the ^{14}C-labeled EO units was approximately 58% by day 28. Biodegradation was sensitive to low temperature (Kravetz et al., 1984). Formation of ^3H_2O decreased from 29% at 25°C to 10% at 12°C and to 2% at 8°C. Lowering the temperature by the same increments decreased the formation of $^{14}CO_2$ from 58% to 50% and to 10%.

Brüschweiler and coworkers (1982; 1983) followed the primary and ultimate degradation of two NPE. Primary degradation of both NPE_{11} and NPE_{23}, as measured by BIAS was approximately 98% complete within 30 days. Ultimate biodegradation as measured by DOC was 82 and 70% for the two compounds, respectively, and ring degradation, measured by UV absorption, was approximately 63% completed in 30 days.

Sato et al. (1963) reported changes in the UV absorption of NPE_{10} upon aeration in activated sludge. After 7 hours, the absorbance had increased and shifted to 270 nm. After 24, 48, and 96 hours, the absorbance at 270-275 nm was 50%, 10-15%, and 0% of the initial absorbance, respectively, representing 100% disappearance of the benzene ring. Frazee et al. (1964) and Osburn and Benedict (1966) reported the disappearance of the UV 275 nm ring band and the aromatic IR 1250/cm frequency band of C_9APE_{10} in river-water dieaway tests. Fuka and Pitter (1978; 1980) and Pitter and Fuka (1979) reported extensive disappearance of the 275 nm band in the Pitter dieaway test (a standardized procedure that measures COD or DOC in an activated-sludge inoculated mineral salts medium). Disappearance of C_9APE_5 - C_9APE_{35} ranged from 72-89% in 7 days. The rate did not appear to be dependent on EO chain length.

The flux of tp-NPE_9 and a mixture of AE at high feed concentrations of 385 and 241 mg/L, respectively, was followed through a continuous-flow, bench-scale reactor seeded with activated sludge from an industrial wastewater treatment plant (Patoczka and Pulliam, 1990). While primary biodegradation was almost 100% complete as measured by CTAS, the BOD, COD, and TOC removals were 98.9%, 74.4%, and 70.6%, respectively. Near the end of the 10-week operation period, the NPE_{0-3} concentrations in the effluent and reactor sludge were <1 and 20 mg/L, respectively. Assuming complete conversion of the feed NPE_9 to NPE_{0-3}, 155-180 mg/L of NPE_{0-3} could have been generated.

Under anaerobic conditions only the polyethoxylate chain of $C_{10-12}APE_9$ was converted to methane (Wagener and Schink, 1987). The alkylphenol chain probably remained unchanged. The authors used anoxic digester sludge in an anaerobic fixed-bed reactor. Ultimate biodegradation as measured by methane production was 45-50%.

3. Other Methods of Removal

In addition to biodegradation, the fate of NP in surface waters as indicated by laboratory studies includes dissipation via volatilization, photolysis, and dilution and adsorption to sediments. Nonylphenol disappeared rapidly from aqueous solution in open tanks, with up to a 90% loss within 48 hr (Ernst et al., 1980). Disappearance was correlated with surface area/volume ratios, aeration, and stirring, suggesting that volatilization is the primary route of dissipation.

The dissipation of NP was studied in stream and pond water incubated in open and closed flasks (Sundaram and Szeto, 1981). The half-life was 2.5 days in open flasks and 16 days in closed flasks. No intermediate biodegradation product was detected in the open flasks, but a transformed, more polar product was detected in the closed flasks. The authors suggested that disappearance of NP in an open system is due to surface volatilization and co-distillation. When sediment was added to the flasks, NP was translocated to the sediment where about 80% was degraded in 70 days.

Tanaka et al. (1991) irradiated 0.25 nmole aqueous solutions of radiolabeled t-OPE_9 with 300 nm sunlight lamps for 30 hours. Approximately 37% of the recovered radioactivity was intact t-OPE_9. Percentages of each of the lower molecular weight homologs, OPE_1-OPE_8, ranged from 0.7 to 1.1%. Polar products including PEG, glycolic aldehydes, and glycolic ethers accounted for 21% of the radioactivity and indicated cleavage at both the carbon-carbon and carbon-oxygen bonds of the polyoxyethylene chain. Because all possible chain lengths of PEG were observed, side-chain cleavage was occurring by a random process at all of the oxygen-ethylene bonds.

Pellzzetti et al. (1989) investigated the photocatalytic degradation of NPE surfactants using TiO_2 particulates as photocatalyst. The degradation pathway and intermediate products were monitored through HPLC, CO_2 evolution, DOC, and particulate organic carbon measurements. Initial hydroxyl radical (OH•) attack on the EO chain and the benzene ring yielded acidic compounds, polyethylene glycols, and ethylene glycols that were further oxidized to CO_2. 4-Nonylphenol was not formed.

Removal of APE from effluents is enhanced by ozonation. Small doses of ozone increased the biodegradability of several NPE by changing the structure of the compounds, making them more amenable to bacteriological breakdown (Narkis et al., 1987). Powdered activated carbon effectively removed NP and dinonyl phenol ethoxylates from aqueous solutions (Narkis and Ben-David, 1983; Weinberg and Narkis, 1987). At a concentration of 10 mg/L surfactant, addition of 40 to 80 mg/L of powdered activated carbon effectively removed the surfactant.

B. FIELD STUDIES

1. Waste Treatment Facilities

Removal at Wastewater Treatment Facilities. The primary factors involved in controlling environmental levels of APE are the efficiency of wastewater treatment plants and the dilution provided by the receiving water. Field studies indicate that APE undergo primary and ultimate degradation at wastewater treatment plants. Alkylphenol polyethoxylate surfactants are biodegraded during aerobic and anaerobic treatment of wastewater to alkylphenols and alkylphenol mono- and diethoxylates as well as to the corresponding carboxylates. Concentrations of NP and NPE have been tracked through several treatment plants. Reported results vary widely from poor treatability in some Swiss treatment plants (Giger et al., 1986) to over 95% NPE removal (Naylor et al., 1992). The reasons for these variations are not readily apparent although Giger found a correlation between NPE treatability and treatment plant nitrification rate. Studies that measured concentrations of nonionics or specific APE and metabolites in influent and effluent streams are summarized in Table 3-1 and briefly discussed here.

During a three-month period, the biodegradability of C_8APE_{10} was studied in an extended aeration-activated sludge treatment plant that served a college campus (Lashen and Lamb, 1967; Lashen and Booman, 1967). The surfactant was spiked into two of three units at concentrations of 5 and 10 mg/L; influent and effluent samples were analyzed by several methods. Primary degradation was over 90% as measured by CTAS and surfactant properties. Occasional TLC analysis results were ~54% less than those measured by CTAS.

Mann and Reid (1971) is one of three references bearing on the effect of winter conditions on APE biodegradation. The other two, Brown et al. (1987) and Naylor et al. (1992), show a much smaller winter operational effect. Mann and Reid (1971) studied the seasonal primary biodegradation of two APE, C_8APE_{8-9} (Nonidet P40) and C_8APE_{14-15} (Nonidet P100), at a trickling filter plant that served a small community in England. The surfactants were introduced by distributing Nonidet-containing dishwashing liquids to the community. In the first trial, degradation of Nonidet P100, as determined by TLC, was between 5% and 25% in March-May. A biodegradability of 64% was recorded in July. Replacement of Nonidet P100 by Nonidet P40 in September resulted in a change in the range of biodegradation to from 70% to nearly 90%. A year-long trial with Nonidet P40 confirmed the seasonal variation: 20-26% in January-May to 80% in late August, September and October as determined by TLC. The authors offered two possible explanations for the variation: (1) the bacteria that are capable of degrading APE in summer are absent or dormant in winter and (2) the rate of bacterial film growth on the filter is affected by seasonal changes and alters the characteristics of the sewage treatment process.

Brown et al. (1987) examined seasonal differences at a small trickling filter plant that serves a population of 6,000 in Germany. Influent concentrations, as measured by BIAS, were similar in March and September, but removal rates were somewhat lower in March (81%)

Table 3-1. Removal at Wastewater Treatment Facilities

Type of Plant[a]/ Location	Surfactant	Influent Concentration (mg/L)	Effluent Concentration (mg/L)	Removal (%)	Measurement Method	Reference
Extended aeration activated sludge/ United States	C_8APE_{10}	5, 10	—	>90 41	CTAS TLC	Lashen and Lamb, 1967; Lashen and Booman, 1967
Trickling filter/ England	C_8APE_{8-9} C_8APE_{14-15}	— —	— —	20-26 (Jan.-May) 80 (Aug.-Oct.) 5-25 (Mar.-May) 64 (July)	TLC TLC TLC TLC	Mann and Reid, 1971
Activated sludge/ Switzerland	NPE_{1-18} NPE_{3-20} NPE_{0-20}	0.8-2.3 0.4-2.2 — — —	0.029-0.369 (NPE_2)[b] — <0.01-0.035 (4-NP) <0.01-0.133 (NPE_1) <0.01-0.070 (NPE_2)	>95 — — — —	HPLC/UV HPLC/UV GC/MS GC/MS GC/MS	Ahel and Giger, 1985b Ahel et al., 1986 Giger et al., 1986; 1987 Schaffner et al., 1982 Stephanou and Giger, 1982
Activated sludge/ Germany	$C_nAPE_{6.4}$[b]	3.1-8.5 0.85-1.63	0.19-0.23 0.12-0.14 0.037-0.047 $(C_nAPE_{2.2})$[b]	96 89	BIAS HPLC/fluorescence	Brown et al., 1986
Trickling filter/ Germany	$C_nAPE_{7.3}$[b] $C_nAPE_{7.6}$[b]	4.0 4.1 — —	0.7 0.5 0.23 $(C_nAPE_{5.3})$[b] 0.04-0.07 (C_nAPE_2) 0.17 $(C_nAPE_{7.1})$[b] 0.01-0.02 (C_nAPE_2)	81 (Mar.) 88 (Sept.) 70 (Mar.) 75 (Sept.)	BIAS[c] BIAS[c] HPLC/fluorescence HPLC/fluorescence	Brown et al., 1987

Table 3-1. (cont.)

Type of Plant[a]/ Location	Surfactant	Influent Concentration (mg/L)	Effluent Concentration (mg/L)	Removal (%)	Measurement Method	Reference
Primary-secondary precipitation - aeration ponds/ Israel	Nonionics Nonionics	2.4 1.2	0.45 0.25	81 79	CTAS CTAS	Zoller, 1985
Not reported/ England	C_nAPE_8	0.126-0.410	0.04-0.23 (C_nAPE_2)[b]	—	HPLC/fluorescence	Holt et al., 1986
Various treatments/ Israel	Nonionics	3.4-10	0.8-6.2	—	CTAS, BIAS	Narkis et al., 1987
Trickling filter/ United States	NP	0.088, 0.038	not detected	—	HPLC/UV	Varma et al., 1987
Activated sludge/ United States	NP NPE_1 NPE_2	— — —	0.002 0.001 0.0002	— — —	HPLC/fluorescence	Radian Corporation, 1989
Activated sludge/ United States	NP NPE_1 NPE_2	— — —	0.009 0.013 0.035	— — —	HPLC/fluorescence	Radian Corporation, 1989
Trickling filter/ United States	NP NPE_1 NPE_2	— — —	0.070 0.080 0.013	— — —	HPLC/fluorescence	Radian Corporation, 1989
Activated sludge (industrial)/ United States	NPE_{1-18}	1.78	0.103 (NPE_{1-18}) 0.0017 (NP)	94	HPLC/fluorescence	Kubeck and Naylor, 1990; Naylor et al., 1992

Table 3-1. (cont.)

Type of Plant[a]/ Location	Surfactant	Influent Concentration (mg/L)	Effluent Concentration (mg/L)	Removal (%)	Measurement Method	Reference
Activated sludge (domestic)/ United States	NPE_{1-18}	2.40	0.071 (NPE_{1-18}) 0.0017 (NP)	97	HPLC/fluorescence	Kubeck and Naylor, 1990; Naylor et al., 1992
Activated sludge (mixed)/ United States	NPE_{1-18} NPE_{1-18}	1.54 1.13	0.043 (NPE_{1-18}) 0.085 (NPE_{1-18})	97.2 (Aug.) 92.5 (Mar.)	HPLC/fluorescence HPLC/fluorescence	Naylor et al., 1992
Activated sludge (wood pulp mill)/ United States	NPE_{1-18}	8.45	0.21 (NPE_{1-18})	97.5	HPLC/fluorescence	Naylor et al., 1992
Activated sludge (wood pulp mill)/ United States	NPE_{1-18}	13.4	2.17 (NPE_{1-18})	84.3	HPLC/fluorescence	Naylor et al., 1992
Activated sludge (chemical)/ United States	NP	0.4-0.8	0.023-0.074 (NP)	92.5	HPLC/fluorescence	Naylor et al., 1992

[a] Non-domestic sources of waste are noted.
[b] Average EO number or major component of the influent or effluent.
[c] APE was 17-22% of the influent and 27-34% of the effluent BIAS measurement.
A dash (—) indicates no data.

compared with September (88%). Removal rates as measured by HPLC were 70% and 75%, respectively. A seasonal study at an activated sludge plant in a small U.S. midwestern city showed a slight drop in extent of NPE removal in winter compared to summer (92.5% vs 97.2%) as measured by HPLC/fluorescence (Naylor et al., 1992). The plant received discharge from a cleaning product manufacturer as well as domestic sewage.

Ahel and Giger and coworkers (Schaffner et al., 1982; Stephanou and Giger, 1982; Ahel and Giger, 1985a; 1985b; Ahel et al. 1986; Giger et al., 1986; 1987; Marcomini et al., 1988) performed a detailed analysis of the concentrations and fluxes of 4-nonylphenol ethoxylates (NPE_3 - NPE_{20}), NPE_1 and NPE_2, NPEC, and NP through several municipal activated-sludge sewage treatment plants in Switzerland. Alkylphenol polyethoxylates with 1-18 ethoxy units were measured in nontreated wastewaters and effluents. Total concentrations, as measured by HPLC/UV absorption, were 0.8-2.3 mg/L in the untreated wastewater and 29-369 µg/L in the treated wastewater (Ahel and Giger, 1985b).

The NPE with 3-20 ethylene oxide units, which were present at total concentrations of 0.4-2.2 mg/L (400-2000 mg/m^3), were >95% removed in the plants (Ahel and Giger, 1985b; Ahel et al., 1986; Giger et al., 1986; 1987). Measurements were based on normal-phase HPLC with UV absorption detection of individual oligomers that were added to give the total concentration (Ahel and Giger, 1985b). Although the range of oligomer distributions in the most commonly-used laundry detergents has a Poisson distribution maximizing at EO_{9-10} with almost no oligomers of EO_{0-2}, nontreated sewage showed a bimodal oligomer distribution with maxima at EO_7 and EO_{1-2}, whereas secondary effluent had a single peak with a maximum at EO_2, with NP also present.

The presence of biotransformation products was recognized and the authors followed their fate through four activated-sludge treatment plants (Ahel et al., 1986; Giger et al., 1986; 1987). The distribution of nonylphenol polyethoxylates, nonylphenol mono- and diethoxylates, nonylphenoxy carboxylic acids, and nonylphenol, measured in primary (untreated) and secondary effluents, decreased from 1-4 µmoles/L to $\leq$0.3 µmoles/L, increased and/or decreased from 0.2-0.3 µmoles/L to 0.06-0.5 µmoles/L, increased (as would be expected from the oxidation of the alcohol moiety) from 0.004-0.05 µmoles/L to 0.2-1.0 µmoles/L, and decreased from 0.1-0.2 µmoles/L to 0.005-0.06 µmoles/L, respectively. The carboxylic acids appeared to be formed under aerobic conditions, resulting in a net production during secondary treatment although tertiary treatment had little effect on their presence. Nonylphenol was strongly accumulated by the sludge. Degradation appeared to be plant specific and was strongly dependent on sludge loading rate and nitrifying conditions in the plant. In a summary paper, Giger and Ahel (1991) estimated that 60-65% of all surfactant-derived NPE compounds (calculated on a molar basis) that have been introduced into sewage treatment plants are discharged into the environment.

Using the GC/MS method described in an earlier paper, Stephanou and Giger (1982) and Schaffner et al. (1982) measured concentrations of nonylphenols and NPE with one and two ethoxylate groups in secondary effluents of the same plants. Total concentrations of the three nonylphenolic compounds ranged from not detectable (<10 µg/L) in three plants operated at low loading conditions to 36-202 µg/L at three other plants. The latter concentrations represented 0.5%-2.3% of the total residual dissolved organic carbon. Nonylphenol ranged from not detected (<10 µg/L) to 35 µg/L, NPE_1 ranged from 24-133 µg/L, and NPE_2 ranged from not detected to 70 µg/L.

The optimized BIAS method as modified by Waters et al. (1986) was used to monitor primary degradation of nonionic surfactants in both an activated sludge plant and a trickling filter plant in Germany (Brown et al., 1986; 1987). The average BIAS removal at the Hochdahl activated sludge plant was 96%. Semi-quantitative results, using an HPLC technique for determination of specific APE, confirmed the high level of removal (89%). HPLC patterns of the APE components showed that the average EO number in the influent was 6.4 and in the effluent was 2.2. Earlier observations at the activated sludge plant measured 60-80% removals (BIAS) (Fischer and Gerike, 1984). The average removal at the Hösel-Dickelsbach trickling filter plant was 81% in March and 88% in September as measured by BIAS and 70 and 75%, respectively, as determined by HPLC. Nonionics had an average EO number of 7.5 in the influent and 6.3 in the effluent. The APE surfactants contributed approximately 20% to the sewage BIAS concentrations at both plants. Concentrations of APE_2 in the effluents were 37-47 µg/L (Hochdahl) and 41-70 µg/L in March and 10-20 µg/L in September (Hösel-Dickelsbach).

In Israel, nonionic surfactants in sewage effluents are relatively high as a result of the widespread use of nonylphenol and dinonylphenol ethoxylates (Narkis et al., 1987). The following concentrations were measured (CTAS or BIAS): 3.4-10.0 mg/L in total raw sewage and 0.8-6.2 mg/L in effluent. Filtration of the raw sewage and effluent through filter paper showed that most of the surfactants were adsorbed onto suspended solids. Concentrations in the filtered sewage were 1.2-5.9 mg/L and in the filtered effluent were 0.7-2.1 mg/L. In another study, typical concentrations of nonionics in sewage influents, as measured by CTAS, were in the range of 1.1-2.2 mg/L; concentrations in treated effluents ranged from 0.25-0.45 mg/L (Zoller, 1985). According to the author, concentrations of nonionics have declined since 1975, corresponding with the upward trend in the use of AE surfactants instead of APE surfactants. Data from two plants are cited in Table 3-1.

Using HPLC with fluorescence detection, Holt et al. (1986) measured APE in the influent and effluent of two plants in England. Total APE in the influent ranged from 126-410 µg/L; total APE in the effluent ranged from 40-228 µg/L. APE with 1 to 19 EO units were detected in both influent and effluent and confirmed by mass spectrometry. C_nAPE_8 was the major component of the influent and C_nAPE_2 was the major component of the effluent, indicating primary, but not ultimate degradation had occurred within the plants.

In a survey of secondary effluents from 10 municipal and industrial wastewater treatment plants discharging into rivers in Illinois, NP was found in one sample (Ellis et al., 1982). Analysis was by GC/mass spectrometry; the concentration and/or limit of detection was not given.

Varma et al. (1987) measured 4-NP as determined by steam distillation and HPLC/UV detection in grab samples from several streams of an advanced wastewater treatment plant in the Washington, DC, area. The plant received primarily domestic waste. Concentrations of two samples were 88 and 38 µg/L (primary influent), 77 and 26 µg/L (primary effluent), 29 and 47 µg/L (trickling filter effluent), 16 and 12 µg/L (effluent from aeration basin), not detected (secondary and final effluents), 17 and 67 µg/L (thickener effluent), 118 and 33 mg/kg (sludge to centrifuge), and 135 and 43 mg/kg (cake from centrifuge). These results show that while NP was effectively removed from the effluent, lipophilic NP adsorbs onto sewage sludge.

The Radian Corporation (1989) measured the concentrations of NP, NPE_1, and NPE_2 in the effluent of two sewage treatment plants on the Alabama River and one on the Saginaw River. Samples were collected during February and March, 1988, when biological activity was at its lowest. The highest effluent concentration of NP (70 µg/L) was present at the trickling filter plant.

Using a limited number of samples, Kubeck and Naylor (1990) and Naylor et al. (1992) assayed wastewater from activated sludge plants for NPE using HPLC with fluorescence detection. Twenty-four hour composite samples of raw wastewater from two treatment plants in North Carolina ranged from 1.6-2.5 mg/L (averages were 1.78 and 2.4 mg/L); effluent concentrations ranged from 0.050-0.1 mg/L, indicating 93-98% removal. The concentration of NP in the effluent ranged from 1-2.5 µg/L. All NPE_{1-18} oligomers were removed to about the same extent with only a slight shift to lower oligomers in the effluent.

A wastewater treatment plant in the midwest that received domestic sewage as well as discharge from a cleaning product manufacturer was sampled in summer and winter (Naylor et al., 1992). Removal of NPE_{1-18} dropped only slightly in winter compared to summer (92.5% vs. 97.2%). Oligomer distribution was skewed toward lower oligomers (NPE_1 and NPE_2), although there was no significant accumulation of these species or NP in the effluents or in digested sludge. Only 0.1% of the influent NPE was present in the anaerobic sludge.

High loadings of surfactant as well as high levels of lignins and other pulp by-products at the wood pulp mill treatment plants placed a heavy load on the plants. The first mill's treatment plant performed as well as the municipal plants, removing 97.5% of NPE. The second plant had a lower removal rate, 84.3%, perhaps due to the higher loading of NPE and a different plant design.

Nonylphenol was measured in the influent and effluent of a treatment plant receiving wastewater from Texaco's chemical plant in Pt. Neches, Texas, which includes a nonylphenol unit. Nonylphenol was removed in the treatment plant to about the same extent (92.5%, average of three days' results) as NPE in the other plants.

Nonylphenol has been found in wastewater effluents or untreated waste from various manufacturing facilities (Etnier, 1985). The following concentrations were listed in this report: 0.013 mg/L in municipal sewage (Lin et al., 1981), 0.05 mg/L in wastewater effluents from a specialty chemical plant (Hites et al., 1979), 0.05 mg/L in untreated waste from a carpet yarn mill and 4.0 mg/L (4-NP) in the treatment pond effluent (Garrison and Hill, 1972), 0.06 mg/L in wastewater effluent from a tire manufacturing plant (Jungclaus et al., 1978), and 2-1617 µg/L from various industrial sources (Shackelford et al., 1983). In another study, alkylphenol-polyethoxy carboxylic acids were identified in treated municipal wastewater, but measurements were not made (Reinhard et al., 1982).

Adsorption to Sludge. Because NP and its mono- and diethoxylate are less water soluble than the higher molecular weight oligomers, they are partially removed from the wastewater stream by sorption to lipophilic flocs of sewage sludge. As a consequence, concentrations of these oligomers are usually higher in sludge than in wastewater. As part of the study of the environmental fate of surfactants, their concentrations in sludge have been determined by several investigators (Table 3-2).

In a series of articles, researchers at the Swiss Federal Institute for Water Resources and Water Pollution Control reported results of their analyses for 4-NP in sludge from more than 30 Swiss wastewater treatment plants (Giger et al., 1984; Schaffner et al., 1984). A steam distillation/solvent extraction method followed by GC analysis was used to determine the NP. Concentrations in anaerobically stabilized samples were higher than in aerobically stabilized sludge, averaging 1010 mg/kg (dry weight) compared to 280 mg/kg for aerobically stabilized sludge. Primary biodegradation products in anaerobically digested sewage sludge from one of the plants, detected by HPLC, contained average concentrations of NP, NPE_1, and NPE_2 of 1200, 220, and 30 mg/kg dry weight (Marcomini and Giger, 1987).

Brunner et al. (1988) investigated the fluxes of NP and NPE_1 and NPE_2 through 29 of the same plants. About 50% of the NPE in the influent was transformed to NP and accumulated in the digested sewage sludge. Nonylphenol was not significantly degraded under anaerobic conditions and accumulated in the digested sludge to concentrations of ~1 g/kg dry matter. Concentrations in aerobically digested sludge averaged 0.3 g/kg dry matter. Mass balances of NP, NPE_1, and NPE_2 were difficult to follow through one of the plants (Marcomini et al., 1988). The loads of these products were higher in the sludge than in the raw sewage. Compared with the input load, the NP enrichment factor in anaerobically digested sludge was 4.7-6.4. The NP was quantitatively (99%) associated with sludge particulate matter. Enrichment of NP, NPE_1, and NPE_2 in the sludges compared to the raw sewage at these and other Swiss treatment plants was interpreted as evidence of their non-biodegradability. This contrasts with recent U.S. data indicating much lower levels ($\leq$ 2.8 mg/kg wet weight, 10 mg/kg dry weight) of these species in sludge and hence greater removal rates (Naylor et al., 1992).

Concentrations of 4-NP in secondary sludge from eight municipal sewage treatment plants in Sweden were measured by Wahlberg et al. (1990). A distinction between aerobic and anaerobic treatments was not made. Waldock and Thain (1991) analyzed effluents and anaerobically digested sewage sludges from treatment plants in the United Kingdom for 4-NP. Concentrations in effluents discharged from several treatment works in the vicinity of the Thames Estuary ranged from <2-21 ng/L. Concentrations in sludges ranged from 30 mg/kg (dry weight) in a sample of primary treated sewage to 4000 mg/kg (dry weight) in a sample of anaerobically treated sewage.

Data from studies on a limited number of sewage treatment plants in the U.S. indicate that NPE metabolites are present in sludge at lower concentrations than at European facilities. Concentrations of NP, NPE_1, and NPE_2 in sludge from different sources within six U.S. plants

Table 3-2. Adsorption of Metabolites to Sludge

Type of Plant or Treatment*/Country	Source of Sample	Number of Plants	Number of Samples	Chemical	Concentration (mg/kg dry weight)	Reference
Various technologies/ Switzerland, Germany	activated sludge mixed, primary, secondary anaerobically stabilized sludge aerobically stabilized sludge	— — 30 8	— — — —	NP NP NP NP	90-150 40-140 450-2530 (range) 1010 (average) 80-500 (range) 280 (average)	Giger et al., 1984 Schaffner et al., 1984
Not given/ Switzerland	digested sludge	1	—	NP NPE_1 NPE_2	1200 220 30	Marcomini and Giger, 1987
Trickling filter/ United States	primary sludge mixed sludge digested sludge	1	—	NP NP NP	23-64 99-102 33-134	Varma et al., 1987
Various technologies/ Switzerland	anaerobically digested sludge aerobically digested sludge	24 5	— —	NP NPE_1 NPE_2 NP NPE_1 NPE_2	1270 183 44 304 362 132	Brunner et al., 1988
Activated sludge/ United States	sludge from aeration basin	1	1	NP NPE_1 NPE_2	27.2 8.8 24.8	Radian Corporation, 1989
Activated sludge/ United States	activated sludge return line	1	1	NP NPE_1 NPE_2	30.9 45.8 51.7	Radian Corporation, 1989
Trickling filter/ United States	raw sludge pump	1	1	NP NPE_1 NPE_2	16.1 3.9 4.4	Radian Corporation, 1989

Table 3-2. (cont.)

Type of Plant or Treatment[a]/Country	Source of Sample	Number of Plants	Number of Samples	Chemical	Concentration (mg/kg dry weight)	Reference
Not given/ Sweden	secondary sludge, aerobic, anaerobic	8	—	NP	26-1100	Wahlberg et al., 1990
Not given/ England	raw sewage digested sludge, aerobic, anaerobic	1 "several"	1 7	NP NP	30 50-4000	Waldock and Thain, 1991
Activated sludge-domestic, chemical/ United States	digested sludge, anaerobic	1	—	NP	10 (Aug) 2.8[b] (Aug.) 1.8[b] (Mar.)	Naylor et al., 1992
Activated sludge-wood pulp mills/ United States	digested sludge	2	—	NP	0.019-0.43[b] (June) 0.74[b] (Sept.)	Naylor et al., 1992

[a]Treatment facilities received primarily domestic wastes unless otherwise noted.
[b]Wet weight.
A dash (—) indicates no data.

that used different treatment technologies ranged from 10-102, 3.9-45.8, and 4.4-51.7 mg/kg dry weight, respectively (Varma et al., 1987; Radian Corporation, 1989; Naylor et al., 1992). In the last study, only 0.1% of the influent NPE was present in the anaerobic sludge.

Based on the data of several investigators (Brüschweiler and Gämperle, 1982; Brüschweiler et al., 1983; and Urano and Saito 1984), Swisher (1987) reported adsorption partition coefficients (P_a = [g of adsorbed substrate/g absorbent]/[g of dissolved substrate/g of solution]) for C_8APE_{11} and NPE_{10} to activated sludge. The P_a were 170-1400 and 7500, respectively. Patoczka and Pulliam (1990) measured the adsorption of NPE_9 and $NPE_{0.3}$ at different concentrations to sludge and graphed adsorption isotherms.

Impact on Waste Treatment Facilities. Salanitro et al. (1988) presented laboratory data that show that NPE, at very high concentrations typical of industrial use, may have an adverse impact on the activated sludge process. NPE_9 (Igepal CO-630) was degraded in bench-scale biotreater units at influent concentrations of 10-40 mg/L. When fed at 80-100 mg/L, substantial BOD breakthrough, loss of nitrification, aerator foaming, and incomplete NPE removal occurred.

The filamentous bacterium *Nocardia* sp. produces a floating foamy scum in aeration tanks. In laboratory experiments, Ho and Jenkins (1991) showed that surfactants such as $NPE_{8.5}$ (Igepal CO-620) can enhance the undesired foaming of *Nocardia*-containing sludge. Significant effects were observed at CTAS concentrations as low as 0.2 mg/L.

2. Soil

As discussed above, APE surfactants may not be fully degraded during sewage treatment, but may be significantly accumulated in the digested sewage sludge. Chemicals in sludge applied to agricultural land or placed in landfills can potentially be taken up by vegetation or migrate to the water table although no evidence of the former has been reported.

At Otis Air Force Base near Falmouth, MA, treated sewage is discharged to sand beds from which it rapidly percolates to the water table (Barber et al., 1988). Several isomers of 4-NP were present in the groundwater below the beds at a total maximum concentration of 790 ng/L. In Canada, OP and OPE_1 were detected in the leachate plumes of two landfills that overlie a sandy bed (Reinhard et al., 1984).

The aerobic degradation of NP, NPE_1, and NPE_2 in a sludge-treated experimental plot was monitored over the period of one year (Marcomini et al., 1989). The initial concentrations of these three metabolites in the soil were 4.7, 1.1, and 0.1 mg/kg (dry weight). One year later the residual concentrations were 0.5, 0.1, and 0.01 mg/kg, respectively. Greater than 80% of the biodegradation took place within the first month. Marcomini et al. (1991) also measured the removal of NP and NPE_1 in sludge-only landfills in Germany. Biodegradation of both compounds was minimal over a 15 year period at a landfill where conditions were anaerobic whereas biodegradation was >90% where conditions were semiaerobic.

Nonylphenol mixed with an insecticide diluent was applied at a rate of 0.47 L/hectare to a forest canopy. Levels of NP in soil, even in open areas, never exceeded the limit of detection (0.1 ppm). Residues persisted in the spruce foliage for about 30 days (Sundaram et al., 1980; Kingsbury et al., 1981).

3. Pesticide Spraying Programs

The environmental fate of NP released during spraying programs as a component of pesticide formulations has also been studied. When NP was applied to a forest ecosystem at a rate of 0.47 L/ha (to simulate aerial spraying of the NP-containing pesticide Metacil), the compound was detected in a stream immediately following application (Sundaram et al., 1980). The highest concentration detected was 9.1 μg/L (ppb) one hour after spraying. Residues declined to trace levels (<2.0 μg/L) after 6 hours and were not detectable (<1 μg/L) after 24 hours. The estimated half-life was 2.5 hr. Residues in a nearly stagnant pool declined from 1100 μg/L 4 hr after spraying to 12 μg/L after 24 hr and were not detectable after 3 days. The authors attributed the rapid dissipation in the stream to dilution by water flow.

C. EFFECT OF CHEMICAL STRUCTURE

Although early data on the biodegradation of APE are conflicting, it is generally agreed that the following factors increase primary biodegradability: a decrease in the number of ethylene oxide groups, an increase in linearity of the alkyl side chain, an increase in length of the alkyl chain, positioning the phenol group at the terminus of the alkyl chain, and *p*-substitution (Swisher, 1987). Although commercial APE all contain branched alkyl chains and are predominantly *p*-substituted, for comparative purposes data on other structures and substitutions are provided in the following discussion.

1. Alkyl-chain structure.

Comparison of the primary biodegradation of branched with corresponding linear APE usually shows the rate for linear APE to be faster (Borstlap and Kortland, 1967; Huddleston and Allred, 1964; Garrison and Matson, 1964; Stead et al., 1972; Sturm, 1973; Ruiz Cruz and Dobarganes Garcia, 1977). (Data on individual studies are provided in the Appendix).

In addition, Saeger et al. (1980) found that the longer chain branched dodecyl Sterox surfactants showed greater biodegradability, as measured by dissolved organic carbon removal in semi-continuous activated sludge assays, than the branched nonyl derivatives. The highly branched Dimersol olefin-derived ethoxylates were degraded to a lesser extent than the less branched Sterox ethoxylates.

Secondary (internal) attachment of the phenol group to the linear alkyl chain also hindered biodegradation as evidenced by the wide divergence between nonrandom linear, random linear, and branched APE in dieaway tests as indicated by TLC (Stead et al., 1972).

According to Smithson (1966), the secondary attachment per se may not be as important as the internal position of the secondary attachment.

Garrison and Matson (1964) compared the biodegradation of linear secondary and branched C_9APE_9 in several types of degradation tests. Both primary and ultimate biodegradation were less extensive for the branched as measured by physical, chemical, and oxygen consumption methods. Oxygen consumption over a 5-day period in a Warburg respirometer was low (0.1 g/g of sample) for the linear secondary compound compared with 0% for the branched compound. Likewise, in river-water dieaway tests, Ruiz Cruz and Dobarganes Garcia (1977) found 89% vs 75% primary biodegradation as measured by CTAS when comparing the same compounds. The tp-C_9APE_9 was faster to degrade than tp-$C_{12}APE_9$, 90% vs 50% in 20 days, as measured by CTAS.

The biodegradability of branched APE may depend on the structure of the individual alkyl components. Giger et al. (1981) separated and identified the isomeric components of technical grade 4-NP and Marlophen 83 (C_9APE) by GC/MS. They identified eight differently branched *p*-substituted structures of the nonyl chain. The same isomers, but with slight differences in relative abundance, were identified in the extract of a secondary sewage effluent.

2. *o*-, *m*-, *p*-Substitution.

The positional attachment of the alkyl group to the phenol ring influences the rate of biodegradation. Using river-water dieaway tests, Blankenship and Piccolini (1963) found a *p*-substituted OPE to degrade faster than the corresponding *o*-substituted OPE, and both degraded faster than a mixture of *o*- and *p*-substituted compounds with secondary attachment of the octyl group.

Marei et al. (1976) synthesized linear primary APE substituted at the *o*-, *m*-, or *p*-position and compared their biodegradability. The rate of biodegradation increased with (1) decreasing EO chain length from 15 to 6.5 units, (2) decreasing alkyl chain length from C_{18} to C_{10} plus EO chain length, and (3) change of substitution from *o*- to *m*- to *p*-.

3. EO chain length.

In addition to the studies cited above, biodegradation rates for branched APE with varying EO chain lengths have been studied. In general, in both primary and ultimate biodegradation studies, rates decreased with increasing EO chain length from 4-5 to 30-50 units (Han 1967; Rudling and Solyom, 1974; Ruiz Cruz and Dobarganes Garcia, 1977; Pitter and Fuka, 1979; Fuka and Pitter, 1980). Saeger et al. (1980), however, found no significant trend with the degree of ethoxylation for the Steroxes ND, NJ, and NK (nonylphenols with 4, 9, and 10 EO units) and Steroxes DF and DJ (dodecylphenols with 6.2 and 10 EO units).

Using river-water dieaway tests, Osburn and Benedict (1966) found that regardless of alkyl chain structure, ethoxy chains greater than 10 units did not readily degrade and suppressed alkyl chain carboxylation. Primary degradation was followed for 36 days by CTAS measurements and IR spectroscopy.

The data of Ruiz Cruz and Dobarganes Garcia (1977) and Pitter and Fuka (1979) for a series of branched C_9 compounds with up to EO_{40} are the most extensive. Primary biodegradation as measured by CTAS decreased from >90% within 7 days for C_9APE_5 to 1% in 17 days for C_9APE_{40}. Above EO_{12}, biodegradation was less than 80% in 17-day tests (Ruiz Cruz and Dobarganes Garcia, 1977). In 7-day tests, ultimate biodegradation in inoculated medium as measured by COD and organic carbon likewise was less than 80% above EO_{10} (Pitter and Fuka, 1979).

D. METABOLIC PATHWAYS OF BIODEGRADATION

The major pathway of primary biodegradation for APE is stepwise oxidation of the EO chain, either by hydrolysis or an oxidative-hydrolytic mechanism. This pathway results in the presence of alkylphenol and alkylphenol mono- and diethoxylate and carboxylate metabolites that have been detected in laboratory studies and in treated wastewater effluents and sludge. Central fission of the aromatic ether bond has not been observed. In most laboratory studies, the highly branched alkyl chains appear to persist unchanged (Lashen et al., 1966; Osburn and Benedict, 1966; Cain, 1981; Swisher, 1987). In two studies (Osburn and Benedict, 1966; Schöberl et al., 1981) a carboxylated group was identified on the branched alkyl chain.

Isolation of primary biodegradation products indicates a single metabolic pathway of degradation. Three types of metabolic byproducts of tripropylene-derived NPE biodegradation have been isolated in river waters, wastewaters, and sludge: (1) NPE with a low number of EO units (NPE_{1-3}) (Giger et al., 1981; Stephanou and Giger, 1982; Reinhard et al., 1982; Brüschweiler et al., 1983; Ahel and Giger, 1985a; Marcomini and Giger, 1987; Kubeck and Naylor, 1990; Radian Corporation, 1990); (2) NP (Giger et al., 1981; Stephanou and Giger, 1982; Brüschweiler et al., 1983; Ahel et al., 1984; Giger et al., 1984; Ahel and Giger, 1985b; Schaffner et al., 1984; Marcomini and Giger, 1987; Varma et al., 1987; Radian Corporation, 1990; Wahlberg et al., 1990; Waldock and Thain, 1991) and (3) nonylphenol carboxylic acids (Ahel et al., 1987). Biotransformation tests under laboratory conditions, using river-water dieaway or activated sludge or primary sewage inoculations, have yielded the same metabolites: (1) APE_{1-3} (Rudling and Solyom, 1974; Brüschweiler et al., 1983; Ball et al., 1989); (2) OP (Ball et al., 1989); and (3) nonyl- or octylphenol carboxylic acids (Yoshimura, 1986; Ball et al., 1989). The proportional composition of octylphenol carboxylic acid homologs in a laboratory biodegradation study was similar to that in wastewater treatment plant effluent (Ball and Reinhard, 1985; Ball et al., 1989). Using the OECD confirmatory test and inoculation with airborne microorganisms, Schöberl et al. (1981) isolated degradation products of br-NPE_9, added at 20 mg/L. They identified two major hydrophobic degradation products: C_9H_{19}-(C_6H_4)-O-$(C_2H_4O)_3$-H at a concentration of 4.1 mg/L and COOH-$(CH_2)_7$-(C_6H_4)-O-$(C_2H_4O)_2$-H at a concentration of 4.5 mg/L. No polydiols were detected. In this test the carboxylated alkyl chain was shortened by one carbon atom.

Based on the fate of APE in these laboratory and field studies, particularly the summary of field studies by Giger et al. (1987) and the laboratory studies of Yoshimura (1986) and Ball et al. (1989), the pathway of primary biodegradation is indicated in Figure 3-1.

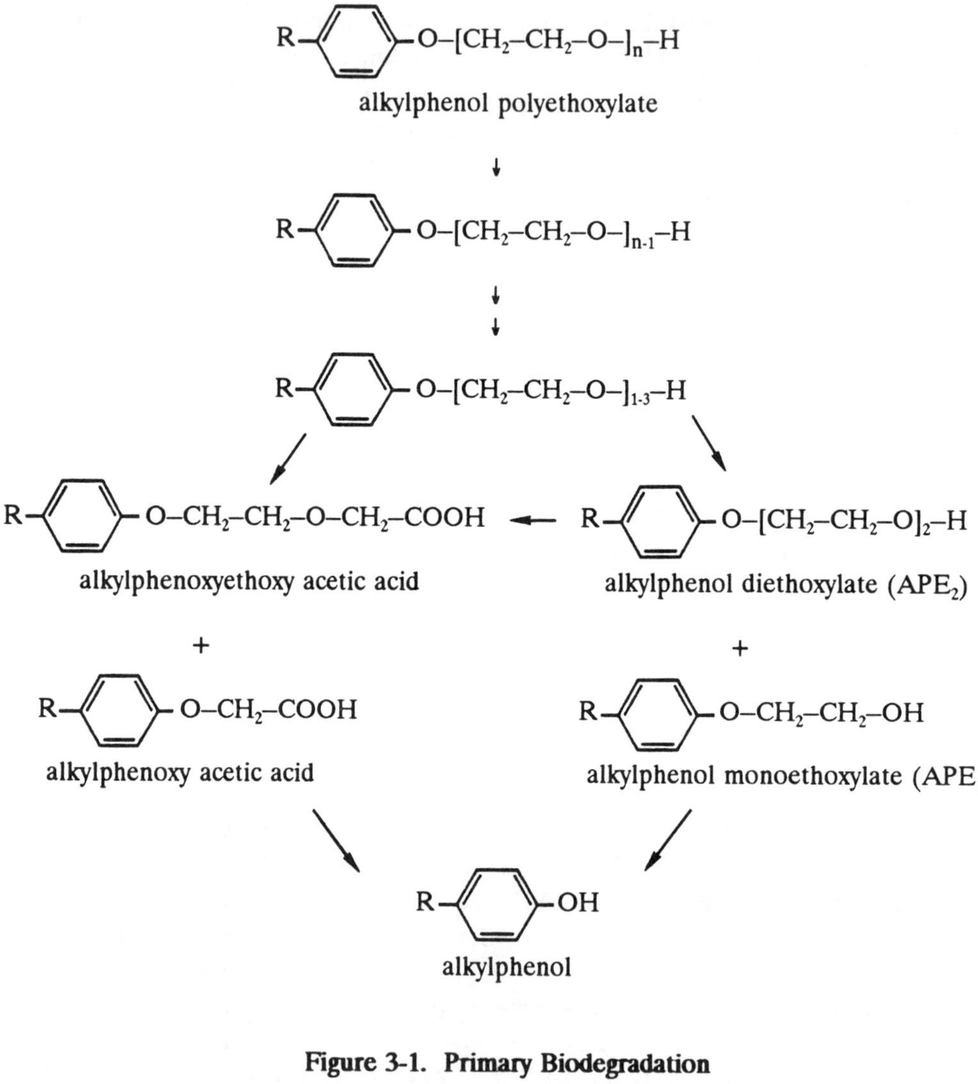

R—⟨phenyl⟩—O–[CH$_2$–CH$_2$–O–]$_n$–H
alkylphenol polyethoxylate

↓

R—⟨phenyl⟩—O–[CH$_2$–CH$_2$–O–]$_{n-1}$–H

↓
↓

R—⟨phenyl⟩—O–[CH$_2$–CH$_2$–O–]$_{1-3}$–H

R—⟨phenyl⟩—O–CH$_2$–CH$_2$–O–CH$_2$–COOH ← R—⟨phenyl⟩—O–[CH$_2$–CH$_2$–O]$_2$–H
alkylphenoxyethoxy acetic acid   alkylphenol diethoxylate (APE$_2$)

+   +

R—⟨phenyl⟩—O–CH$_2$–COOH   R—⟨phenyl⟩—O–CH$_2$–CH$_2$–OH
alkylphenoxy acetic acid   alkylphenol monoethoxylate (APE$_1$)

R—⟨phenyl⟩—OH
alkylphenol

Figure 3-1. Primary Biodegradation

Under both aerobic and anaerobic conditions, the transformation of APE$_n$ (where n=≥3) by ether cleavage was rapid. Although terminal ether cleavage may be preceded by alcohol oxidation, no acid intermediates of the higher molecular weight homologs were detected

(Ball et al., 1989). When biodegradation reached APE_2, there was accumulation of this intermediate up to day 36 in the aerobic system. At this point, terminal alcohol oxidation resulted in the accumulation of some acid intermediates (up to day 36) and only traces of alkylphenol. Under anaerobic conditions, there was near quantitative conversion to APE_1 which was gradually transformed to alkylphenol. Alkylphenol was still present after 190 days. Terminal oxidation of the alcohol group was slow, but carboxylic intermediates persisted for 190 days. When the carboxylic acids were used as the starting compounds, their transformation was slow, and in the anaerobic system, still incomplete after 120 days. GC/MS chromatograms revealed no biotransformation of the alkyl group (designated by R in the figure).

Ultimate biodegradation of the lower molecular weight APE metabolites proceeds by slow oxidation to alkylphenols followed by breakage of the phenol ring. Most of the studies did not indicate aromatic ring degradation. However, Swisher (1987) reports several mechanisms of microbial aromatic ring degradation, the most common being formation of catechol from phenol followed by ring scission between or adjacent to the two hydroxyl groups.

No information on the ultimate biodegradation of the branched alkyl chain moiety was located. Studies in which radioactive tracers were used were limited to radiolabels on the phenol moiety and the EO units. However, Swisher (1987) reports that branched fatty acid chains can be oxidized by a combination of ß- and α-oxidation.

REFERENCES

Ahel, M. and W. Giger. 1985a. Determination of alkylphenols and alkylphenol mono- and diethoxylates in environmental samples by high-performance liquid chromatography. Anal. Chem. 57:1577-1583.

Ahel, M. and W. Giger. 1985b. Determination of nonionic surfactants of the alkylphenol polyethoxylate type by high-performance liquid chromatography. Anal. Chem. 57:2584-2590.

Ahel, M., T. Conrad and W. Giger. 1987. Persistent organic chemicals in sewage effluents. 3. Determination of nonylphenoxy carboxylic acids by high-resolution gas chromatography/mass spectrometry and high-performance liquid chromatography. Environ. Sci. Technol. 21:697-703.

Ahel, M., W. Giger, and M. Koch. 1986. Behavior of nonionic surfactants in biological waste water treatment. Commun. Eur. Communities Report 414-428.

Ahel, M., W. Giger, E. Molnar-Kubica and C. Schaffner. 1984. Analysis of organic micropollutants in surface waters of the Glatt Valley, Switzerland. In: Analysis of Organic Micropollutants in Water, G. Angeletti and A. Bjorseth, eds. Reidel Publishing Company, Dordrecht, Holland.

APHA (American Public Health Association). 1989. *Standard Methods for the Examination of Water and Wastewater*, 17th ed., APHA, Washington, DC.

ASTM (American Society for Testing and Materials. 1991a. E1297 Standard Test Method for Biodegradation by a Shake-Flask Die-Away Method. *1991 Annual Book of ASTM Standards*, Philadelphia, PA, Vol. 11.04, pp. 911-915.

ASTM (American Society for Testing and Materials. 1991b. D2667 Standard Test Method for Biodegradability of Alkylbenzene Sulfonate. *1991 Annual Book of ASTM Standards*, Philadelphia, PA, Vol. 15.04, pp. 267-272.

Ball, H.A. and M. Reinhard. 1985. In: Water chlorination: chemistry, environmental impact, and health effects (Vol. 5), R.L. Jolley et al., Eds. Lewis Publishers, Chelsea MI, pp. 1505-1514.

Ball, H.A., M. Reinhard and P.L. McCarty. 1989. Biotransformation of halogenated and nonhalogenated octylphenol polyethoxylate residues under aerobic and anaerobic conditions. Environ. Sci. Technol. 23:951-961.

Barber, L.B., II, E.M. Thurman and M.P. Schroeder. 1988. Long-term fate of organic micropollutants in sewage-contaminated groundwater. Environ. Sci. Technol. 22:205-211.

Birch, R.R. 1984. Biodegradation of nonionic surfactants. J. Am. Oil. Chem. Soc. 61:340-343.

Birch, R.R. 1991a. Prediction of the fate of detergent chemicals during sewage treatment. J. Chem. Tech. Biotechnol. 50:411-422.

Birch, R.R. 1991b. Recent developments in the biodegradability testing of nonionic surfactants. Riv. Ital. Sost. Grasse 68:433-437.

Blankenship, F.A. and V.M. Piccolini. 1963. Biodegradation of nonionics. Soap Chem. Specialties. 39(12):75-78,181.

Borstlap, C. and C. Kortland. 1967. Biodegradability of nonionic surfactants under aerobic conditions. Fette-Seifen-Anstrichmittel 69:736-738.

Brown, D., H. de Henau, J.T. Garrigan, P. Gerike, M. Holt, E. Keck, E. Kunkel, E. Matthijs, J. Waters and R. J. Watkinson. 1986. Removal of nonionics in a sewage treatment plant. Removal of domestic detergent nonionic surfactants in an activated sludge sewage treatment plant. Tenside 23:190-195.

Brown, D., H. de Henau, J.T. Garrigan, P. Gerike, M. Holt, E. Kunkel, E. Matthijs, J. Waters and R.J. Watkinson. 1987. Removal of nonionics in sewage treatment plants. II. Removal of domestic detergent nonionic surfactants in a trickling filter sewage treatment plant. Tenside 24:14-19.

Brunner, P.H., S. Capri, A. Marcomini and W. Giger. 1988. Occurrence and behaviour of linear alkylbenzenesulphonates, nonylphenol, nonylphenol mono- and nonylphenol diethoxylates in sewage and sewage sludge treatment. Water Res. 22:1465-1472.

Brüschweiler, H. and H. Gämperle. 1982. Primary and ultimate biodegradation of alkylphenol ethoxylates. J. Com. Esp. Deterg. pp. 55-71.

Brüschweiler, H., H. Gämperle and F. Schwager. 1983. Primary degradation, ultimate degradation and intermediate degradation products of alkylphenol ethoxylates. Tenside 20:317-324.

Cain, R.B. 1981. Microbial degradation of surfactants and builder components. In: Microbial Degradation of Xenobiotics and Recalcitrant Compounds, T. Leisinger et al., eds., Academic Press, London, pp. 325-370.

Ellis, D.D., C.M. Jone, R.A. Larson and D.J. Schaeffer. 1982. Organic constituents of mutagenic secondary effluents from wastewater treatment plants. Arch. Environ. Contam. Toxicol. 11:373-382.

Ernst, B., G. Julien, K. Doe and R. Parker. 1980. Environmental investigations of the 1980 spruce budworm spray program in New Brunswick. EPS-5-AR-81-3. Surveillance Report, Canada Environmental Protection Service. (Cited in Etnier, 1985).

Etnier, E.L. 1985. Chemical Hazard Information Profile: Nonylphenol. Draft Report. Office of Toxic Substances, U.S. Environmental Protection Agency.

Fischer, W.K. and P. Gerike. 1984. A surfactant balance for the input to a municipal sewage treatment plant. Tenside 21:71-73.

Frazee, C.D., Q.W. Osburn and R.O. Crisler. 1964. Application of infrared spectroscopy to surfactant degradation studies. J. Am. Oil Chem. Soc. 41:808-812.

Fuka, T. and P. Pitter. 1978. Relation between molecular structure and biodegradability of organic compounds. VII. Biodegradability of nonionic sulfated surfactants. Sb. VSChT. F22:51-73.

Fuka, T. and P. Pitter. 1980. Relation between molecular structure and biodegradability of organic compounds. VII. Biodegradability of nonionic sulfated surfactants. Sb. VSChT F22:51-73.

Garrison, A.W. and D.W. Hill. 1972. Organic pollutants from mill persist in downstream waters. Am. Dyestuff Rep. 61:21-25.

Garrison, L.J. and R.D. Matson. 1964. A comparison by Warburg respirometry and die-away studies of the degradability of select nonionic surfactants. J. Am. Oil Chem. Soc. 41:799-804.

Giger, W and M. Ahel. 1991. Behaviour of nonylphenol polyethoxylates and their metabolites in mechanical-biological sewage treatment. In: Swedish EPA Seminar on Nonylphenol Ethoxylates/Nonylphenol held in Saltsjobaden, Sweden, February 6-8, 1991, pp. 87-103.

Giger, W., E. Stephanou and C. Schaffner. 1981. Persistent organic chemicals in sewage effluents: 1. Identification of nonylphenols and nonylphenolethoxylates by glass capillary gas chromatography/mass spectrometry. Chemosphere 10:1253-1263.

Giger, W., P.H. Brunner and C. Schaffner. 1984. 4-Nonylphenol in sewage sludge: Accumulation of toxic metabolites from nonionic surfactants. Science 225:623-625.

Giger, W., M. Ahel and M. Koch. 1986. Behaviour of alkylphenol polyethoxylate surfactants in mechanical-biological sewage treatment. Vom Wasser 67:69-81.

Giger, W., M. Ahel, M. Koch, H.U. Laubscher, C. Schaffner and J. Schneider. 1987. Behaviour of alkylphenol polyethoxylate surfactants and of nitrilotriacetate in sewage treatment. Wat. Sci. Tech. 19:449-460.

Gledhill, W.E. 1975. Screening test for assessment of ultimate biodegradability: linear alkyl benzene sulfonate. Appl. Microbiol. 30:922-929.

Han, K.W. 1967. Determination of biodegradability of nonionic surfactants by sulfation and methylene blue extraction. Tenside 4:43-45.

Hites, R.A., G.A. Jungclaus, V. Lopez-Avila and L.S.Sheldon. 1979. Potentially toxic organic compounds in industrial wastewaters and river systems: two case studies. In: Monitoring toxic substances. D. Scheutzle (ed.), American Chemical Society, Washington, DC, pp. 63-90.

Ho, C.F. and D. Jenkins. 1991. The effect of surfactants on *Nocardia* foaming in activated sludge. Water Sci. Tech. 23:879-887.

Holt, M.S., E.H. McKerrell, J. Perry and R.J. Watkinson. 1986. Determination of alkylphenol ethoxylates in environmental samples by high-performance liquid chromatography coupled to fluorescence detection. J. Chromatog. 362:419-424.

Huddleston, R.L. and R.C. Allred. 1964. Effect of structure on biodegradation of nonionic surfactants. Proceedings of the Fourth International Congress on Surface-Active Substances held in Brussels, Belgium. Gordon and Breach, Science Publishers, New York, 3:871-882.

Hughes, A.I., D.R. Peterson and R.K. Markarian. 1989. Comparative biodegradability of linear and branched alcohol ethoxylates. Presented at the American Oil Chemists' Society Annual Meeting, May 3-7, Cincinnati, OH.

Jungclaus, G.A., V. Lopez-Avila and R.A. Hites. 1978. Organic compounds in an industrial wastewater: a case study of their environmental impact. Environ. Sci. Technol. 12:88-96.

Kingsbury, P.D., B.B. McLeod and R.L. Millikin. 1981. The environmental impact of nonylphenol and the Matacil^R formulation. Part 2. Terrestrial ecosystems. Report FPM-X-36, Forest Pest Management Institute, Sault Ste. Marie, Ontario, Canada.

Kravetz, L. H., Chung, K.F. Guin, W.T. Shebs, L.S. Smith and H. Stupel. 1982. Ultimate biodegradation of an alcohol ethoxylate and a nonylphenol ethoxylate under realistic conditions. Soap Cosmet. Chem. Spec. 58:34-42, 102B.

Kravetz, L., H. Chung, K.F. Guin, W.T. Shebs and L.S. Smith. 1984. Primary and ultimate biodegradation of an alcohol ethoxylate and an alkylphenol ethoxylate under average winter conditions in the USA. Shell Chemical Company Technical Bulletin SC:779-83; Tenside 21:1-6.

Kravetz, L., J.P. Salanitro, P.B. Dorn and K.F. Guin. 1991. Influence of hydrophobe type and extent of branching on environmental response factors of nonionic surfactants. J. Am. Oil Chem. Soc. 68:610-618.

Kubeck, E. and C.G. Naylor. 1990. Trace analysis of alkylphenol ethoxylates. J. Am. Oil Chem. Soc. 67:400-405.

Lashen, E.S., F.A. Blankenship, K.A. Booman and J. Dupré. 1966. Biodegradation studies on a p-tert-octylphenoxypolyethoxyethanol. J. Am. Oil Chem. Soc. 43:371-376.

Lashen, E.S. and K.A. Booman. 1967. Biodegradability and treatability of alkylphenol ethoxylates: A class of nonionic surfactants. Water Sew. Works 114:R155-R163.

Lashen, E.S. and J.C. Lamb. 1967. III. Biodegradation of a nonionic detergent. Water Wastes Eng. 4:55-59.

Lin, D.C.K., R.G. Melton, F.C. Kopfler and S.V. Lucas. 1981. Glass capillary gas chromatographic/mass spectrometric analysis of organic concentrates from drinking and advanced waste treatment waters. In: L.H. Keith, ed., Advances in the Identification and Analysis of Organic Pollutants in Water, Vol. 2, Ann Arbor Science Publications, Ann Arbor, MI, pp. 891, 895. (Cited in Etnier, 1985).

Mann, A.H. and V.W. Reid. 1971. Biodegradation of synthetic detergents: evaluation by community trials. Part 2. Alcohol and alkylphenol ethoxylates. J. Am. Oil Chem. Soc. 48:794-797.

Marcomini, A. and W. Giger. 1987. Simultaneous determination of linear alkylbenzenesulfonates, alkylphenol polyethoxylates, and nonylphenol by high-performance liquid chromatography. Anal. Chem. 59:1709-1715.

Marcomini, A., S. Capri, P.H. Brunner and W. Giger. 1988. Mass fluxes of linear alkylbenzenesulphonates, nonylphenol, nonylphenol mono- and diethoxylate through a sewage treatment plant. Commun. Eur. Comm. Rep. 11350:266-277.

Marcomini, A., P.D. Capel, T. Lichtensteiger, P.H. Brunner and W. Giger. 1989. Behavior of aromatic surfactants and PCBs in sludge-treated soil and landfills. J. Environ. Qual. 18:523-528.

Marcomini, A., F. Cecchi and A. Sfriso. 1991. Analytical extraction and environmental removal of alkylbenzene sulphonates, nonylphenol and nonylphenol monoethoxylate from dated sludge-only landfills. Environ. Toxicol. 12:1047-1054.

Marei, A., T.M. Kassem and B.A. Gebril. 1976. Alkylphenol ethoxylate surfactants. Indian J. Technol. 14:447-452.

Narkis, N. and B. Ben-David. 1983. Evaluation of non-ionic surfactants removal by adsorption. In: Dev. Ecol. Environ. Qual, Proc. Int. Meet. ISR:431-442.

Narkis, N., B. Ben-David and M. Schneider-Rotel. 1987. Non-ionic surfactants: interactions with ozone. Tenside 24:200-205.

Naylor, C.G., J.P. Mieure, I. Morici, R.R. Romano. 1992. Alkylphenol ethoxylates in the environment. Presented at the Third CESIO International Surfactants Congress held in London, U.K., June 1992.

Neufahrt, A. K. Hoffman, G. Täuber and Z. Damo. 1987. Biodegradation of nonylphenol ethoxylate (NPE) and certain environmental effects of its catabolites. Unpublished paper.

Osburn, Q.W. and J.H. Benedict. 1966. Polyethoxylated alkyl phenols: Relationship of structure to biodegradation mechanism. J. Am. Oil Chem. Soc. 43:141-146.

Patoczka, J. and G.W. Pulliam. 1990. Biodegradation and secondary effluent toxicity of ethoxylated surfactants. Water Res. 24:965-972.

Pellzzetti, E., C. Minero, V. Maurino, A. Sciafani, H. Hidaka and N. Serpone. 1989. Photocatalytic degradation of nonylphenol ethoxylated surfactants. Environ. Sci. Technol. 23:1380-1385.

Pitter, P. and T. Fuka. 1979. Biodegradation of non-sulfated and sulfated nonylphenol ethoxylate surfactants. Env. Protect. Eng. 5:47-56.

Radian Corporation. 1989. Environmental sampling and analysis for nonylphenol ethoxylate species in the Alabama and Saginaw Rivers. Report prepared for the Alkylphenol and Ethoxylates Panel of the Chemical Manufacturers Association by Radian Corporation, Austin, TX.

Radian Corporation. 1990. Nonylphenol and nonylphenol ethoxylates in river water and bottom sediments. Final Report. Prepared for The Alkylphenol and Ethoxylates Panel of the Chemical Manufacturers Association by Radian Corporation, Austin, TX.

Reinhard, M., W. Goodman and K.E. Mortelmans. 1982. Occurrence of brominated alkylphenol polyethoxy carboxylates in mutagenic wastewater concentrates. Environ. Sci. Technol. 16:351-363.

Reinhard, M., N.L. Goodman and J.F. Barker. 1984. Occurrence and distribution of organic chemicals in two landfill leachate plumes. Environ. Sci. Technol. 18:953-961.

Rudling, L. and P. Solyom. 1974. The investigation of biodegradability of branched nonyl phenol ethoxylates. Water Res. 8:114-119.

Ruiz Cruz, J. and M.C. Dobarganes Garcia. 1977. Pollution of natural waters by synthetic detergents. XII. Relation between structure and biodegradation of nonionic surfactants in river water. Grasas Aceites 28:325-331.

Saeger, V.W., R.G. Kuehnel, C. Linck and W.E. Gledhill. 1980. Biodegradation of Sterox and Dimersol olefin-derived alkylphenol ethoxylates. Monsanto Industrial Chemical Company Report No. ES-80-SS-46, 6 pp.

Salanitro, , J.P., G.C. Langston, P.B. Dorn and L. Kravetz. 1988. Activated sludge treatment of ethoxylate surfactants at high industrial use concentrations. Presented at the International Conference on Water and Wastewater Microbiology, Newport Beach, California, February 8-11.

Sato, M., K. Hashimoto and M. Kobayashi. 1963. Microbial degradation of nonionic synthetic detergent, poly EO nonylphenol ethers. Water Treat. Eng. 4:31-36.

Schaffner, C., E. Stephanou and W. Giger. 1982. Determination of nonylphenols and nonylphenolethoxylates in secondary sewage effluents. Comm. Eur. Communities Rep.

Schaffner, C., P.H. Brunner and W. Giger. 1984. 4-Nonylphenol, a highly concentrated degradation product on nonionic surfactants in sewage sludge. In: Proceedings of the Third International Symposium on Processing and Use of Sewage Sludge. Reidel Publishing Company, Dordrecht, Holland.

Schick, M.J. (Ed.). 1967. *Nonionic Surfactants*, Volume 1, Surfactant Science Series. Marcel Dekker, Inc., NY.

Schöberl, P., E. Kunkel and K. Espeter. 1981. Comparative investigations on the microbial metabolism of a nonylphenol and an oxoalcohol ethoxylate. Tenside 18:64-72.

SDA (Soap and Detergent Association). 1965. A procedure and standards for the determination of the biodegradability of alkyl benzene sulfonate and linear alkylate sulfonate. J. Am. Oil Chem. Soc. 42:986-933.

Shackelford, W.M., D.M. Cline, L. Faas and G. Kurth. 1983. Evaluation of automated spectrum matching for survey identification of wastewater components by gas chromatography-mass spectrometry. PB83-182931. U.S. Environmental Research Laboratory. Anal Chimn Acta 146:15-27. (Cited in Etnier, 1985).

Smithson, L.H. 1966. Properties of ethoxylate derivatives of nonrandom alkylphenols. J. Am. Oil Chem. Soc. 43:568-571.

Stead, J.B., A.T. Pugh, I.I. Kaduji and R.A. Morland. 1972. A comparison of biodegradability of some alkylphenol ethoxylates using three methods of detection. Proceedings of the Sixth International Congress on Surface-Active Substances, Zurich, Switzerland. Carl Hanser Verlag, Munich, Germany, 3:721-734.

Steber, J. and P. Wierich. 1987. The anaerobic degradation of detergent range fatty alcohol ethoxylates. Studies with ^{14}C-labelled model surfactants. Water Res. 21:661-667.

Stephanou, E. and W. Giger. 1982. Persistent organic chemicals in sewage effluents. II. Quantitative determinations of nonylphenols and nonylphenol ethoxylates by glass capillary gas chromatography. Env. Sci. Technol. 16:800-805.

Sturm, R.N. 1973. Biodegradability of nonionic surfactants: screening test for predicting rate and ultimate biodegradation. J. Am. Oil Chem. Soc. 50: 159-167.

Sundaram, K.M.S. and S. Szeto. 1981. The dissipation of nonylphenol in stream and pond water under simulated field conditions. J. Environ. Sci. Health B16:767-776.

Sundaram, K.M.S., S. Szeto, R. Hindle and D. MacTavish. 1980. Residues of nonylphenol in spruce foliage, forest soil, stream water and sediment after its aerial application. J. Environ. Sci. Health B15:403-419.

Swisher, R.D. 1987. *Surfactant Biodegradation*, 2nd. Ed. Surfactant Science Series, Vol. 18. Marcel Dekker, Inc., New York. pp. 47-146.

Tanaka, F.S., R.G. Wien and R.G. Zaylskie. 1991. Photolytic degradation of a homogeneous Triton X nonionic surfactant: nonaethoxylated *p*-(1,1,3,3-tetramethylbutyl)phenol. J. Agric. Food Chem. 39:2046-2052.

Urano, K. and M. Saito. 1984. Adsorption of surfactants on microbiologies. Chemosphere 13:285-292.

Varma, M.M., D. Patel, L. Wan, J.H. Johnson, and J.N. Cannon. 1987. Measurement of 4-nonylphenol in water and wastewater effluents. PB89-171086/WEP, NTIS, Springfield Virginia.

Wagener, S. and B. Schink. 1987. Anaerobic degradation of nonionic and anionic surfactants in enrichment cultures and fixed-bed reactors. Water Res. 21:615-622.

Wahlberg, C., L. Renberg and U. Wideqvist. 1990. Determination of nonylphenol and nonylphenol ethoxylates as their pentafluorobenzoates in water, sewage sludge and biota. Chemosphere 20:179-195.

Waldock, M.J. and J.E. Thain. 1991. Environmental concentrations of 4-nonylphenol following dumping of anaerobically digested sewage sludges: A preliminary study of occurrence and acute toxicity. Unpublished paper, Ministry of Agriculture, Fisheries and Food, Fisheries Laboratory, Burnham-on-Crouch, Essex, Britain.

Waters, J., T.J. Garrigan and A.M. Paulson. 1986. Investigations into the scope and limitations of the bismuth active substances procedure (Wickbold) for the determination of nonionic surfactants in environmental samples. Water Res. 20:247-253.

Weinberg, H. and N. Narkis. 1987. Physico-chemical treatments for the complete removal of non-ionic surfactants from effluents. Environ. Poll. 45:245-260.

Yoshimura, K. 1986. Biodegradation and fish toxicity of nonionic surfactants. J. Am. Oil Chem. Soc. 63:1590-1596.

Zoller, U. 1985. The "hard" and "soft" surfactant profile of Israel municipal wastewaters. J. Am. Oil Chem. Soc. 62:1006-1009.

IV. ENVIRONMENTAL TOXICOLOGY

A. AQUATIC TOXICITY

The U.S. EPA prescribes test guidelines for chemicals subject to environmental effects test regulations under the Toxic Substances Control Act (TSCA) (Fed. Reg. 50:39321-39374). The tests include (1) acute and chronic toxicity tests using freshwater algae, aquatic plants, daphnids, fish, oysters, and shrimp and (2) fish bioconcentration tests. EPA-designated species were used in most of the following studies.

The toxicity of APE and some of their metabolites have been evaluated for a variety of species under a wide range of conditions. The results suggest that concentrations of APE and their metabolites, such as NP, in U.S. river waters and sediments do not approach acutely toxic (LC_{50}) levels. Sublethal effects for sensitive aquatic organisms may occur at concentrations as low as 0.5 mg/L of APE and 0.0067 mg/L of NP.

1. Acute Toxicity to Fish

Lethality data for freshwater and marine fish are summarized in Table 4-1. For APE, 48-hr and 96-hr LC_{50} values ranged from 1.3 mg/L for C_9APE_4 to >1000 mg/L for C_9APE_{30} (Macek and Krzeminski, 1975), with most toxicity values for commercially available APE between 4 and 14 mg/L. When metabolites of APE such as NP were tested, they were found to be more toxic than the parent compounds. For all species tested, the 48-hr and 96-hr LC_{50} values of NP were ≤ 3.0 mg/L and ranged from 0.13 mg/L (average measured concentration) for juvenile Atlantic salmon (*Salmo salar*) in flow-through tests (McLeese et al., 1980b) to 3.0 mg/L for the saltwater cod (*Gadus morhua*) (test conditions not given) (Swedmark, 1968). The LC_{50} values for other alkylphenols ranged from 0.19 mg/L for *p*-hexylphenol (McLeese et al., 1980b) to 5.14 mg/L for *p*-tert-butylphenol (Holcombe et al., 1984). When tested on the killifish (*Oryzias latipe*), the toxicities of the nonylphenol ethoxycarboxylic acids, with LC_{50} values of 8.9 and 9.6 mg/L, were similar to that of many APE (Yoshimura, 1986).

In general, toxicity decreased with increasing EO chain length. For a series of Igepal surfactants, toxicity to bluegill sunfish (*Lepomis macrochirus*) decreased from 2.4-2.8 mg/L for C_9APE_5 to >1000 mg/L for C_9APE_{30} (Macek and Krzeminski, 1975). Toxicities were comparable for similar molecular weight products acquired from different manufacturing sources. The same trend was present for an octylphenol series of ethoxylates. This trend was most obvious for the data of Yoshimura (1986) who tested a series of nine C_9APE, including synthesized biodegradation intermediates, on the Japanese killifish. Toxicity decreased from 1.4 mg/L for NP (NPE_0) to 110 mg/L for $NPE_{16.6}$. The author found that the source of the water had a small effect on the test results, with toxicity slightly reduced in river water compared to laboratory water.

Table 4-1. Acute Toxicity of Alkylphenol Ethoxylates and Alkylphenols to Fish

Species/ Common Name	Surfactant/ Trade Name	LC_{50} (mg/L)	Test Duration	Reference
Agonus cataphractus hook-nose	nonylphenol	0.51	96-hr	Waldock and Thain, 1991
Carassius auratus goldfish	C_9APE_{10}	13.8	48-hr	Unilever Research Laboratories, 1977
Carassius auratus goldfish	C_9AE_9	18	48-hr	Tomiyama, 1974
Carassius auratus goldfish	C_9APE_{10}	5.4	48-hr	Kurata et al., 1977
Carassius auratus goldfish	C_9APE_{9-10}	6.9	6-hr	Reiff et al., 1979
Cyprinodon variegatus sheepshead minnow	br-4-nonylphenol	0.31^a	96-hr	CMA, 1990a
Gadus morhua cod	nonylphenol	3.0	96-hr	Swedmark, 1968
Gadus morhua cod	C_9APE_{10}	2.5-6.8	96-hr	Swedmark et al., 1971; Swedmark et al., 1976
Gasterosteus aculeatus stickleback	4-nonylphenol	0.4	96-hr	Granmo et al., 1991
Lebistes reticulatus guppy	C_9APE_n (Lissapol NXA) C_nAPE_n (Hostapal)	$>7^b$ $>4-6^b$	— —	Madai and An der Lan, 1964

Table 4-1. (cont.)

Species/ Common Name	Surfactant/ Trade Name	LC_{50} (mg/L)	Test Duration	Reference
Lebistes reticulatus guppy	br-C_9AE_{11}	52-64[c]	24-hr	Van Emden et al., 1974
Lepomis macrochirus bluegill sunfish	C_9APE_4 (Surfonic N-40)	1.3	96-hr	Macek and Krzeminski, 1975
	$C_9APE_{9.5}$ (Surfonic N-95)	7.6	96-hr	
	C_9APE_5 (Igepal CO-520)	2.4-2.8	96-hr	
	C_9APE_9 (Igepal CO-630)	7.9	96-hr	
	C_9APE_{30} (Igepal CO-880)	>1000	96-hr	
	C_8APE_5 (Triton X-45)	2.8-3.2	96-hr	
	C_8APE_{10} (Triton X-100)	12.0	96-hr	
	C_8APE_{30} (Triton X-305)	531	96-hr	
Leuciscus idus ide or golden orfe	C_9APE_{10}	7.4, 9.5[d]	48-hr	Unilever Research Laboratories, 1977
Leuciscus idus ide or golden orfe	C_9APE_9	7	48-hr	Fischer, 1973
Leuciscus idus ide or golden orfe	C_nAPE_n	3.7->10	48-hr	Hamburger et al., 1977
Leuciscus idus ide or golden orfe	C_9APE_{9-10}	4.9, 7.0, 11.2	48-hr	Reiff et al., 1979

Table 4-1. (cont.)

Species/ Common Name	Surfactant/ Trade Name	LC_{50} (mg/L)	Test Duration	Reference
Oryzias latipse killifish	nonylphenol[e]	1.4	48-hr	Yoshimura, 1986
	C_9APE_1	3.0	48-hr	
	$C_9APE_{3.3}$	2.5	48-hr	
	C_9APE_5	3.6	48-hr	
	$C_9APE_{6.4}$	5.4	48-hr	
	$C_9APE_{8.4}$	11.6	48-hr	
	$C_9APE_{8.9}$	11.2–14	48-hr	
	C_9APE_{13}	48	48-hr	
	$C_9APE_{16.6}$	110	48-hr	
	C_9-C_6H_n-$OCH_2CH_2OCH_2COONa$	8.9	48-hr	
	C_9-C_6H_n-OCH_2COONa	9.6	48-hr	
Phoxinus phoxinus minnow	C_9APE_{10}	8.8	48-hr	Unilever Research Laboratories, 1977
Phoxinus laevis minnow	APE	65	48-hr	Hamburger et al. 1977
Pimephales promelas fathead minnow	nonylphenol[f]	0.135[a]	96-hr	Holcombe et al., 1984
	4-tert-butylphenol	5.14[a]	96-hr	
	4-tert-pentylphenol	2.50[a]	96-hr	
Pimephales promelas fathead minnow	C_9APE_9	1.6	96-hr	Shell Chemical Company, 1987; Salanitro et al., 1988
Pimephales promelas fathead minnow	C_9APE_7	3.2	96-hr	Markarian et al., 1989
Pimephales promelas fathead minnow	br-C_9APE_9	4.6[a]	96-hr	Kravetz et al., 1991 Dorn et al., 1993

Table 4-1. (cont.)

Species/ Common Name	Surfactant/ Trade Name	LC_{50} (mg/L)	Test Duration	Reference
Pimephales promelas fathead minnow	nonylphenol	0.3	96-hr	Monsanto Chemical Co., 1985a
Pleuronectes flesus flounder	C_9APE_{10}	3.0	96-hr	Swedmark et al., 1971
Rasbora heteromorpha harlequin fish	C_9APE_{10}	11.3	48-hr	Unilever Research Laboratories, 1977
Rasbora heteromorpha harlequin fish	C_9APE_{9-10}	8.6	96-hr	Reiff et al., 1979
Salmo gairdneri[g] rainbow trout	$C_9APE_{9.5}$ $C_9APE_{9.5}$ (biodegradation products) C_9APE_{10}	7.5-12.5 4.5-5.9 7.5-11.7[h]	96-hr 96-hr 96-hr	Unilever Research Laboratories, 1977
Salmo gairdneri rainbow trout	C_9APE_{10}	2.5-62[i]	3-hr	Marchetti, 1965
Salmo gairdneri rainbow trout	C_9APE_8	4.7	96-hr	Calamari and Marchetti, 1973
Salmo gairdneri rainbow trout	C_nAPE_n	2-6.3	48-hr	Hamburger et al., 1977
Salmo gairdneri rainbow trout [fingerlings]	nonylphenol	0.230	96-hr	Armstrong and Kingsbury, 1979
Salmo gairdneri rainbow trout [embryos; juveniles]	nonylphenol	0.48 0.6-0.9	24-hr 96-hr	Ernst et al., 1980

Table 4-1. (cont.)

Species/ Common Name	Surfactant/ Trade Name	LC_{50} (mg/L)	Test Duration	Reference
Salmo salar[j] Atlantic salmon	p-sec-butylphenol p-hexylphenol p-nonylphenol	0.74[a] 0.19[a] 0.13-0.19[a]	96-hr 96-hr 96-hr	McLeese et al, 1980b; 1981
Salmo trutta brown trout	C_9APE_{10}	2.7	48-hr	Unilever Research Laboratories, 1977
Salmo trutta brown trout	C_9APE_{9-10}	1.0	96-hr	Reiff et al, 1979
Salvelinus fontinalis brook trout	nonylphenol	0.145	96-hr	Armstrong and Kingsbury, 1979

[a] Measured concentration.
[b] Deleterious effect; no other information provided.
[c] LC_{100}.
[d] Variations in water hardness.
[e] Tween 80 used as dispersant.
[f] Composition: 91% 4-nonylphenol, 4% 2-nonylphenol, 5% 2,4-dinonylphenol.
[g] Renamed *Oncorhynchus mykiss*.
[h] Variation in temperature (10 and 15°C), water hardness (20-300 mg/L CaCO$_3$), and fish size (3-10 cm).
[i] Results varied with age which ranged from newly hatched alvin (62 mg/L) to 210-day old fingerlings (8.0 mg/L).
[j] Tested at 10°C.
A dash (—) indicates no data.

Age of the test species and experimental conditions affect toxicity. Tolerance to C_9APE_{10} varied among alevin, fry, and fingerling rainbow trout (*Oncorhynchus mykiss*) (Marchetti, 1965). Alevins with completely absorbed yolk sacs were more sensitive than younger alevins; tolerance increased in the fry and fingerlings. In another study, young guppies were slightly more sensitive than adults (Van Emden et al., 1974). Eggs and larvae of cod were more sensitive to NPE_{10} than adults (Swedmark et al., 1971). Survival rates and percentage of normal development for cod decreased compared to controls at >0.2 mg/L.

When toxicity tests were conducted over a period of time under static conditions, toxicity decreased slightly with increasing time, presumably due to the gradual biodegradation of the surfactant. The first 24 hours was the period of greatest toxicity; frequent changing of the test solutions during 96-hour static tests increased the toxicity of APE (Calamari and Marchetti, 1973; Arthur D. Little, 1977; Unilever Research Laboratories, 1977). Kurata et al. (1977) exposed goldfish (*Carassius auratus*) to C_9APE_{10} which had been maintained in river water for four days, and determined an LC_{50} of 3.7 mg/L. The LC_{50} for the intact surfactant (starting at day 0) was 5.4 mg/L, indicating little biodegradation during the four-day period. McLeese et al. (1980b) measured the concentration of NP during 48-hr toxicity tests with salmon. By the end of the test, the concentration had decreased by 30%.

To determine whether toxic effects persist after primary biodegradation, several investigators have coupled biodegradation tests to aquatic toxicity tests. The bioassays were carried out by exposing the organisms to the biodegrading medium or effluent and comparing survival times to those in known concentrations of the starting materials. In a laboratory river-water dieaway test, rainbow trout were exposed to biodegrading C_8APE_9 at an initial concentration of 20 mg/L (Reiff, 1976). Ten to eleven weeks were required for the surfactant to be rendered nontoxic to the trout, at which time the concentration as indicated by BIAS was less than the LC_{50}.

Yoshimura (1986) tested the toxicity to Japanese killifish (*Oryzias latipse*) of biodegrading C_9APE_9 (initial concentration 20 mg/L) in a river-water dieaway system. Although no peaks of C_9APE_9 or its biodegradation intermediates were present on HPLC chromatograms after 10 days, 50% of the tested fish died at this time (96-hr test). The author speculates that carboxylate metabolites may be responsible for the toxicity remaining after the APE disappeared. All fish survived when tested after 14 days of biodegradation.

In a recent study, fathead minnows (*Pimephales promelas*) were subjected to effluent from activated sludge biotreater units degrading 50 mg/L br-C_9APE_9 (1.7 methyl branches/hydrophobe) at 25°C. Acute toxicity was still present (20-40% of the LC_{50}), indicating the presence of the intact surfactant or its metabolites (Kravetz et al., 1991; Dorn et al., 1993).

Symptoms of fish exposed to lethal concentrations of APE surfactants were not described. Behavioral and physiological changes noted in fathead minnows exposed to the metabolites, NP, 4-tert-butylphenol, and 4-tert-pentylphenol, during 96-hr acute toxicity tests included narcosis, loss of equilibrium (4-tert-pentylphenol only) and spinal deformities. In addition to these effects, fish exposed to NP suffered hemorrhaged body areas and their bodies were swollen with fluid (Holcombe et al., 1984).

Differences between nontoxic and acutely toxic concentrations appear to be very small, i.e. lethality curves are very steep, though data are limited. For C_9APE_9, Fischer (1973) cited 48-hr LC_0, LC_{50}, and LC_{100} values of 5, 7, and 10 mg/L, respectively, for the ide or golden orfe (*Leuciscus idus*). The 14-day LC_{50} of C_9APE_8 for rainbow trout (*Salmo gairdneri*) of 4.3 mg/L was close to the 96-hr LC_{50}, 4.7 mg/L (Calamari and Marchetti, 1973).

2. Acute Toxicity to Aquatic Invertebrates

Toxicity data on aquatic invertebrates are summarized in Table 4-2. It can be observed that LC_{50} values for APE vary widely among species and that NP is more toxic to all tested species. For NPE with 7, 9, or 10 EO units, LC_{50} values for *Daphnia magna/pulex* ranged from 2.9 mg/L in a 48-hr study (Salanitro et al., 1988) to 44.2 mg/L in a 24-hr study (Unilever Research Laboratories, 1977). Nonylphenol was much more toxic to this species, with an LC_{50} of <0.5 mg/L. Other crustaceans and bivalves exhibited similar sensitivities to APE, with 96-hr LC_{50} values of 2.9 to >100 mg/L and <5 to >100 mg/L, respectively.

As noted in the previous section on fish, toxicity was related to chemical structure. Using *Daphnia* sp., Janicke et al. (1969) tested a series of technical products containing mixtures with different EO chain lengths. The "lethal threshold" decreased with increasing EO chain length from 4 to 30 units. In addition, Hall et al. (1989) found that EO chain length was the best predictor of toxicity to mysid shrimp, *Mysidopsis bahia*. Of several aromatic or aliphatic and linear or branched APE, low solubility surfactants with low EO molar ratios were the most toxic surfactants tested.

Maxwell and Piper (1968) studied the capacity of several series of surfactants with varying alkyl chain structures and EO chain lengths to prevent emergence of mosquito (*Aedes aegypti*) pupae. LC_{50} values ranged from 1 to >400 mg/L. Dinonylphenol and tridecylphenol ethoxylates were the most toxic, with values of 1-5 mg/L, regardless of the length of the EO chain.

For a series of alkylphenols with the alkyl chain ranging in carbon number from 4 to 12, toxicity to shrimp (*Crangon septemspinosa*) increased with increasing molecular weight which also correlated with increasing log K_{ow} (McLeese et al., 1981).

Early developmental stages of aquatic invertebrates were less tolerant than adults. Eggs and larvae of mussels (*Mytilus edulis*) were exposed to NPE_{10} by Swedmark et al. (1971). While mortality of adults held at 5 mg/L for 16 days was less than 50%, at 2 mg/L embryos did not develop beyond the blastula stage and at 1 mg/L they did not develop beyond the veliger larval stage. Likewise, larvae of both the spider crab (*Hyas araneus*) and the barnacle (*Balanus balanoides*) were also more sensitive than adults (Swedmark et al., 1971). *Mysidopsis bahia* that were less than one day old were extremely sensitive to br-4-NP, with a 96-hr LC_{50} of 43 μg/L (CMA, 1990b). The no-observed-effect concentration (NOEC) was 18 μg/L.

Following biodegradation of NPE_9 in an activated sludge reactor, the effluent toxicity for mysid shrimp (*Mysidopsis bahia*) was 17% that of the 48-hr LC_{50} of the intact surfactant (Patoczka and Pulliam, 1990).

Table 4-2. Acute Toxicity of Alkylphenol Ethoxylates and Alkylphenols to Aquatic Invertebrates

Species/Common Name	Surfactant/Trade Name	LC_{50} (mg/L)	Test Duration	Reference
Arthropoda				
Daphnia magna water flea	C_9APE_{10}	44.2	24-hr	Unilever Research Laboratories, 1977
Daphnia magna water flea	4-nonylphenol	0.18[a]	24-hr	Bringmann and Kühn, 1982
Daphnia magna water flea	nonylphenol	0.44	48-hr	Monsanto, 1985b
Daphnia magna water flea	C_9APE_7	4.1	96-hr	Markarian et al., 1989
Daphnia magna water flea	br-C_9APE_9	14.0[b]	48-hr	Kravetz et al., 1991; Dorn et al., 1993
Daphnia magna water flea	nonylphenol	0.19[a,b]	48-hr	Comber et al., 1993
Daphnia pulex water flea	nonylphenol	0.14-0.19	48-hr	Ernst et al., 1980
Daphnia pulex water flea	C_9APE_{10}	12.5	48-hr	Burlington Research, Inc., 1985; Moore et al., 1987
Daphnia pulex water flea	br-C_9APE_9	2.9	48-hr	Shell Chemical Company, 1987; Salanitro et al., 1988

Table 4-2. (cont.)

Species/ Common Name	Surfactant/ Trade Name	LC_{50} (mg/L)	Test Duration	Reference
Daphnia sp. water flea	C_9APE_4 C_9APE_6 C_9APE_7 C_9APE_{10} C_9APE_{20} C_9APE_{30}	5^c 5^c 10^c 10^c 1000^c 10000^c	— — — — — —	Janicke et al., 1969
Ceriodaphnia dubia cladoceran	$NPE_1 + NPE_2$ nonylphenol	1.04 0.47	48-hr 48-hr	Ankley et al., 1990
Gammarus pulex amphipod	C_9APE_n (Lissapol NXA)	>1-2	—	Madai and An der Lan, 1964
Nitocra spinipes crustacean	nonylphenol	0.118; 0.139	—	Bergström, 1984
Leander adspersus decapod	C_9APE_{10}	>100 (6-8°C) 10-50 (15-17°C)	96-hr 96-hr	Swedmark et al., 1971
Leander squilla decapod	C_9APE_{10}	>100	96-hr	Swedmark et al., 1971
Mysidopsis bahia mysid shrimp	4-t-octylphenol	47-113^d	96-hr	Cripe et al., 1989
Mysidopsis bahia mysid shrimp (<24 hrs old)	br-4-nonylphenol	0.043^b	96-hr	CMA, 1990b
Mysidopsis bahia mysid shrimp	br-C_9APE_9	1.23	48-hr	Patoczka and Pulliam, 1990

Table 4-2. (cont.)

Species/ Common Name	Surfactant/ Trade Name	LC$_{50}$ (mg/L)	Test Duration	Reference
Mysidopsis bahia mysid shrimp	n-C$_9$APE$_{1.5}$	1.66-3.34	48-hr	Hall et al., 1989
	n-C$_9$APE$_9$	1.23-1.89	48-hr	
	n-C$_9$APE$_{50}$	4148	48-hr	
	tp-NPE$_{1.5}$	0.11	48-hr	
	tp-NPE$_9$	0.71-2.2	48-hr	
	tp-NPE$_{15}$	2.57	48-hr	
	tp-NPE$_{40}$	100	48-hr	
	tp-NPE$_{50}$	4110	48-hr	
	OPE$_{1.5}$	6.51-7.07	48-hr	
	OPE$_5$	1.83	48-hr	
Crangon crangon brown shrimp	C$_9$APE$_{12}$	89.5	48-hr	Portmann and Wilson, 1971
Crangon crangon brown shrimp	4-nonylphenol	0.6	96-hr	Granmo et al., 1991
Crangon crangon brown shrimp	nonylphenol	0.42	96-hr	Waldock and Thain, 1991
Crangon septemspinosa sand shrimp	o-sec-butylphenol	1.3	96-hr	McLeese et al., 1980b; 1981
	p-sec-butylphenol	1.8	96-hr	
	o-tert-butylphenol	2.4	96-hr	
	m-tert-butylphenol	5.2	96-hr	
	p-tert-pentylphenol	1.7	96-hr	
	p-hexylphenol	0.9	96-hr	
	p-heptylphenol	0.6	96-hr	
	p-tert-octylphenol	1.1	96-hr	
	p-nonylphenol	0.3;0.4	96-hr	
	p-dodecylphenol	0.15	96-hr	

Table 4-2. (cont.)

Species/Common Name	Surfactant/Trade Name	LC_{50} (mg/L)	Test Duration	Reference
Pandalus montagui pink shrimp	C_9APE_{12}	19.3	48-hr	Portmann and Wilson, 1971
Balanus balanoides barnacle (adults; nauplii)	C_9APE_{10}	<25; 1.5	96-hr	Swedmark et al., 1971
Carcinus maenas shore crab	C_9APE_{10}	>100	96-hr	Swedmark et al., 1971
Carcinus maenas shore crab	C_9APE_{12}	>100	48-hr	Portmann and Wilson, 1971
Eupagurus bernhardus hermit crab	C_9APE_{10}	>100	96-hr	Swedmark et al., 1971
Hyas araneus spider crab (adults; larvae)	C_9APE_{10}	>1070; 10	96-hr	Swedmark et al., 1971
Homarus americanus lobster	nonylphenol	0.2	96-hr	McLeese et al., 1980b
Aedes aegypti mosquito	C_9APE_{11}	500	24-hr	Van Emden et al., 1974
Culex pipiens mosquito	various nonionics	1 - >400	—	Maxwell and Piper, 1968
Mollusca				
Cardium edule cockle (adult; juvenile)	C_9APE_{10}	5(6-8°C) <<10 (15-17°C)	96-hr 96-hr	Swedmark et al., 1971

Table 4-2. (cont.)

Species/Common Name	Surfactant/Trade Name	LC_{50} (mg/L)	Test Duration	Reference
Cardium edule cockle	C_9APE_{12}	92.5	48-hr	Swedmark et al., 1971
Mya arenaria clam	C_9APE_{10}	18 (6-8°C) <10 (15-17°C)	96-hr	Portmann and Wilson. 1971
Mya arenaria clam	nonylphenol	>1.0	360-hr	McLeese et al., 1980b
Mytilus edulis mussel	C_9APE_{10}	12 (6-8°C) <10 (15-17°C)	96-hr	Swedmark et al., 1971; Swedmark et al., 1976
Mytilus edulis mussel	nonylphenol	3.0	96-hr	Granmo et al., 1989; 1991
Pecten maximus scallop	C_9APE_{10}	<<5.0	96-hr	Swedmark et al., 1971
Pecten opercularis scallop	C_9APE_{10}	<<10	96-hr	Swedmark et al., 1971
Biomphalaria glabrata snail	C_9APE_{11}	23 (LC_{100})	24-hr	Van Emden et al., 1974

[a]EC_{50} (immobilization).
[b]Measured concentration.
[c]Toxicity threshold.
[d]Toxicity inversely related to food availability.
A dash (—) indicates no data.

3. Toxicity to Algae

The concept of toxicity is difficult to apply to algal cultures because of the rapid, often logarithmic, growth following removal to fresh media. Thus the endpoint in studies of the toxicity of chemicals to algae is usually growth inhibition. Occasionally, acute or longer exposures of algae to certain concentrations of surfactants may stimulate cell division and growth. Furthermore, because algae may reproduce within the acute exposure periods (48- or 96-hr periods) applied to other organisms, these studies would be classified as chronic for algae. Because of the growth rather than mortality endpoint and the difficulty in conducting acute studies, all studies on algae are contained in a single table (Table 4-3).

The following studies demonstrate that most species of algae, when tested individually, are not negatively affected by APE concentrations of ≤ 10 mg/L. On the other hand, a few species of green algae are sensitive to concentrations as low as 0.21 mg/L. Changes in the community structure of natural assemblages of algae, based on common indices of species diversity and similarity, were affected at 8.4 mg/L. Algae were sensitive to NP, with changes in growth and cell ultrastructure at 0.5 mg/L. Variability among the studies may be due to many factors, the most important of which may be the source or strain of the test species.

Ukeles (1965) tested a series of branched NP surfactants on 12 species of marine phytoplankton. The Igepal series contained 9, 9.5, 10.5, 15, 20, and 30 EO units. Growth for all species was inhibited to some degree at 10-1000 mg/L, with variable, but less inhibition with increasing EO units. Growth of the most sensitive species, *Stichococcus* sp. and *Nannochloris* sp., was completely inhibited at ≥ 100 mg/L. The least sensitive species was *Protococcus* sp. for which growth generally equaled that of controls at all concentrations. The authors suggested that thickness and composition of the cell wall may be factors that influence the sensitivity of algae to surfactants.

Davis and Gloyna (1967) tested the toxicity of $C_9APE_{9.5}$ on several species of green and blue-green algae. At a concentration of 20 mg/L, growth of *Oscillatoria borneti*, *O. chalybia*, *O. formosa*, *Gloccapsa alpicola*, *Scenedesmus obliquus*, and *Ulothrix fimbriata* was appreciably inhibited. Growth of *Anabaena variabilis*, *O. tenuis*, *Ankistrodesmus braunii*, *Chlorella pyrenoidosa*, and *C. vulgaris* was markedly inhibited. Batch additions of 50 mg/L of the chemical to pond water had a stimulatory effect on the growth of blue-green algae but an inhibitory effect on the green alga, *Euglena* sp.

The toxicity of $C_8APE_{9.5}$ (Triton X-100) to the diatom *Nitzschia holsatica* was studied at two temperatures (Nyberg, 1976). At 25°C, growth, compared to that of controls, was normal up to 10 mg/L, depressed up to 45% at 15 mg/L, and completely inhibited at 25 mg/L. At 15°C, the temperature of Finnish waters in summer, growth was depressed up to 57% at 5 mg/L and completely inhibited at 15 mg/L.

Additional studies were conducted using single species. For the green alga, *Selenastrum capricornutum* the LC_{50} value was >1000 mg/L for C_9APE_7 (Markarian et al., 1989; 1990) and 48-hr EC_{50} values were 20 mg/L for C_9APE_8 (Emulgen 910) and 50 mg/L for C_9APE_9 (Emulgen

Table 4-3. Effects on Growth of Algae

Test Species	Surfactant	Effect	Concentration (mg/L)	Reference
12 species of marine phytoplankton	br-C_9APE_3 - br-C_9APE_{30}	14-day growth inhibition	10-1000	Ukeles, 1965
green and blue-green algae	$C_9APE_{9.5}$	growth inhibition	20	Davis and Gloyna, 1967
Chlamydomonas reinhardii	nonylphenol	cell membrane disruption	0.5-0.7	Weinberger and Rea, 1981
Chlamydomonas reinhardii	nonylphenol	inhibition of photosynthesis	0.75	Moody et al., 1983
Chlorella pyrenoidosa	nonylphenol	growth depression 24-hr LC_{50} LC_{100}	0.025-7.5 1.5 25	Weinberger and Rea, 1981; 1982
Chlorella fusca	C_8APE_{10} (Triton X-100)	no effect, 14-day growth	131, 263, 525	Wong, 1985
Microcystis aeruginosa	C_8APE_{10}	growth (96-hr EC_{50})	7.4	Lewis and Hamm, 1986
Nitzschia actinastroides	$C_8APE_{9.5}$ (Triton X-100)	5-day growth inhibition	10-15	Nyberg, 1985
Nitzschia holsatica	$C_8APE_{9.5}$ (Triton X-100)	no effect 5-day growth (EC_{45}) 5-day growth (EC_{100}) 5-day growth (EC_{57}) 5-day growth (EC_{100})	10 (25°C) 15 (25°C) 25 (25°C) 5 (15°C) 15 (15°C)	Nyberg, 1976
Phorphyridium purpureum	$C_8APE_{9.5}$ (Triton X-100)	5-day growth inhibition	5-10	Nyberg, 1985
Poterioochromonas malhamensis	C_8APE_{10} C_8APE_{40}	lethality lethality	124 17,784	Röderer, 1987

Table 4-3 (cont.)

Test Species	Surfactant	Effect	Concentration (mg/L)	Reference
Scendesmus sp.	C_9APE_4 C_9APE_6 C_9APE_7 C_9APE_{10} C_9APE_{20} C_9APE_{30}	"lethal threshold" "lethal threshold" "lethal threshold" "lethal threshold" "lethal threshold" "lethal threshold"	6 10 16 31 125 5000	Janicke et al., 1969
Selenastrum capricornutum	C_9APE_8 (Emulgen 910) C_9APE_9 (Emulgen 909)	growth (48-hr EC_{50}) growth (48-hr EC_{50})	20 50	Yamane et al., 1984
Selenastrum capricornutum	C_8APE_{10}	growth (96-hr EC_{50})	0.21	Lewis and Hamm, 1986
Selenastrum capricornutum	NPE_6 (Synperonic NP6) NPE_9 (Synperonic NP9) NPE_{30} (Synperonic NP30) $OPE_{9.5}$ (Triton X-100)	3-week growth, slight decrease 3-week growth rate increase 3-week growth rate decrease (~50%) 3-week growth rate increase 3-week growth, slight decrease 3-week growth rate increase	100 200, 300, 400, 500 100 - 500 100 - 500 100 300 - 500	Nyberg, 1988
Selenastrum capricornutum	C_9APE_7	growth (96-hr EC_{50})	>1000	Markarian et al., 1989; 1990
Selenastrum capricornutum	br-4-nonylphenol	growth (96-hr EC_{50})	0.41[a]	CMA, 1990c
Selenastrum capricornutum	br-NPE_9	growth (96-hr EC_{50}) 96-hr NOEC 96-hr LOEC MATC[b]	12 8 16 11.3	Dorn et al., 1993
Skelatonema costatum	br-4-nonylphenol	growth (96-hr EC_{50})	0.027[a]	CMA, 1990d

[a] Measured concentration. [b] MATC or Chronic Value is the geometric mean of the NOEC and LOEC.

909) (Yamane et al., 1984). In another study, the 96-hr EC_{50} was 12 mg/L for NPE_9 (Dorn et al., 1993). In contrast, the 96-hr EC_{50} value for C_8APE_{10}, reported by Lewis and Hamm (1986), was 0.21 mg/L. In a study lasting 3 weeks, NPE_6, NPE_9, and $NPE_{9.5}$ at a concentration of 100 mg/L slightly inhibited growth, but growth with NPE_{30} was nearly twice that of the control culture (Nyberg, 1988). At higher concentrations, growth in the other surfactants was variable among concentrations, but continued lower than that of the controls for only NPE_9. No measurements of the percent of the surfactants remaining or degradation products during the course of the study were made.

Wong (1985) cultured the green alga, *Chlorella fusca*, in growth media containing 0.2, 0.4, or 0.8 mM (131, 263, or 523 mg/L) C_8APE_{10} (Triton X-100) for 14 days. Specific growth rates were not different from that of the control culture; however, the maximum yield of cells was significantly reduced ($P<0.05$) at 0.4 and 0.8 mM. When metals and organic substances present as contaminants in natural lakes were added to the growth medium, there was a synergistic toxic effect with the addition of metals and an ameliorating effect with the addition of several organic chemicals.

Lewis (1986) studied the short-term *in situ* effects of C_8APE_{11} on the community structure of enclosed natural assemblages of phytoplankton in Acton Lake, Ohio, during the summer. Concentrations of less than 3.2 mg/L had no significant effect on the community; however, both species number and a similarity index were significantly reduced at concentrations of ≥ 8.4 mg/L. Between concentrations of 0.26 and 15.2 mg/L, relative abundances of the Chlorophyta (green), Chrysophyta (yellow), and Cyanophyta (blue-green) algae were similar, with the blue-greens progressively increasing and then declining at 40 mg/L. According to the author, the ability of single species short-term toxicity tests performed in the laboratory to predict effects in the field was dependent upon the surfactant, the species, and the parameter measured.

In a continuation of the above study, the effects of the same surfactant on lake community photosynthesis were determined using a ^{14}C technique (Lewis and Hamm, 1986). The mean 3-hr EC_{50} for photosynthesis was 28.7 mg/L, with a range of 2.5 to 101.5 mg/L. The range was largely due to monthly changes in water temperature and phytoplankton community structure. When compared to predictions from laboratory single-species tests for toxicity, the concentrations that produced effects in the laboratory were lower than or similar to concentrations that produced effects in the field. Lewis (1990) reviewed studies on the chronic toxicities of surfactants to algae. He concluded that the toxicity of most surfactants to natural assemblages of algae under natural conditions is less than that predicted from laboratory tests.

Azov et al. (1982) studied the effect of a nonylphenolethoxylate detergent (structure not given) on algal production in pilot-scale high-rate-oxidation ponds. These ponds combine wastewater treatment with algal biomass production; the latter is used as animal feed. At concentrations above 60 mg/L, algal production as measured by optical density and chlorophyll a decreased within 5 days; at 100 mg/L, cell lysis was observed.

The effects of NP on the ultrastructure, photosynthetic activity, and growth of several species of algae have been investigated. Concentrations of br-4-NP that depressed growth of

Selenastrum capricornutum and *Skelatonema costatum* were 0.41 and 0.027 mg/L, respectively (CMA, 1990c; 1990d). At concentrations of 0.5 to 0.7 mg/L in the nutrient media, the flagellae and ultrastructural architecture of the cell membranes of *Chlamydomonas reinhardii* were disrupted (Weinberger and Rea, 1981). Using fluorometric determinations, at 1 hour the photosynthetic activity of *C. reinhardii* was inhibited 55% at 0.5 mg/L NP and 100% at 0.75 mg/L and 1.0 mg/L (Moody et al., 1983).

Weinberger and Rea (1981, 1982) exposed cultures of *Chlorella pyrenoidosa* to concentrations of NP of 0.025-25 mg/L. At concentrations of 0.025, 0.25, 2.5, and 7.5 mg/L, the exponential growth was depressed while at 25 mg/L, a 100% kill was achieved. From these values a 24-hr LC_{50} of 1.5 mg/L was calculated.

4. Sublethal Effects

The sublethal effects of C_9APE_{10} on the activity of saltwater fish and invertebrates were described by Swedmark et al. (1971). At <1 mg/L, cod (*Gadus morhua*) maintained normal behavior for several months. At 5 and 10 mg/L, an initial period of normal behavior was followed by increased swimming activity and subsequent loss of equilibrium. Breathing rate and opercular movements increased in frequency. At 10 mg/L the bivalve (*Mytilus edulis*) was unable to form byssal threads, and within 36 hr the ability to close the valve was inhibited. At 5 mg/L the siphon retraction ability was gradually inactivated. At 0.5 mg/L juvenile *M. edulis* had reduced and irregular heart beats, and at 4 mg/L the burrowing activity of *Cardium edule*, *Astarte montagui*, and *A. sulcata* ceased.

Weinberger and Rea (1981) described the behavior of fish and snails (scientific names not provided) exposed to 0.5 mg/L NP in microcosms containing natural lake water and algae. Guppies reacted immediately with an initial startle reaction followed by slightly disorientated swimming behavior and reduced feeding. Two of the six guppies died within 24 hr and the others recovered after 36 hr. The snails dropped from the inner surface of the tank and did not emerge for 8 hr. Following this, five of the six snails returned to their normal feeding behavior. At a concentration of 0.056 mg/L NP, *Mytilus edulis* showed sublethal effects of decreased byssal strength and change of scope for growth (energy available above that needed for maintenance) (Granmo et al., 1989; 1991). At a concentration of 0.1 mg/L, no byssus thread formation occurred. Fertilization and early development were not affected at 0.2 mg/L, the highest concentration tested.

During flow-through toxicity tests with *Pimephales promelas*, exposure to $>LC_{50}$ of NP (0.135 mg/L) resulted in lethargic, but stimuli-reactive behavior in surviving fish (Holcombe et al., 1984). At concentrations as low as 0.098 mg/L, some loss of equilibrium was observed.

5. Chronic Toxicity

Chronic toxicity tests using br-NPE_9 were conducted with *Selenastrum capricornutum* (Table 4-3), *Daphnia magna*, and *Pimephales promelas* (Table 4-4). The 96-hr EC_{50} for growth of *Selenastrum capricornutum* was 12 mg/L; the 96-hr NOEC was 8 mg/L. In a 7-day chronic test, *Daphnia magna* were able to survive and grow at a NOEC of 10.0 mg/L. The 7-day NOEC for fathead minnows was 1.0 mg/L and the lowest-observed-effects-concentration (LOEC) was 2.0 mg/L (Kravetz et al., 1991; Dorn et al., 1993). Other studies in Table 4-4 were conducted with NP. The most sensitive species appeared to be juvenile mysid shrimp, with a LOEC for 28-day growth of 6.7 µg/L and a maximum acceptable toxicant concentration (MATC) of 5.1 µg/L (CMA, 1991a).

Several aquatic organisms were maintained for six months in the effluents from biotreater units that received C_9APE_{10} and its metabolites (Neufahrt et al., 1987). Effluent, which contained 460 µg/L C_9 polyethoxylates, 2 µg/L NP, 1 µg/L C_9APE_1, and 3 µg/L C_9APE_2, was diluted 1:5 with water to simulate receiving water. There were no long-term reproductive effects on *Daphnia magna* or the snail *Planorbis corneus* or growth effects on the aquatic vascular plant *Lemna minor*. The number of young produced by the guppy (*Lebistes reticulatus*) was smaller than that of controls, but the difference was not evaluated. Cod (*Gadus morhua*) could be maintained in flow-through aquaria for several months at concentrations of ≤ 1 mg/L of C_9APE_{10} (Swedmark et al., 1971).

6. Structure-Activity Relationship

The log K_{ow} is a measurement of the tendency of a chemical to partition from water into a nonpolar solvent such as octanol. This tendency is also referred to as hydrophobicity or lipophilicity. Chemicals with a high partition coefficient tend to bioaccumulate in organisms in contact with a solution of the chemical. This observation may not be relevant for nonionic surfactants which probably accumulate at the octanol-water interface with portions of the molecule in each layer. The majority of studies that addressed structure-activity relationships, however, reported a relationship between EO chain length (which is related to water solubility) of similar C-chain surfactants and toxicity to aquatic organisms. In general, toxicity to aquatic organisms increased with decreasing EO chain length.

The toxicities of APE with a range of EO units were tested in several species of aquatic organisms (Tables 4-1, 4-2, and 4-3). In all cases, toxicity increased (LC_{50} values decreased) with decreasing EO chain length. This was true for *Daphnia* sp., tested with C_9APE_4 - C_9APE_{30} (Janicke et al., 1969); the mysid crustacean *Mysidopsis bahia* tested with n-$C_9APE_{1.5}$ - n-C_9APE_{50} and tp-$NPE_{1.5}$ - tp-NPE_{50} (Hall et al., 1989); the bluegill sunfish *Lepomis macrochirus*, tested with C_9APE_4 - C_9APE_{30} (Macek and Krzeminski, 1975); and the Japanese killifish *Oryzias latipes*, tested with C_9APE_0 - $C_9APE_{16.6}$ (Yoshimura, 1986). The data of Yoshimura (1986), who tested a series of nine C_9APE, including synthesized biodegradation intermediates, most clearly show this trend (Figure 4-1). Toxicity decreased from 1.4 mg/L for NP (C_9APE_0) to 110 mg/L for $C_9APE_{16.6}$. In another study, however, toxicities of APE to mosquito pupae appeared to be less at intermediate chain lengths of 4 to 15 units compared to longer and shorter lengths (Maxwell and Piper, 1968), but the results were complicated by similar toxicities for many of the compounds which probably reflected the broad spectrum of chain lengths in a single

Table 4.4. Chronic Effects on Aquatic Organisms[a]

Species/Common Name	Chemical	Parameter Measured	NOEC (mg/L)	LOEC (mg/L)	MATC[b] (mg/L)	LC$_{50}$ (mg/L)	Reference
Daphnia magna water flea	br-NPE$_9$	7-day growth 7-day mortality (LC$_{50}$)	10[c] 10[c]	>10[c] 20[c]	>10[c] 14[c]	9.0[c]	Kravetz et al., 1991; Dorn et al., 1993
Daphnia magna water flea	NP	21-day survival, growth, reproduction	0.024[c]				Comber et al., 1993
Mysidopsis bahia mysid shrimp (juveniles)	br-4-NP	length of F$_1$ at 28 days survival of F$_1$ at 29 days number of young/female sublethal effects 28-day lethality	0.0039[c] 0.0067[c] 0.0067[c] 0.021[c]	0.0067[c] 0.0091[c] 0.0091[c]	0.0051[c] 0.0078[c] 0.0078[c]	>0.021[c]	CMA, 1991a
Crangon crangon brown shrimp	NP	7-day lethality				0.34	Waldock and Thain, 1991
Anodonta cataractae freshwater clam	NP	6-day lethal threshold				5.0	McLeese et al., 1980b
Mya arenaria clam	NP	15-day lethal threshold				>1	McLeese et al., 1980b
Mytilus edulis mussel	br-4-NP	15-day lethality 35-day lethality 15- and 30-day byssus strength, scope for growth 35-day fertilization	0.2	0.056		0.5 0.14	Granmo et al., 1989; 1991
Agonus cataphractus hook nose	NP	7-day lethality				0.36	Waldock and Thain, 1991
Gadus morhua cod	NPE$_{10}$	several months	≤1				Swedmark et al, 1971
Pimephales promelas fathead minnow	br-NPE$_9$	7-day growth 7-day mortality (LC$_{50}$)	1.0[c] 1.8[c]	2.0[c] 2.0[c]	1.4[c] 1.4[c]	2.9[c]	Kravetz et al., 1991; Dorn et al., 1993

Table 4-4. (cont.)

Species/ Common Name	Chemical	Parameter Measured	NOEC (mg/L)	LOEC (mg/L)	MATC[b] (mg/L)	LC$_{50}$ (mg/L)	Reference
Pimephales promelas fathead minnow	br-4-NP	length at 28 days survival at 33 days	0.023 0.0074	0.014	0.010		CMA, 1991b
Rana catesbiana bullfrog (tadpole)	br-4-NP	30-day sublethal effects 30-day lethality	155 mg/kg[d]	390 mg/kg[d]	250 mg/kg[d]	260 mg/kg[c]	CMA, 1992

[a] Data for algae are located in Table 4-3.
[b] MATC or Chronic Value is calculated as the geometric mean of the NOEC and LOEC.
[c] Measured concentration.
[d] Nonylphenol was added to sediment; values are mg/kg dry weight of sediment; organic carbon content of sediment was 0.05%; water was supplied on a flow-through basis.

product. For algae, toxicity, as measured by growth inhibition, was variable but also decreased with increasing EO chain length as reported by Ukeles (1965) for 12 species of marine phytoplankton, by Janicke et al. (1969) for *Scendesmus* sp., and by Nyberg (1988) for *Selenastrum capricornutum*.

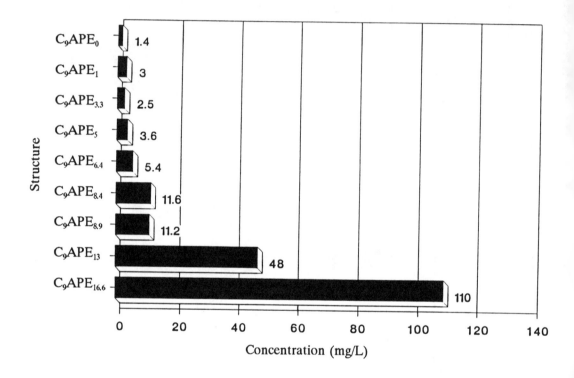

Figure 4-1. **Relationship between Toxicity and Ethylene Oxide Chain Length.** Data are 48-hr LC_{50} Values for Japanese killifish (*Oryzias latipse*). Source: Yoshimura, 1986

For a series of alkylphenols, potential degradation products of APE, toxicity increased with increasing molecular weight which correlated with increasing log K_{ow}. 4-Nonylphenol and 4-dodecylphenol, with log K_{ow} values of 4.2 and 5.5, respectively, were more toxic than several butylphenol isomers with log K_{ow} values of 2.1-2.8. Ninety-six-hr LC_{50} values for shrimp and salmon ranged from 0.13-0.30 mg/L for the former compounds compared to 0.77-5.2 mg/L for the butylphenols (McLeese et al., 1981). Ahel (1987) additionally directly measured the log K_{ow} of several short chain length APE. Log K_{ow} values for OP, NP, NPE_1, NPE_2, and NPE_3 were 4.12, 4.48, 4.17, 4.21, and 4.20, respectively.

7. Effects of Environmental Variables on Toxicity

Uptake of surfactants may be more rapid and reach higher tissue concentrations at higher temperatures than at low temperatures, thus affecting toxicity. The surfactant C_9APE_{10} was taken up by cod (*Gadus morhua*) more rapidly and reached higher concentrations in several tissues at 18°C compared to 11°C (Granmo and Kohlberg, 1976).

Water temperature can affect the toxicity of APE. For cod, toxicity of C_9APE_{10} was greater at 15 to 17°C than at 6 to 8°C, the latter being the normal winter temperature for this species (Swedmark et al., 1971). The toxicity of C_9APE_{10} increased with an increase in temperature for several other marine species (Swedmark et al., 1971). For flounder (*Pleuronectes flesus*), clam (*Mya arenaria*), mussel (*Mytilus edulis*), cockle (*Cardium edule*), decapod (*Leander adspersus*), and barnacle (*Balanus balanoides*), 96-hr LC_{50} values were generally higher at winter temperatures of 6 to 8°C (5 to >100 mg/L) then at 15 to 17°C (2.5 to 50 mg/L). Golden orfe (*Leuciscus idus*) were also less sensitive at the lower temperature of 15°C ("critical concentration" range of 12-20 mg/L) than at 20°C ("critical concentration" range of 6-11 mg/L) (Fischer and Gode, 1978).

Changes in temperature affected the response of individual species of algae and phytoplankton communities to toxicants. Growth of the untreated diatom *Nitzschia holsatica* was slower at 15°C than at 25°C and the inhibitory effect of $C_8APE_{9.5}$ was more marked at the lower temperature (Nyberg, 1976). However, the inhibitory effect of C_8APE_{11} and other surfactants on the photosynthetic activity of lake phytoplankton increased with increasing water temperature which ranged from 17°C in May to 28.3°C in July (Lewis and Hamm, 1986).

8. Bioaccumulation

The relatively high log K_{ow} values of >3 for NP and the lower molecular weight APE (measured by Ahel, 1987; McLeese et al., 1981; and Caux et al., 1988) indicate a potential to bioaccumulate in tissues. However, while uptake is rapid, excretion of APE and the lower molecular homologs is rapid following removal to clean water. The bioconcentration factor (BCF) refers to the concentration of a chemical in an aquatic organism divided by the concentration in water. None of the following laboratory uptake/excretion studies determined the formation of metabolites. Studies were also conducted under field conditions.

Exposure of cod (*Gadus morhua*) to 5 mg/L of C_9APE_{10} labeled in the EO chain with ^{14}C resulted in rapid uptake through the gills and distribution to the tissues, reaching an equilibrium after eight hours (Granmo and Kollberg, 1976). At this time concentrations in tissues were: gills and blood, 100 ppm; liver and kidney, 500 ppm; and gall bladder, 4000 ppm. Except for the gall bladder, 60% of the radioactivity was eliminated from the tissues within 24 hours.

For a series of alkylphenols, equilibrium BCF in juvenile Atlantic salmon (*Salmo salar*), calculated from a one-compartment model and taking into account the exponential decrease of the concentrations in water, increased with increasing log K_{ow} values (McLeese et al., 1981). Bioconcentration factors, as determined from uptake/excretion, for *p*-sec-butylphenol, *p*-hexylphenol, *p*-NP, and *p*-dodecylphenol were 37, 350, 10 and 280 (results of two studies with *p*-NP), and 6000, respectively. Excretion half-lives were 4, 10, 0.3 and 4, and 690 days, respectively. For *p*-NP, the whole-body concentration averaged 235 µg/g wet weight after 24 hours of exposure; during this time the concentration in the water averaged 0.31 mg/L.

Mussels (*Mytilus edulis*) were exposed for four days in static tests to NP at nominal concentrations of 0.10 and 1.13 mg/L (McLeese et al., 1980a). The concentrations in the water declined by 25% over the four-day exposure period. Maximum concentrations in the mussels were reached between one and two days and declined by four days. Estimated BCF ranged between 1.4 and 13. At an exposure concentration of 1.13 mg/L, uptake was highest during day 1, 10.2 µg/g (wet weight), and declined to 1.2 µg/g by day 4. By day 8 post-treatment, NP was below the limit of detection of <0.3 µg/g. The calculated rate constants indicated low uptake and high excretion rates. The half-life was 0.3 days.

The CMA Panel studied the bioaccumulation of NP during 20-day exposures of the fathead minnow (*Pimephales promelas*) at two exposure concentrations, 5 and 25 µg/L (Naylor et al., 1992). Bioconcentration factors at the two exposures were 270 and 350, respectively. The times for 90% steady-state conditions were 4.5 and 4.1 days, respectively. Upon removal to clean water, half-times to depuration were 1.4 and 1.2 days, respectively.

Recent studies reported by Ekelund et al. (1990) and Granmo et al. (1991) measured bioaccumulation of 4-NP in three marine animals. They reported higher BCF values than other investigators, possibly because they ensured that exposure concentrations remained constant. The BCF values in shrimp (*Crangon crangon*), mussels (*Mytilus edulis*), and sticklebacks (*Gasterosteus aculeatus*) were 100, 3430, and 1250, respectively.

Four studies addressed uptake under field conditions. *Mytilus edulis* placed in cages near the wastewater outlet of a facility that manufactured NPE contained concentrations of 4-NP of 0.20 to 0.40 µg/g wet weight (Wahlberg et al., 1990). Concentrations of NPE_1 ranged from 0.08 to 0.28 µg/g. Concentrations in the water and sediment were not measured.

Granmo et al. (1991) placed *Mytilus edulis* in cages at several stations in an area receiving wastewater from a chemical plant on the west coast of Sweden. Fresh-weight tissue concentrations of NP ranged from <0.1 µg/g to 0.4 µg/g and were related to proximity to the plant. The NP concentration in the water was assumed to be below the limit of detection. Individual concentrations of NPE_1, NPE_2, and NPE_3 were less than that of NP. During the same study, mussels were also placed in known concentrations of the wastewater. In this case the estimated BCF for NP was 340.

Nonylphenol, NPE_1, and NPE_2 were determined in several freshwater organisms from surface waters of the Glatt Valley, Switzerland (Ahel et al., 1993). High concentrations occurred in the macroalga, *Cladophora glomerata*, 38.0, 4.7, and 4.3 mg/kg dry weight for NP, NPE_1 and NPE_2, respectively, where respective water concentrations were 3.9, 24, and 9.4 µg/L. Based on the dry weight concentrations, BCF were 10,000, 200, and 500, respectively. Dry weight concentrations for the respective compounds in two species of aquatic plants were lower: 4.2, 0.9, and 0.6 mg/kg in *Fontinalis antipyretica* and 2.5, 1.1, and 1.9 mg/kg in *Potamogeton crispus*. Concentrations varied among plant samples taken from different areas of the rivers and among seasons. The concentrations of NP, NPE_1, and NPE_2 in various tissues of several species of fish were <0.03-1.6, 0.06-7.0, and 0.03-3.1 mg/kg dry weight, respectively, resulting in BCF of 13-410, 3-300, and 3-330, respectively. Concentrations tended to be low in the edible parts of the fish. Concentrations in a single mallard duck captured on the bank of the river were similar to those of fish.

Marcomini et al. (1990) analyzed NPE metabolites (NP, NPE_1, and NPE_2) in macroalga (*Ulva rigida*), water, and marine sediments of the Venice lagoon, Italy. The mean concentration of total NPE in the water was 1.8 $\mu g/L$; of this, the sum of NP, NPE_1, and NPE_2 constituted <10% (<0.18 $\mu g/L$). The total concentration of these three metabolites in algae samples was 0.049 $\mu g/g$ dry weight. The total concentration in decomposing macroalgae overlying the sediment was 0.25 $\mu g/g$ dry weight (0.0325 $\mu g/g$ wet weight). The mean concentration of the three metabolites in sediment was 49 $\mu g/g$ dry weight. The concentrations in sediment samples overlain by decomposing macroalgae were higher than those in sediments free from macroalgae.

9. Comparison of Toxicity Data with Environmental Concentrations

Naylor et al. (1992) reviewed the risk of adverse environmental impact of alkylphenols and their ethoxylates. Based on (1) extensive biodegradation/removal at wastewater treatment plants, (2) low concentrations in 30 U.S. rivers at locations most likely to contain significant amounts, (3) a substantial margin of safety between "worst case" river water concentrations and sublethal toxic effects, and (4) the absence of significant accumulation of NP and NPE in the water columns, interstitial water, sediment, and fish, the risk of adverse environmental impact is low.

B. EFFECTS ON MICROORGANISMS

A limited number of studies on soil microorganisms indicate that effects are species- and acclimation-specific; the lowest concentration to inhibit growth of bacteria is ≥ 20 mg/L in relatively uncontaminated areas and ≥ 50 mg/L in areas exposed to chemicals.

When an unidentified octylphenol ethoxylate was applied to natural soil and surface water bacteria that had been plated in Petri dishes, response appeared to depend on previous exposure to chemicals (Hartmann, 1966). In soil samples taken from the surface of a forest, a 10 mg/L solution had little effect on bacterial colonies; at ≥ 20 mg/L the number of colonies was 50% less than that of the control group. Bacteria from soil samples taken at depths of 4 to 15 cm showed sensitivity with increasing depth, with 50 to 100% loss of colonies at 20 mg/L. Bacteria plated from soils from a grassy field and surface water were little affected by concentrations up to 100 mg/L.

Vandoni and Goldberg-Federico (1973) found that a solution of 1000 mg/L of octylphenol ethoxylate (0.1% EO by weight) had a slight inhibitory effect on the nitrification process in two soils. A concentration of 50 mg/L C_8APE_5 (Triton X-45) inhibited the growth of bacteria and various fungi cultured in agar, but had no affect on yeast (Hislop et al., 1977). A concentration of 5 mg/L had no effect on bacteria and fungi.

The surfactants C_8APE_5 (Triton X-45) and C_8APE_{10} (Triton X-100) inhibited the growth of *Staphylococcus aureus* at concentrations of 0.0585 µM (24 µg/L) and 0.159 µM (103 µg/L), respectively (Lamikanra and Allwood, 1976). At higher concentrations, growth was completely inhibited over the six-hour observation period. Toxicity was related to absorption by the cells. Triton X-45 was more rapidly taken up by the cells than Triton X-100.

Janicke et al. (1969) reported the "toxicity threshold" for a series of NPE for *Pseudomonas* sp. Ethoxylates and threshold concentrations were C_9APE_4, 50 mg/L; C_9APE_6, 500 mg/L; C_9APE_7, 63-500 mg/L; and C_9APE_{10}, C_9APE_{20}, and C_9APE_{30}, 1000 mg/L.

Hislop et al. (1977) also tested the toxicity of several Triton surfactants (C_8APE_5-C_8APE_{70}) on apple scab (*Venturia inaequalis*). Dipping apple leaves into a 50,000 mg/L solution inhibited the release of ascospores from 97.3% for C_8APE_5 to 67.4% for C_8APE_{12-13}, indicating an inverse relationship between ethoxylate chain length and toxicity. However, C_8APE_{70} was similar in toxicity to C_8APE_5. In another study, neither C_8APE_{10} (Triton X-100) nor C_9APE_6 (Triton N-57) sprayed on dormant Cortland apple trees had an effect on powdery mildew, apple scab, or fruit set (Spotts and Feree, 1979).

The effect of NPE containing 4, 6, 9, 10, 11, 15, 23, and 30 EO units on the growth of the soil bacteria, *Bacillus megaterium*, *B. cereus*, *B. polymyxa*, *B. subtilis*, *Pseudomonas fluorescens*, and *Azobacter chroococcum*, was investigated by the agar diffusion method (Cserhati et al., 1991). With the exception of *Azobacter chroococcum*, which was insensitive to each surfactant at each concentration, additions of the surfactants at concentrations of 20-800 mg/L inhibited or restricted growth in the agar zone immediately surrounding the surfactant. The growth of *Bacillus megaterium* and *B. cereus* was stimulated in the outer zones of the plate where the concentration of the diffusing surfactant was presumably lower than in the center of the plate. Toxicity decreased with increasing EO chain length.

For the marine luminescent bacterium, *Photobacterium phosphoreum*, the 5-min LC_{50} of NPE_9 was 60.6 mg/L (Dorn et al., 1993).

Two studies addressed the effects of NP. Knie et al. (1983) studied the effect of 4-NP on a population of *Pseudomonas putida*. They reported an EC_{50} of >10 mg/L. 4-Nonylphenol inhibited germination of *Bacillus megaterium* (99% inhibition at 2 hr) at 32 mg/L, but a saturated solution of NP proved ineffective in inhibiting sporostatic activity after 24 hr incubation (Lewis and Jurd, 1972).

C. EFFECTS ON HIGHER PLANTS

Because they improve wetting, penetration, absorption, and water solubility characteristics, surfactants are used in formulations of foliar-applied agrochemicals as stabilizing, emulsifying, and dispersing agents. APE and NP have been applied to higher plants as part of pesticide formulations. The lowest concentration of APE found to inhibit growth of young terrestrial and aquatic plants or trees was 10 µg/L, with most growth-inhibiting concentrations for mature plants several orders of magnitude higher. Nonylphenol, at a concentration of 20 mg/L, inhibited the growth of white birch seedlings and, at a concentration of 0.5 mg/L, was lethal to the aquatic vascular plant *Lemna minor*. APE are rapidly taken up by plants and metabolized to polar metabolites.

Damage to apple and plum tree leaves following immersion in a series of iso-octylcresol/ethylene oxide condensates (C_8 with 5, 8, 10, 15, or 20 EO units) at two concentrations, 0.05% and 0.5% (500 and 5,000 mg/L), was assessed (Furmidge, 1959a; 1959b). Compared to ionic materials, the nonionic surfactants were practically non-phytotoxic. Maximum wetting occurred with the lower ethoxylate homologs.

Single spray treatments of C_9APE_5 (Triton N-57) or C_8APE_{10} (Triton X-100) at concentrations of 10,000-50,000 mg/L to various fruit crops affected bud opening (Spotts and Ferree, 1979). Treatments of 30,000 and 50,000 mg/L of either surfactant to apple trees (*Malus domestica*) produced a 2-5 day delay in bud break and bud kills of 30-50%. The same concentrations were lethal to the buds of Concord grape vines and C_9APE_5 produced a bud delay in Aurore grapes. At concentrations of 10,000-50,000 mg/L, both compounds caused a 67-100% bud kill in peach trees (*Prunus persica*). Neither of the surfactants at any of the concentrations tested had an effect on pear buds (*Pyrus* sp.).

Solutions of 1.0% and 0.10% v/v (10,000 and 1,000 mg/L) of several OPE surfactants were applied to the leaves of sweet corn, cucumber, cotton, lima bean, tobacco, and tomato; injury was rated one week later (Gast and Early, 1956). Applications of 1.0% solutions produced severe injury while solutions of 0.10% had no effect. C_8APE_{10} (Triton X-100), C_8APE_5 (Triton X-45), and $C_8APE_{10.5}$ (Igepal CA-710) were the most phytotoxic. Triton X-155 was practically nontoxic.

A concentration of 25 mg/L of C_8APE_{10} (Triton X-100) in the culture medium significantly reduced pollen germination of the bitter gourd, *Momordica charantia* (Regupathy and Subramaniam, 1974). At a concentration of 100 mg/L, 22 nonionic surfactants repressed the elongation of the primary root of cucumber seedlings (Parr and Norman, 1964). Triton X-100 at a concentration of 0.1% v/v (1,000 mg/L) for 4 hours was toxic to the roots of pea (*Pisum sativum*) seedlings (Nethery, 1967). C_8APE_{10} (Triton X-100) protected beet cells from plasmolysis at concentrations above 70-100 mg/L (Haapala, 1970). Below this concentration, retention of solutes was enhanced.

The growth rate of cress seedlings grown on filter paper in 5000 or 50,000 mg/L solutions of a series of C_8APE_n (Triton X) surfactants was reduced (Hislop et al., 1977). No effects on barley or pea seedlings were observed when a concentration of 50,000 mg/L of C_8APE_5 was sprayed on the potting soil used to grow the plants.

Horowitz and Givelberg (1979) exposed the roots and leaves of seedling sorghum (*Sorghum bicolor*) to various concentrations of Agral 90™, a 92% active alkylphenol-ethylene oxide condensate. When roots were exposed to a concentration of 10 mg/L for 8 days, significant growth reduction in the roots and shoots occurred. Exposure to 10,000 mg/L resulted in leakage of amino acids and inorganic ions, and within three days, wilting of the test plants.

Effects of applications of solutions ranging in concentration from 0.01 to 1.0% v/v (100 to 10,000 mg/L) of C_8APE_n (Triton X surfactants with EO units of 3, 5, 9.5, 30, and 40) to the adaxial surface of 10-day-old cowpea (*Vigna unguiculata*) were studied (Lownds and Bukovac, 1988). Toxicity generally decreased with increasing EO chain length. For $C_8APE_{9.5}$, a concentration of 0.01% was not phytotoxic, but at higher concentrations damage, including sunken, discolored areas characterized by loss of structural integrity and necrosis, occurred. For a given surfactant, phytotoxicity increased with increasing concentration, volume of application, and temperature, and decreased with increasing humidity.

Knoche et al. (1992) studied the phytotoxicity of two homologous series of APE (OPE_5, $OPE_{7.5}$, $OPE_{9.5}$, OPE_{16}, OPE_{30} and NPE_7, NPE_{10}, NPE_{14}, NPE_{20}, NPE_{30}) to brussels sprout (*Brassica oleracea*). Toxicity was determined by visual observation and measurement of ethylene production. Surfactants with medium EO chain lengths, $OPE_{9.5}$ and NPE_{10}, were the most phytotoxic. No difference was found between OPE and NPE surfactants.

C_8APE_n (Triton X) surfactants were also tested on the aquatic vascular plant, duckweed (*Lemna minor*) (Caux et al., 1988). The authors monitored growth, frond florescence and chlorophyll content, conductivity of the test media, and specific ion leakage. Solutions of 1 mg/L of each surfactant and a 10 mg/L solution of C_8APE_{7-8} (Triton X-114) were without effects. A 10 mg/L solution of the other Triton surfactants (C_8APE_1, C_8APE_3, C_8APE_{9-10}) depressed frond development by 25 to 50%. At 50 mg/L, the two lower homologs were about twice as phytotoxic as the higher ethoxylates. The toxicity of the surfactants to *L. minor* was inversely related to EO chain length and directly related to log K_{ow}.

Concentrations of 0.5-5 mg/L NP had a small effect on the growth (frond multiplication) of cultures of *Lemna minor* when treated once and observed for six days (Prasad, 1989). Under the same conditions, concentrations of 2.5-25 mg/L were toxic to another macrophyte, *Salvinia molesta*; frond production was reduced by day 3 and treated cultures were dead by day 9. When cultures of *Lemna minor* were treated with fresh solutions of NP every day for 4 days, concentrations of 0.125 and 0.250 mg/L produced significant effects on growth beginning at 24 hours post-treatment. Concentrations of ≥ 0.50 mg/L caused considerable bleaching, chlorosis, and mortality of cultures. Studies with ^{14}C-NP indicated absorption into the cells. Ultrastructural examination revealed disruption of the chloroplast membrane.

Nonylphenol at a concentration of 100 mg/L significantly depressed seed germination of jackpine (*Pinus banksiana*) and paper birch (*Betula papyrifera*). Seedling growth was decreased in jack pine at 100 mg/L and in white birch at 20 mg/l, the only level tested (Weinberger and Vladut, 1981).

A group of investigators studied the metabolic fate of APE in higher plants. Frear et al. (1977) exposed the excised tissues of soybean, corn, wheat, pea, and barley seedlings to ^{14}C-labeled t-C_8APE_6 or t-C_8APE_9. The terminal end of the ethoxylate chain was rapidly conjugated with glucosides or fatty acid esters, mainly of palmitic and linoleic acids. Stolzenberg and Olson (1977; 1978) exposed the excised leaves of barley and rice to aqueous solutions of the same labeled compounds. Within two days the surfactants were largely converted to polar metabolites; minor amounts of C_8 with shortened ethoxylate chains were formed.

Stolzenberg et al. (1982) also studied the metabolic fate of radiolabeled C_8APE_{10} (Triton X-100) in the leaves of barley plants (*Hordeum vulgare*). Following growth of the excised leaves in surfactant solutions or application to the foliar surface of intact plants, the surfactant was rapidly taken up and metabolized; there was little translocation from the treated leaves of intact plants. Metabolites were primarily polar; hydrolysis released the parent compound and hydroxylated and unidentified metabolites. In an earlier paper (Stolzenberg and Olson, 1977), transpiration and photosynthesis of the excised plants continued during growth in solutions of 150 mg/L; the rate was at 75% that of controls when surfactant residues in the tissues reached 150 mg/g fresh weight.

D. EFFECTS ON BIRDS AND WILDLIFE

No studies were found that addressed the effects of APE on birds or wildlife. Kingsbury et al. (1981) applied NP to forests at a rate equivalent to the quantity applied in allowable maximum seasonal sprayings of Matacil. Nonylphenol did not have any significant (level of significance not specified) effect on forest songbird populations or the ability of songbirds to defend their breeding territories. Terrestrial invertebrates, likewise, did not appear to be affected. According to Hislop et al. (1977), concentrations lower than 50,000 mg/L had no effects on earthworms (*Lumbricus terrestris*).

E. MODE OF ACTION

Many nonionic surfactants interact with proteins, changing the shape and activity of enzymatic proteins and solubilizing structural proteins which results in changes in cell permeability (Helenius and Simons, 1975; Swisher, 1987). In cell physiology studies, C_8APE_{10} (Triton X-100) among other surfactants has been used to solubilize membrane proteins without loss of their biological activity. Large amounts of C_8APE_{10} remain bound to the proteins (Helenius and Simons, 1975).

REFERENCES

Ahel, M. 1987. Biogeochemical behaviour of alkylphenol polyethoxylates in the aquatic environment. Ph.D. Dissertation, University of Zagreb, Yugoslavia, 200 pp.

Ahel, M., J. McEvoy and W. Giger. 1993. Bioaccumulation of the lipophilic metabolites of nonionic surfactants in freshwater organisms. Environ. Pollut. 79:243-248.

Ankley, G.T., G.S. Peterson, M.T. Lukasewycz and D.A. Jenner. 1990. Characteristics of surfactants in toxicity identification evaluations. Chemosphere 21:3-12.

Armstrong, J.A. and P.D. Kingsbury. 1979. Interim Progress Report, Forest Pesticide Management Institute, Canadian Forestry Service, 63 pp. (Cited in McLeese et al., 1981).

Arthur D. Little. 1977. Human Safety and Environmental Aspects of Major Surfactants. Arthur D. Little, Cambridge, MA.

Azov, Y, G. Shelef and N. Narkis. 1982. Effect of hard detergents on algae in a high-rate-oxidation pond. Appl. Environ. Microbiol. 43:491-492.

Bergstöm, B. 1984. Unpublished data cited in Wahlberg et al., 1990.

Bringmann, G. and R. Kühn. 1982. Results of toxic action of water pollutants on *Daphnia magna* Straus tested by an improved standardized procedure. Zeitsch. Wasser Abwasser Forschung. 15:1-6.

Burlington Research, Inc. 1985. Acute and chronic bioassays of industrial surfactants. Final Report, 40 pp.

Calamari, D. and R. Marchetti. 1973. The toxicity of metals and surfactants to rainbow trout (*Salmo gairdneri* Rich.). Water Res. 7:1453-1464.

Caux, P.Y., P. Weinberger and D.B. Carlisle. 1988. A physiological study of the effects of Triton surfactants on *Lemna minor* L. Environ. Toxicol. Chem. 7:671-676.

CMA (Chemical Manufacturers Association). 1990a. Acute flow through toxicity of nonylphenol to the sheepshead minnow, *Cyprinodon variegatus*. Unpublished report, Alkylphenol & Ethoxylates Panel, CMA, Washington, DC.

CMA (Chemical Manufacturers Association). 1990b. Acute flow through toxicity of nonylphenol to the mysid, *Mysidopsis bahia*. Unpublished report, Alkylphenol & Ethoxylates Panel, CMA, Washington, DC.

CMA (Chemical Manufacturers Association). 1990c. Acute static toxicity of nonylphenol to the freshwater alga, *Selenastrum capricornutum*. Unpublished report, Alkylphenol & Ethoxylates Panel, CMA, Washington, DC.

CMA (Chemical Manufacturers Association). 1990d. Acute static toxicity of nonylphenol to the marine alga, *Skeletonema costatum*. Unpublished report, Alkylphenol & Ethoxylates Panel, CMA, Washington, DC.

CMA (Chemical Manufacturers Association). 1991a. Chronic toxicity of nonylphenol to the mysid, *Mysidopsis bahia*. Unpublished report, Alkylphenol & Ethoxylates Panel, CMA, Washington, DC.

CMA (Chemical Manufacturers Association). 1991b. Early life stage toxicity of nonylphenol to the fathead minnow, *Pimephales promelas*. Unpublished report, Alkylphenol & Ethoxylates Panel, CMA, Washington, DC.

CMA (Chemical Manufacturers Association). 1992. Toxicity of nonylphenol to the tadpole *Rana catesbiana*. Unpublished report, Alkylphenol & Ethoxylates Panel, CMA, Washington, DC.

Comber, M.H.I., T.D. Williams and K.M. Stewart. 1993. The effects of nonylphenol on *Daphnia magna*. Water Res. 27:273-276.

Cripe, G.M., A. Ingley-Guezou, L.R. Goodman and J. Forester. 1989. Effect of food availability on the acute toxicity of four chemicals to *Mysidopsis bahia* (Mysidacea) in static exposures. Environ. Toxicol. Chem. 8:333-338.

Cserhati, T., Z. Illes and I. Nemes. 1991. Effect of non-ionic tensides on the growth of some soil bacteria. Appl. Microbiol. Biotechnol. 35:115-118.

Davis, E.M. and E.F. Gloyna. 1967. Biodegradability of nonionic and anionic surfactant by blue-green algae. Environmental Health Engineering Research Laboratory, University of Texas, Center for Research in Water Resources.

Dorn, P.B., J.P Salanitro, S.H. Evans and L. Kravetz. 1993. Assessing the aquatic hazard of some linear and branched nonionic surfactants by biodegradation and toxicity. Environ. Toxicol. Chem. 12:1751-1762.

Ekelund, R., A. Bergman, A. Granmo and M. Berggren. 1990. Bioaccumulation of 4-nonylphenol in marine animals - A re-evaluation. Environ. Pollut. 64:107-120.

Ernst, B., G. Julien, K. Doe and R. Parker. 1980. Environmental investigations of the 1980 spruce budworm spray program in New Brunswick. EPS-5-AR-81-3. Surveillance Report, Canada Environmental Protection Service.

Etnier, E.L. 1985. Chemical Hazard Information Profile: Nonylphenol. Draft Report. Office of Toxic Substances, U.S. Environmental Protection Agency.

Fischer, W.K. 1973. (Personal communication cited in Gloxhuber, 1974).

Fischer, W.K., and P. Gode. 1978. Comparative investigation of various methods of examining fish toxicity, with special consideration of the German golden orfe test and the iso zebra fish test. A.F. Wasser Abwasser Forsch. 11:99-105. (Cited in Goyer et al., 1981).

Frear, D.S., H.R. Swanson and G.E. Stolzenberg. 1977. Polyethoxylated alkylphenol detergents: *in vitro* plant metabolism studies. 174th National Meeting of the American Chemical Society, September, 1977 (Abstract).

Furmidge, C.G.L. 1959a. Physico-chemical studies on agricultural sprays. II. The phytotoxicity of surface-active agents on leaves of apple and plum trees. J. Sci. Food Agric. 10:274-282.

Furmidge, C.G.L. 1959b. Physico-chemical studies on agricultural sprays. III. Variation of phytotoxicity with the chemical structure of surface-active agents. J. Sci. Food Agric. 10:419-425.

Gast, R. and J. Early. 1956. Phytoxicity of solvents and emulsifiers used in insecticide formulations. Agric. Chem. 11:42-45, 136-137, 139.

Gloxhuber, C. 1974. Toxicological properties of surfactants. Arch. Toxicol. 32:245-270.

Goyer, M.M. J.H. Perwak, A. Sivak and P.S. Thayer. 1981. Human Safety and Environmental Aspects of Major Surfactants (Supplement). Arthur D. Little, Cambridge, MA.

Granmo, A. and S. Kollberg. 1976. Uptake pathways and elimination of a nonionic surfactant in cod (*Gadus morhua* L.). Water Res. 10:189-194.

Granmo, A., R. Eklund, K. Magnusson and M. Berggren. 1989. Lethal and sublethal toxicity of 4-nonylphenol to the common mussel (*Mytilus edulis* L.). Environ. Pollut. 59:115-127.

Granmo, A., R. Eklund, M. Berggren and K. Magnusson. 1991. Toxicity of 4-nonylphenol to aquatic organisms and potential for bioaccumulation. In: Swedish EPA Seminar on Nonylphenol Ethoxylates/Nonylphenol held in Saltsjobaden, Sweden, February 6-8, 1991, pp. 53-75.

Haapala, E. 1970. The effect of a nonionic detergent on some plant cells. Physiol. Plant 23:187-201.

Hall, W.S., J.B. Patoczka, R.J. Mirenda, B.A. Porter and E. Miller. 1989. Acute toxicity of industrial surfactants to *Mysidopsis bahia*. Arch. Environ. Contam. Toxicol. 18:765-772.

Hamburger, B., H. Haberling and H.R. Hitz. 1977. Comparative tests on toxicity to fish using minnows, trout, and golden orfe. Arch. Fischerewiss. 28:45-55. (English Abstract) (Cited in Goyer et al., 1981).

Hartmann, L. 1966. Effect of surfactants on soil bacteria. Bull. Environ. Contam. Toxicol. 1:219-224.

Helenius, A. and K. Simons. 1975. Solubilization of membranes by detergents. Biochim. Biophy. Acta 415:29-79.

Hislop, E.C., V.M. Barnaby and R.T. Burchill. 1977. Aspects of the biological activity of surfactants that are potential eradicants of apple mildew. Ann. Appl. Biol. 87:29-39.

Holcombe, G.W., G.L. Phipps, M.L. Knuth and T. Felhaber. 1984. The acute toxicity of selected substituted phenols, benzenes and benzoic acid esters to fathead minnows *Pimephales promelas*. Environ. Pollut (Series A) 35:367-381.

Horowitz, M. and A. Givelberg. 1979. Toxic effects of surfactants applied to plant roots. Pestic. Sci. 10:547-557.

Janicke, W., G. Bringmann and R. Kühn. 1969. Aquatic toxicological studies of the harmful action of ethoxylate nonionic surfactants. Gesund. Ing. 90:133-138.

Kingsbury, P.D., B.B. McLeod and R.L. Millikin. 1981. The environmental impact of nonylphenol and the Matacil formulation. Part 2. Terrestrial ecosystems. Report FPM-X, Forest Pesticide Management Institute, ISS FPM-X-36. (Cited in Etnier, 1985).

Knoche, M., G. Noga and F. Lenz. 1992. Surfactant-induced phytotoxicity: evidence for interaction with epicuticular wax fine structure. Crop Protect. 11:51-56.

Knie, V.J., A. Halke, I. Juhnke and W. Schiller. 1983. Ergebnisse der untersuchungen von chemischen stoffen mit vier biotests. Deutsche Gewasser. Mitt. 27:77-79.

Kravetz, L., J.P. Salanitro, P.B. Dorn and K.F. Guin. 1991. Influence of hydrophobe type and extent of branching on environmental response factors of nonionic surfactants. J. AM. Oil Chem. Soc. 68:610-618.

Kurata, N., K. Koshida and T. Fujii. 1977. Biodegradation of surfactants in river water and their toxicity to fish. Yukagaku 26:115-118.

Lamikanra, A. and M.C. Allwood. 1976. The antibacterial activity of non-ionic surface-active agents. Microbios Lett. 1:97-101.

Lewis, J.C. and L. Jurd. 1972. Sporostatic action of cinnamylphenols and related compounds on *Bacillus megaterium*. Spores 5:384-389.

Lewis, M.A. 1986. Comparison of the effects of surfactants on freshwater phytoplankton communities in experimental enclosures and on algal population growth in the laboratory. Environ. Toxicol. Chem. 5:319-332.

Lewis, M.A. 1990. Chronic toxicities of surfactants and detergent builders to algae: A review and risk assessment. Ecotoxicol. Environ. Safety 20:123-140.

Lewis, M.A. and B.G. Hamm. 1986. Environmental modification of the photosynthetic response of lake plankton to surfactants and significance to a laboratory-field comparison. Water Res. 20:1575-1582.

Lownds, N.K. and M.J. Bukovac. 1988. Studies on octylphenoxy surfactants. V. Toxicity to cowpea leaves and effects of spray application factors. J. Am. Soc. Hort. Sci. 113:205-210.

Macek, K.J. and Krzeminski. 1975. Susceptibility of bluegill sunfish (*Lepomis macrochirus*) to nonionic surfactants. Bull. Environ. Contam. Toxicol. 13:377-384.

Madai, J. and H. An der Lan. 1964. Zur wirkung einiger detergentien auf sübwasser-organismen. Wasser Abwasser. 4:168-183. (Cited in Arthur D. Little, 1977).

Marchetti, R, 1965. The toxicity of nonyl phenol ethoxylate to the developmental stages of rainbow trout, *Salmo gairdneri* Richardson. Ann. Appl. Biol. 55:425-430.

Marcomini, A., B. Pavoni, A. Sfriso and A.A. Orio. 1990. Persistent metabolites of alkylphenol polyethoxylates in the marine environment. Marine Chem. 29:307-323.

Markarian, R.K., M.L. Hinman and M.E. Targia. 1990. Acute and chronic toxicity of selected nonionic surfactants. Presented at the Eleventh Annual Meeting of the Society of Environmental Toxicology and Chemistry, Arlington, VA. (Poster abstract).

Markarian, R.K., K.W. Pontasch, D.R. Peterson and A.I. Hughes. 1989. Review and analysis of environmental data on Exxon surfactants and related compounds. Technical Report. Exxon Biomedical Sciences, Inc.

Maxwell, K.E. and W.D. Piper. 1968. Molecular structure of nonionic surfactants in relation to laboratory insecticidal activity. J. Econ. Entomol. 61:1633-1636.

McLeese, D.W., D.B. Sergent, C.D. Metcalfe, V. Zitko and L.E. Burridge. 1980a. Uptake and excretion of aminocarb, nonylphenol, and pesticide diluent 585 by mussels (*Mytilus edulis*). Bull. Environ. Contam. Toxicol. 24:575-581.

McLeese, D.W., V. Zitko, C.D. Metcalfe and D.B. Sergeant. 1980b. Lethality of aminocarb and the components of the aminocarb formulation to juvenile Atlantic salmon, marine invertebrates and a freshwater clam. Chemosphere 9:79-82.

McLeese, D.W., V. Zitko, D.B. Sergeant, L. Burridge and C.D. Metcalfe. 1981. Lethality and accumulation of alkylphenols in aquatic fauna. Chemosphere 10:723-730.

Monsanto Industrial Chemicals Co. 1985a. (Cited in Etnier, 1985).

Monsanto Industrial Chemicals Co. 1985b. (Cited in Etnier, 1985).

Moody, R.P., P. Weinberger and R. Greenhalgh. 1983. Algal fluorometric determination of the potential phytotoxicity of environmental pollutants. In: Aquatic Toxicology, J.O. Nriagu, ed. John Wiley & Sons, New York, pp. 503-512.

Moore, S.B., R.A. Diehl, J.M. Barnhardt and G.B. Avery. 1987. Aquatic toxicity of textile surfactants. Text. Chem. Col. 19:29-32.

Naylor, C.G., J.P. Mieure, I. Morici, R.R. Romano. 1992. Alkylphenol ethoxylates in the environment. Presented at the Third CESIO International Surfactants Congress, June 1992, London, U.K.

Nethery, A.A. 1967. Inhibition of mitosis by surfactants. Cytologia 32:321-327.

Neufahrt, A., K. Hofmann, G. Täuber and Z. Damo. 1987. Biodegradation of nonylphenol-ethoxylate (NPE) and certain environmental effects of its catabolites. Hoechst Aktiengesellschaft, unpublished paper.

Nyberg, H. 1976. The effects of some detergents on the growth of *Nitzschia holsatica* Hust. (Diatomeae). Ann. Bot. Fennici 13:65-68.

Nyberg, H. 1985. Physiological effects of four detergents on the algae *Nitzschia actinastroides* and *Porphyridium purpureum*. Publication 12, Department of Botany, University of Helsinki. (Cited in Lewis, 1990).

Nyberg, H. 1988. Growth of *Selenastrum capricornutum* in the presence of synthetic surfactants. Water Res. 22:217-223.

Parr, J.F. and A.G. Norman. 1964. Effects of nonionic surfactants on root growth and cation uptake. Plant Physiol. 39:502-507.

Patoczka, J. and G.W. Pulliam. 1990. Biodegradation and secondary effluent toxicity of ethoxylated surfactants. Water Res. 24:965-972.

Portmann, J.E. and K.W. Wilson. 1971. Shellfish Information Leaflet. Ministry of Agriculture, Fisheries, and Food (U.K.). (Cited in McLeese et al., 1980b).

Prasad, R. 1989. Effects of nonylphenol adjuvant on macrophytes. Adjuvants Agrochem. 1:51-61.

Regupathy, A. and T.R. Subramaniam. 1974. Effect of surfactants on the *in vitro* germination of pollen of bitter-gourd (*Momordica charantia* L.). J. Hort. Sci. 49:197-198.

Reiff, B. 1976. Proceedings of the Seventh International Congress on Surface-Active Substances, Moscow, USSR, September, 1976. (Cited in Scharer et al., 1979).

Reiff, B., R. Lloyd, M.J. How, D. Brown and A.S. Alabaster. 1979. The acute toxicity of eleven detergents to fish: Results of an interlaboratory exercise. Water Res. 13:207-210.

Röderer, G. 1987. Toxic effects of tetraethyl lead and its derivatives on the chrysophyte *Poterioochromonas malhamensis*. VIII. Comparative studies with surfactants. Arch. Environ. Contam. Toxicol. 16:291-301.

Salanitro, J.P., G.C. Langston, P.B. Dorn and L. Kravetz. 1988. Activated sludge treatment of ethoxylate surfactants at high industrial use concentrations. Presented at the International Conference on Water and Wastewater Microbiology, Newport Beach, CA., February 8-11.

Scharer, D., L. Kravetz and J.B. Carr. 1979. Biodegradation of nonionic surfactants. J. Pulp Paper Ind. 62:8 pp.

Shell Chemical Company. 1987. Neodol Product Guide for Alcohols, Ethoxylates and Ethoxysulfates. SC:7-87.

Spotts, R.A. and D.C. Feree. 1979. Effect of a dormant application of surfactants on bud development and disease control in selected deciduous fruit plants. Hort. Sci. 14:38-39.

Stolzenberg, G.E. and P.A. Olson. 1977. Behavior and fate of ethoxylated alkyl-phenol detergents in barley plants. Abstract, 173rd ACS Meeting, New Orleans, March 20-25.

Stolzenberg, G.E. and P.A. Olson. 1978. Nonionic surfactant behavior and fate in rice plants: hexaethoxylated octylphenol. 175th National Meeting of the American Chemical Society, March 1978 (Abstract).

Stolzenberg, G.E., P.A. Olson, R.G. Zaylskie and E.R. Mansager. 1982. Behavior and fate of ethoxylated alkylphenol surfactant in barley plants. J. Agric. Food Chem. 30:637-644.

Swedmark, M. 1968. Resistens hos fisk mot glykol, tensider och en vanlig tensidravara. Vatten 5:430-443. (Cited in Granmo et al., 1991).

Swedmark, M., B. Braaten, E. Emanuelsson and A. Granmo. 1971. Biological effects of surface active agents on marine animals. Marine Biol. 9:183-201.

Swedmark, M., A. Granmo and S. Kollberg. 1976. Toxicity testing at Kristinberg Marine Biology Station. In: Pollutants in the Aquatic Environment. FAO/SIDA/TF 108, Suppl. 1. (Cited in Goyer et al., 1981).

Swisher, R.D. 1987. *Surfactant Biodegradation*, 2nd ed., Surfactant Science Series: Vol. 18. Marcel Dekker, NY.

Tomiyama, S. 1974. Effects of surfactants on fish. Bull. Jpn. Soc. Sci. Fish. 40:1291-1296.

Ukeles, R. 1965. Inhibition of unicellular algae by synthetic surface-active agents. J. Phycol. 1:102-110.

Unilever Research Laboratories. 1977. (Unpublished data cited in Arthur D. Little, 1977).

Vandoni, M.V. and L. Goldberg-Federico. 1973. Sul comportamento dei detergenti di sintest nel terreno agrario--Nota VIII. Riv. Ital. Sost. Grasse 50:185-192. (Cited in Arthur D. Little, 1977).

Van Emden, H.M., C.C.M. Kroon, E.N. Schoeman and H.A. Van Seventer. 1974. The toxicity of some detergents tested on *Aedes aegypti* L., *Lebistes reticulatus* Peters, and *Biomphalaria glabrata* Say. Environ. Pollut. 6:297-308.

Wahlberg, C., L. Renberg and U. Wideqvist. 1990. Determination of nonylphenol and nonylphenol ethoxylates as their pentafluorobenzoates in water, sewage sludge and biota. Chemosphere 20:179-195.

Waldock, M.J. and J.E. Thain. 1991. Environmental concentrations of 4-nonylphenol following dumping of anaerobically digested sewage sludges: A preliminary study of occurrence and acute toxicity. Unpublished paper, Ministry of Agriculture, Fisheries and Food, Fisheries Laboratory, Burnham-on-Crouch, Essex, Britain.

Weinberger, R. and M. Rea. 1981. Nonylphenol: A perturbant additive to an aquatic ecosystem. In: Bermington, N. et al. (Eds.), Proc. 7th Annual Aquatic Toxicity Workshop, November 5-7, Montreal, Quebec. Can. Techn. Rep. Fish Aquatic Sci. No. 990, pp. 371-380.

Weinberger, R. and M. Rea. 1982. Effects of aminocarb and its formulation adjuncts on the growth of *Chlorella pyrenoidosa* Chick. Environ. Exper. Bot. 22:491-496.

Weinberger, R. and R. Vladut. 1981. Comparative toxic effects of some xenobiotics on the germination and early seedling growth of jack pine (*Pinus banksiana* Lamb.) and white birch (*Betula papyrifera* Marsh.). Can. J. For. Res. 11:796-804.

Wong, S.L. 1985. Algal assay evaluation of trace contaminants in surface water using the nonionic surfactant, Triton X-100. Aquatic Toxicol. 6:115-131.

Yamane, A.N., M. Okada and R. Sudo. 1984. The growth inhibition of planktonic algae due to surfactants used in washing agents. Water Res. 18:1101-1105.

Yoshimura, K. 1986. Biodegradation and fish toxicity of nonionic surfactants. J. Am. Oil Chem. Soc. 63:1590-1596.

V. HUMAN SAFETY

An evaluation of the data up to 1981 dealing with various measures of mammalian toxicity as indicators of potential human safety as well as actual human exposure indicate that APE do not represent a hazard to human health (Arthur D. Little, 1977; Goyer et al., 1981). The data reviewed below indicate that for both acute and chronic exposures, APE exhibit a low order of toxicity to mammalian species by the oral, dermal, or inhalation route of intake. Alkylphenol ethoxylates have not been found to cause reproductive, genotoxic, or carcinogenic effects, although cardiotoxicity in dogs as evidenced by focal myocardial necrosis was demonstrated with APE_{15} - APE_{20}.

A. ANIMAL STUDIES

1. Acute Exposures

Oral. Studies of various APE surfactants, administered by gavage to rats, rabbits, guinea pigs, and mice indicate that these materials are practically nontoxic to slightly toxic according to the ranking of Gosselin et al. (1984). The acute oral LD_{50} for the rat ranges from 1410 mg/kg (C_9APE_9) to >28,000 mg/kg (C_8APE_{40}) (Table 5-1). Although APE with an average EO chain length of 9-10 units exhibited the greatest toxicity in some of the studies (Finnegan and Dienna, 1953; CIR Expert Panel, 1983), the data from other studies were too scattered to make conclusions concerning toxicity and chemical structure. Data were similarly scattered for the rabbit and guinea pig. At ≥ 30 EO units, the materials were essentially nontoxic. The length and degree of branching of the alkyl chain did not appear to influence toxicity.

Thompson and Gibson (1984) found a difference between the sexes in sensitivity to acute doses of C_8APE_1 (Triton X-15). They administered the chemical to four groups of five male and five female unfasted rats. The LD_{50} for female rats was less than half that of male rats. Other studies used either one sex (Larson et al., 1963; Smyth and Calandra, 1969) or sexual differences were not apparent (Monsanto Chemical Company, 1975).

Toxic symptoms of rats dying of acute doses included lethargy or depression, diarrhea, tremors, and coma. Most deaths occurred within 24 hours. Autopsies revealed congestion and discoloration of the liver, congestion and hemorrhage of the lungs, pale or mottled kidneys, and inflammation and congestion of the gastric mucosa. In rabbits, tremors, salivation, diarrhea, lethargy, and liver congestion were observed. Animals that survived the high doses and were sacrificed seven days later showed congestion of the lungs and slight discoloration or congestion of the liver. At low doses the viscera appeared normal (Monsanto Chemical Company, 1959; 1972; Larson et al., 1963; Smyth and Calandra, 1969).

Dermal. The dermal LD_{50} for rabbits following exposure to a series of C_9APE of 4 to 13 EO units ranged from 2000 to >10,000 mg/kg (Table 5-2). The toxicity decreased as the number of ethylene oxide units increased (Monsanto Chemical Company, 1975). In this study,

Table 5-1. Acute Oral Toxicity of Alkylphenols and Alkylphenol Ethoxylates to Mammals

Species	Chemical	LD_{50} (mg/kg)	Reference
Octylphenols			
Rat	o-Octylphenol	2800	RTECs, 1987
	p-t-Octylphenol	2160	RTECs, 1987
Mouse	p-t-Octylphenol	3210	RTECs, 1987
Octylphenol Ethoxylates			
Rat	p-t-C_8APE_1	7200	Finnegan and Dienna, 1953
	p-t-C_8APE_3	3900	
	p-t-C_8APE_4	3700	
	p-t-$C_8APE_{9.7}$	1700	
	p-t-$C_8APE_{12.5}$	1800	
Rat	p-t-C_8APE_{16}	2700	Larson et al., 1963
	p-t-C_8APE_{20}	3600	
	p-t-C_8APE_{30}	21,200	
	p-t-C_8APE_{40} (Triton X-405)	>28,000	
Rat	C_8APE_8 (Igepal CA-630)	4250^a	Schick, 1967
	Triton CF-10	2800	
Rat	C_8APE_1 (Triton X-15)	11,600 (♂) 4190 (♀)	Thompson and Gibson, 1984
Guinea pig	C_8APE_8 (Igepal CA-630)	1650^a	Schick, 1967
Nonylphenol			
Rat	Nonylphenol (mixed isomers)	1620^a	Smyth et al., 1969
Rat	Nonylphenol	580	Texaco Chemical Company, 1985a
Rat	Nonylphenol	1300	Monsanto Chemical Company, 1985
Nonylphenol Ethoxylates			
Rat	C_9APE_{9-10}	1600	Olson et al., 1962
Rat	C_9APE_9	2600	Smyth and Calandra, 1969

Table 5-1. (cont.)

Species	Chemical	LD$_{50}$ (mg/kg)	Reference
Rat	C$_9$APE$_4$ (Igepal C0-430)	5000[a]	Schick, 1967
	C$_9$APE$_4$ (Tergitol NP 14)	4290[a]	
	C$_9$APE$_7$ (Tergitol NP 27)	3670[a]	
	C$_9$APE$_{8-9}$ (Igepal CO-630)	3000	
	C$_9$APE$_9$ (Tergitol TP 9)	2600[a]	
	C$_9$APE$_{10.5}$ (Tergitol NPX)	2500[a]	
	C$_9$APE$_{13.5}$ (Igepal CO-730)	2500	
	C$_9$APE$_{15}$ (Tergitol NP 35)	4000	
	C$_9$APE$_{20}$ (Igepal CO-880)	>16,000	
	C$_9$APE$_{20}$ (Tergitol NP 40)	15,900	
	Igepal DJ-970	31,500	
	Surfonic N-90	2580	
Rat	br-C$_9$APE$_4$ (Sterox ND)	4800	Monsanto Chemical Company, 1975
	br-C$_9$APE$_5$ (Sterox NE)	3250	
	br-C$_9$APE$_7$ (Sterox NG)	3600	
	br-C$_9$APE$_9$ (Sterox NJ)	5600	
	br-C$_9$APE$_{10.3}$ (Sterox NK)	4800	
	br-C$_9$APE$_{12}$ (Sterox NL)	2170	
	br-C$_9$APE$_{13}$ (Sterox NM)	5600	
Rat	br-C$_9$APE$_2$	3550	CIR Expert Panel, 1983
	br-C$_9$APE$_4$	7400	
	br-C$_9$APE$_4$	4300	
	br-C$_9$APE$_6$	1980	
	br-C$_9$APE$_7$	3670[a]	
	br-C$_9$APE$_9$	1410-3000[a,b]	

Table 5-1. (cont.)

Species	Chemical	LD_{50} (mg/kg)	Reference
Rat (cont.)	br-C_9APE_{10} br-C_9APE_{12} br-C_9APE_{13} br-C_9APE_{15} br-C_9APE_{40}	1300 5100[a] 3730[a] 2500 not determined[c]	CIR Expert Panel, 1983 (cont.)
Rat	C_9APE_4 (Surfonic N-40) $C_9APE_{9.5}$ (Surfonic N-95 C_9APE_{12} (Surfonic N-120)	>5000 3300 3900	Texaco Chemical Company, 1991a Texaco Chemical Company, 1991b Texaco Chemical Company, 1991c
Rabbit	C_9APE_{12} (Sterox NL)	>871<1050[d]	Monsanto Chemical Company, 1959
Rabbit	br-C_9APE_9 br-C_9APE_9	4400 620[a]	CIR Expert Panel, 1983
Guinea pig	C_9APE_4 (Igepal CO-430) C_9APE_{8-9} (Igepal CO-630)	5000[a] 2000	Schick, 1967
Guinea pig	br-C_9APE_9 br-C_9APE_9	2000 840[a]	CIR Expert Panel, 1983
Mouse	br-C_9APE_9	4290[a]	CIR Expert Panel, 1983
Dodecylphenol			
Rat	Dodecylphenol	2140[a]	Smyth et al., 1962
Dodecylphenol Ethoxylates			
Rat	br-$C_{12}APE_{6.2}$ (Sterox DF) br-$C_{12}APE_{11}$ (Sterox DJ) n-$C_{12}APE_{11}$ (Sterox MJ-b)	4800 3300 2510-3550	Monsanto Chemical Company, 1964, 1975

[a] μL/kg (the specific gravity of most APE = ~1).
[b] Range of eight studies.
[c] Doses ranged from 4 to 64 mL/kg; product was characterized as relatively harmless.
[d] Minimum lethal dose.

the protocol consisted of application of the undiluted compound to the closely clipped, intact skin of New Zealand rabbits. The treated areas were covered with plastic and the animals were placed in wooden stocks for 8 hours, after which time they were placed in individual cages. No deaths occurred in this study.

In data compiled by the CIR Expert Panel (1983), a series of undiluted C_9APE with 4, 7, 9, 10, or 13 EO units was applied under occlusion to the shaved, abraded backs of rabbits. The animals were observed for 14 days. The LD_{50} values ranged from 1800 μL/kg (1800 mg/kg)[1] for C_9APE_7 to 4400 mg/kg for C_9APE_9. When applied as a 50% solution (diluent not defined), C_9APE_{40} had an LD_{50} of >10,000 mg/kg. Toxic effects included erythema and necrosis of the skin at the site of application, diarrhea, lung congestion and hemorrhages, liver congestion, and mottled kidneys.

Inhalation. Six male rats were exposed for six hours to the concentrated vapor of C_9APE_7 (Sterox NG) formed by passing a stream of air through 112 grams of the compound contained in a 500-mL Erlenmeyer flask. Upon removal from the chamber, no symptoms of irritation or toxicity were observed and respiration was normal. No mortalities occurred during a 10-day observation period. The animals were sacrificed after 10 days at which time the viscera appeared normal (Monsanto Chemical Company, 1972).

Data submitted to the CIR Expert Panel (1983) included one report on inhalation toxicity. In this study, groups of six male rats were placed in chambers and exposed to 1% aqueous aerosol dispersions of C_9APE_4 or C_9APE_7 or the concentrated vapor of C_9APE_9, formed at ambient temperature, for eight hours. Concentrations for the first two compounds were 0.0213 and 0.025 mL/L. The animals were observed for 14 days during which time weight gains were normal and there were no mortalities. Six male rats exposed for four hours to the vapor of C_9APE_9 formed at 179°C (decomposition products) also suffered no mortalities.

Smyth et al. (1962; 1969) list four hours as the maximum time to death for rats inhaling a concentrated vapor of NP.

Other Routes. In order to evaluate the direct effects of surfactants on the lungs and to assess the potential hazards of aspiration, 10 male and 10 female Sprague-Dawley rats were administered NPE_9 or NPE_4 by endotracheal injection (Myers et al., 1988; Tyler et al. 1988). NPE_9 was more toxic than NPE_4, with a minimum lethal dose of 0.02 mL/kg (1.0% dilution in saline) for the former compared to a minimum lethal dose of 0.64 mL/kg (32% emulsion in saline) for the latter. A dose of 0.01 mL/kg of NPE_9 resulted in no deaths or signs of toxicity and weight gains after the first day were similar to controls. After 14 days, absolute and relative lung weights and the incidence of alveolar histiocytosis were increased in females. Female rats dosed with 0.32 mL/kg NPE_4 did not gain weight by 14 days post treatment; microscopically,

---

[1]The specific gravity of most APE ≈ 1.

Table 5-2. Acute Dermal Toxicity of Nonylphenol and Alkylphenol Ethoxylates to Rabbits

Chemical	LD_{50} (mg/kg)	Reference
Nonylphenol	2,031 >2,000<3,160	Smyth et al., 1969 Monsanto Chemical Co., 1985
C_9APE_4 (Sterox ND)	>2,000 2,500[a]	Monsanto Chemical Co., 1975 CIR Expert Panel, 1983
C_9APE_4 (Surfonic N-40)	>3,000	Texaco Chemical Company, 1992a
C_9APE_5 (Sterox NE)	>3,160	Monsanto Chemical Co., 1975
C_9APE_7 (Sterox NG)	>3,160 1,800[a]	Monsanto Chemical Co., 1975 CIR Expert Panel, 1983
C_9APE_9 (Sterox NJ)	>5,010 4,400[a] 2,830[a]	Monsanto Chemical Co., 1975 CIR Expert Panel, 1983 CIR Expert Panel, 1983
$C_9APE_{9.5}$ (Surfonic N-95)	>3,000	Texaco Chemical Company, 1992b
C_9APE_{10} (Sterox NK)	>2,000 2,000[a]	Monsanto Chemical Co., 1975 CIR Expert Panel, 1983
C_9APE_{12} (Sterox NL)	>10,000	Monsanto Chemical Co., 1975
C_9APE_{12} (Surfonic N-120)	>3,000	Texaco Chemical Company, 1992c
C_9APE_{13} (Sterox NM)	>7,940 3,970[a]	Monsanto Chemical Co., 1975 CIR Expert Panel, 1983
C_9APE_{40}[b]	>10,000	Monsanto Chemical Co., 1975
$C_{12}APE_6$ (Sterox DF)	>3,160	Monsanto Chemical Co., 1975
$C_{12}APE_{11}$ (Sterox DJ)	>1,260	Monsanto Chemical Co., 1975

[a] $\mu L/kg$.
[b] Applied as 50% solution.

the only lesion present was alveolar histiocytosis in the females. The authors concluded that significant aspiration hazards exist for ethoxylated surfactants.

Intratracheal instillation may not be the best method for predicting aspiration potential. However, Zerkle et al. (1987) state that ethoxylates can be naturally emetic in man, making aspiration a possible route of lung exposure. Osterberg et al. (1976) appraised existing methodology for aspiration toxicity testing and concluded that the Gerarde (1963) technique in which the material is placed in the mouth is superior for predicting aspiration hazard and toxicity. The Gerarde technique allowed a larger dose to enter the lungs and was consistent in dose-response. Low viscosity and low surface tension increased the aspiration hazard of chemicals.

Female mice were injected with C_9APE_9 intraperitoneally (undiluted), subcutaneously (undiluted), or intravenously (1% solution in saline). The respective LD_{50} values were 210 mg/kg, 1000 mg/kg, and 44 mg/kg (CIR Expert Panel, 1983).

Ariyoshi et al. (1990) studied the effects of a single intraperitoneal injection of C_9APE_n (Emulgen 913) on the hemoproteins and heme-metabolizing enzymes in the livers of rats. A dose of 50 mg/kg increased heme oxygenase activity and depressed microsomal total heme and cytochromes P-450 and b_5. There were no changes in several other enzymes including the rate-limiting enzyme in the heme biosynthetic pathway. The relevance of the intraperitoneal route of administration and the enzyme modulating effects to human health is unknown.

2. Subchronic Exposures

Oral. The inclusion of polyethylene glycol monoisooctylphenol (EO content not stated) in the diet of male Osborne-Mendel rats for 16 weeks at levels of 2% or 4% (1000 or 2000 mg/kg/day) resulted in significant reductions in body weight gain. At $\leq 1\%$ (500 mg/kg/day) in the diet, the weight gain was not significantly different from that of controls. Two of five rats on the 4% dosage and 5/5 rats on an 8% (4000 mg/kg/day) dosage died within the first week of the experiment (Fitzhugh and Nelson, 1948).

Larson et al. (1963) placed 15 male and 15 female albino rats on diets containing 5% (2500 mg/kg/day) C_8APE_{40} (Triton X-405) for three months. No effects on weight gain, hematologic indices, organ to body weight ratios, or tissues were noted. Beagle dogs treated with 0.35% or 5% (88 or 1250 mg/kg/day) in the diet were similarly unaffected.

In their analysis of the toxicology of APE surfactants, Smyth and Calandra (1969) conducted 90-day subchronic feeding studies with several species at dose levels of 40 to 5000 mg/kg/day (concentrations of 0.01-1.0% of APE in the diet). Ingestion of APE with 20 or more EO units resulted in little toxicity to rats. Ingestion of APE of less than 20 EO units caused growth retardation, attributed to the unpalatability of the food, and increased absolute and relative liver weights for the lower homologs at 0.04 or 0.2% (20 or 100 mg/kg/day) in the diet. In several additional feeding studies with rats using C_9APE_9, the following effects were reported: retardation of weight gain at 0.64% (320 mg/kg/day) and higher in the diet (attributed to poor palatability) and reversible cellular changes in the liver and kidney such as cloudy swelling, focal necrosis, and lipid deposition at 0.25 and 1.25% (125 and 625 mg/kg/day) in the diet. In one study, no effects were observed at 0.1% (50 mg/kg/day) in the diet, whereas in another study there was a reduced weight gain at 0.01% (5 mg/kg/day) in the diet. No effects were seen with $\leq$0.3% (150 mg/kg/day) C_9APE_{40} (Smyth and Calandra, 1969) or 5% (2500 mg/kg/day) C_8APE_{40} (Larson et al., 1963) in the diet for 90 days. In summarizing their data, Smyth and Calandra (1969) reported no-effect levels on growth and liver weight in the 90-day feeding studies with rats and dogs (Table 5-3).

Table 5-3. No-Observed Effect Level in 90-Day Feeding Studies

Chemical	Species	Dose (mg/kg/day)[a]
C_8APE_9	Rat Dog	40 —[b]
C_8APE_{40}	Rat Dog	5% in diet (2500) 5% in diet (1250)
C_9APE_4	Rat Dog	40 40
C_9APE_6	Rat Dog	—[b] 40
C_9APE_9	Rat Dog	10 0.04% in diet (10)
C_9APE_{15}	Rat Dog	40 40
C_9APE_{20}	Rat Dog	1000 —[b]
C_9APE_{30}	Rat Dog	5000 1000
C_9APE_{40}	Rat	0.3% in diet (150)
$C_{12}APE_6$	Rat	—[b]
$C_{12}APE_9$	Rat	—[b]
$C_{12}APE_{40}$	Rat	5000

[a]Percent in diet converted to mg/kg/day using food factors of 0.05 for rats and 0.025 for dogs; mg/kg/day = percent x 10,000 x food factor (U.S. EPA, 1985).
[b]Retarded growth and/or liver weight changes were present at 40 mg/kg/day, the lowest dose tested.
Source: Smyth and Calandra, 1969.

Smyth and Calandra (1969) also addressed the effect of chemical structure on toxicity in a similar subchronic feeding study with dogs. Feeding of 40, 200, 1000 or 5000 mg/kg/day of several APE surfactants with 4 to 30 EO units resulted in toxic effects for only C_9APE_{20}. At 40 mg/kg/day there was microscopic evidence of cardiotoxicity as seen by focal myocardial necrosis. At 1000 mg/kg/day, six of eight dogs died and focal myocardial necrosis could be grossly observed. A dose of 5000 mg/kg/day of this same surfactant had no effect on rats. Both dogs and guinea pigs showed evidence of cardiac lesions, but rabbits, rats, and cats did not.

In order to further study the cardiotoxicity of APE in dogs and eliminate the possibility that an impurity was present in the commercial product previously studied, one male and one female Beagle dog were fed different commercial products of various chain lengths (Smyth and Calandra, 1969). At a dose of 1000 mg/kg/day, the lesions were present by the 14th day. Regardless of the alkyl group, cardiotoxicity was evident for all materials having an average EO chain length of 15-20 units (Table 5-4).

Table 5-4. Cardiotoxicity in Dogs Fed APE of Various Chain Lengths

Chemical	Dose (mg/kg/day)	Cardiotoxicity
C_9APE_{12}	1000	0
C_9APE_{15}[a]	1000	0
	200	0
	1000	+
$C_9APE_{17.5}$	1000	+
C_8APE_{20}	1000	+
C_9APE_{20}[a]	1000	+
	1000	+
$C_{12}APE_{20}$	1000	+
C_9APE_{25}	1000	0

[a]Products from different manufacturers tested.
Source: Smyth and Calandra, 1969

Dermal. Several octylphenol ethoxylates with ethylene oxide units ranging from 1 to 13 were applied to rabbit skin at concentrations up to 1%, 5 days/week, for 4 weeks (Finnegan and Dienna, 1953). Local irritant effects were noted, but there were no systemic effects.

Brown (1969) examined the effects of isooctylphenol ethoxylates on oxygen consumption and enzyme activity of skin. Application of a 2% aqueous solution (20,000 mg/L) of $C_8APE_{8.9}$ or C_8APE_{15} for one to four weeks had no effect on oxygen consumption in excised skins of hairless mice. In skin which had been painted for 4 weeks with a 2% aqueous solution of an isooctylphenol ethoxylate, there was an increase in activity of the enzymes phosphogluconate dehydrogenase, glucose-6-phosphate dehydrogenase, monoamine oxidase, and succinic dehydrogenase. This was not observed in skin treated with C_8APE_{8-9}.

3. Chronic Exposures

Oral. Thirty male and 30 female Wistar rats fed C_8APE_{40} at dietary concentrations of 0, 0.035, 0.35, or 1.4% (0, 17.5, 175, or 700 mg/kg/day) for two years showed no adverse effects on growth and survival, food consumption, hematologic values, urinary concentrations of reducing substances and protein, organ to body weight ratios, or incidence of pathologic lesions (Larson et al., 1963).

In two-year feeding studies, C_9APE_4 at doses of 200 and 40 mg/kg/day produced no significant effects on rats and dogs, respectively (Smyth and Calandra, 1969). C_9APE_9 at doses of 140 and 30 mg/kg/day in the diet of rats and dogs, respectively, also produced no significant toxicological effects. Parameters considered were body and organ weights and histopathology of 28 tissues. A dose of 1000 mg/kg/day of C_9APE_4 resulted in reduced body weights and enlarged livers in rats and reduced weight, emesis, and increased serum alkaline phosphatase in dogs. At a dose of 88 mg/kg/day, C_9APE_9 produced increased liver to body weight ratios in dogs. These effects were attributed to reduced food intake as demonstrated by paired feeding studies.

4. Dermal and Ocular Irritation

Dermal. The primary skin irritation potential of APE has been tested using primarily the patch test method of Draize et al. (1944). In this test, 0.5 mL of the undiluted or diluted test material is applied to the shaved backs of albino rabbits. Usually four sites, two intact and two abraded are prepared. Patches are taped over the exposure sites and the trunk of the animal is wrapped with plastic. After 24 hours, the wrappings are removed and the sites are scored for irritation. The backs are then washed and the sites are re-scored after 72 hours. The sites are scored on a basis of weighted scores of 0 (non-irritating) to 4 (severe reaction) for both erythema (reddening) and edema (swelling) for a total score, referred to as the Primary Irritation Index, of up to 8.

Although results among studies are conflicting for some surfactants, NPE with 2 to 9 EO units were generally moderately to severely irritating when applied undiluted to rabbit skin, whereas NPE with EO ≥ 10 were non- to mildly irritating (Table 5-5). The dodecyl APE were both moderately irritating. Effects were similar for intact and abraded skin. Applications of NPE_9 at concentrations of $\leq 25\%$ resulted in very slight to slight erythema.

Undiluted NP was severely irritating. A volume of 0.5 mL administered once to the shaved, intact skin of albino rabbits and wiped off after 4 hours was considered corrosive when scored 48 hours after application (Texaco Chemical Company, 1985b). When massaged into shaved rabbit skin daily for 30 days, a 10% solution of NP in water (0.5 mL/day) induced

Table 5-5. Rabbit Skin Irritation

Surfactant/ Trade Name	Concentration [a,b]	Irritation Category (Score)[c]	Reference
NPE_2	undiluted	moderate (2/8)	CIR Expert Panel, 1983
NPE_4 (Sterox ND)	undiluted	severe (7.5/8)	Monsanto Chemical Company, 1975
NPE_4	undiluted	primary (5.6/8)	CIR Expert Panel, 1983
NPE_4	undiluted (0.01 mL)	non-irritating	CIR Expert Panel, 1983
NPE_4 (Surfonic N-40)	undiluted	slightly irritating (1.8/8.0)	Texaco Chemical Company, 1992d
NPE_5 (Sterox NE)	undiluted	severe (7.5/8)	Monsanto Chemical Company, 1975
NPE_6	undiluted	severe	CIR Expert Panel, 1983
NPE_7 (Sterox NG)	undiluted	severe (7.5/8)	Monsanto Chemical Company, 1975
NPE_7	undiluted (0.01 mL)	non-irritating	CIR Expert Panel, 1983
NPE_9 (Sterox NJ)	undiluted	severe (6.7/8)	Monsanto Chemical Company, 1975
NPE_9	undiluted	non-irritating	CIR Expert Panel, 1983
NPE_9	undiluted (0.01 mL)	minimal (2/10)	CIR Expert Panel, 1983
$NPE_{9.5}$	1% (5 mL)[d] 5% (5 mL)[d] 25% (5 mL)[d]	very slight very slight slight erythema	Olson et al., 1962
$NPE_{9.5}$ (Surfonic N-95)	undiluted	slightly irritating (1.04/8.0)	Texaco Chemical Company, 1991d
NPE_{10} (Sterox NK)	undiluted	mild (2.5/8)	Monsanto Chemical Company, 1975
NPE_{10}	undiluted (0.01 mL)	non-irritating	CIR Expert Panel, 1983
NPE_{12} (Sterox NL)	undiluted	mild (2.0/8)	Monsanto Chemical Company, 1975
NPE_{12}	undiluted	slight (0.75/8)	CIR Expert Panel, 1983
NPE_{12} (Surfonic N-120)	undiluted	non-irritating (0.0/8.0)	Texaco Chemical Company, 1992e

Table 5-5. (cont.)

Surfactant/ Trade Name	Concentration [a,b]	Irritation Category (Score)[c]	Reference
NPE_{13} (Sterox NM)	undiluted	non-irritating (0.7/8)	Monsanto Chemical Company, 1975
NPE_{13}	undiluted (0.01 mL)	marked (3/10)	CIR Expert Panel, 1983
NPE_{15}	undiluted	slight (0.45/8)	CIR Expert Panel, 1983
NPE_{30}	70%	mild (1.83/8)	CIR Expert Panel, 1983
NPE_{40}	undiluted	mild (1.46/8)	CIR Expert Panel, 1983
NPE_{40}	undiluted (0.01 mL)	minimal (2/10)	CIR Expert Panel, 1983
$C_{12}APE_{6.2}$ (Sterox DF)	undiluted	moderate (4.5/8)	Monsanto Chemical Company, 1975
$C_{12}APE_{11}$ (Sterox DJ)	undiluted	moderate (4.4/8)	Monsanto Chemical Company, 1975
Nonylphenol	undiluted	severe (8/8)	Monsanto Chemical Company, 1985; Texaco Chemical Company, 1985c

[a]Volume of 0.5 mL unless otherwise specified.
[b]Dilutions are aqueous solutions.
[c]Score/total possible score.
[d]Applied to intact (10 applications/14 days) and abraded (1 application daily for 3 days) skin.

changes in the epithelial cells, including increased layers and numbers of cells in the superficial dermis (Rantuccio et al., 1984).

Of several classes of surfactants studied by Brown (1969), APE were ranked as the least irritating in rabbits, guinea pigs, and hairless mice. Brown (1971) cautioned that animal tests should be used with caution in predicting irritant effects in humans.

Primary Eye Irritation. The potential of APE to irritate the eye has been assessed using the standard Draize test (Draize, 1959; Draize et al., 1944) and a scoring system of 0 to 110 points with the albino rabbit as the animal model. A volume of 0.1 mL of the test material is introduced into the conjunctival sac of one eye while the other eye serves as the control. The unwashed eye is then examined for ocular injury at one hour and at daily intervals up to 14 days after instillation. The cornea, iris, and conjunctiva are graded on the basis of 0 (no visible response) to 4 (severe response). A solution of ophthalmic fluorescein is used for detection of corneal lesions.

Undiluted NPE were generally moderately to severely irritating to the eyes (Table 5-6). A series of C_9APE and $C_{12}APE$ surfactants, tested undiluted in the rabbit eye, were severely irritating, producing ulceration of the cornea in at least two of six animals after the 10th day (Monsanto Chemical Company, 1975). Ten percent solutions of the compounds were moderately to severely irritating, with C_9APE_5 and $C_9APE_{9.2}$ producing ulceration. These lesions may or may not be reversible.

In one study (Finnegan and Dienna, 1953), effects were related to EO chain length. Lowest concentrations of p-t-C_8APE that resulted in eye mucosa irritation decreased with increasing EO chain length up to p-t-$C_8APE_{12.5}$. In addition, NPE_{40} was non-irritating. Rinsing the eyes immediately after instillation reduced the irritation index (Olson et al., 1962; Gershbein and McDonald, 1977).

Based on the Draize method, Texaco Chemical Company (1985a) and Monsanto Chemical Company (1985) list an eye irritation index for NP of 57/110 and 58/110, respectively. Based on a scale of 1 to 10, Smyth et al. (1969) assigned a score of 10 to NP, indicating a severe burn from 0.5 mL of a 1% solution in water or propylene glycol. According to the Texaco Chemical Company (1985a), NP is extremely irritating to the eyes and may cause permanent eye injury.

5. Skin Sensitization

No studies on an allergic response to APE surfactants were reported. Skin sensitization was not observed in guinea pigs treated with NP in a modified Buehler test (Texaco Chemical Company, 1985c).

6. Carcinogenicity

No carcinogenic effects were noted in chronic (two-year) feeding studies in which rats were administered C_8APE_{40} at doses up to 700 mg/kg/day in the diet (Larson et al., 1963), rats and dogs were administered C_9APE_4 at doses up to 1000 mg/kg/day, and rats and dogs were administered C_9APE_9 at concentrations up to 140 and 88 mg/kg/day, respectively (Smyth and Calandra, 1969). (See Chronic Exposures - Section A.3 for noncarcinogenic effects).

The tumor-promoting activity of multiple applications of a 22% solution of octylphenol [4-(1,1,2,2-tetramethylbutyl)phenol] applied in benzene to the skin of female Sutter mice following a single initiating dose of dimethylbenzanthracene (DMBA) was reported by Boutwell and Bosch (1959). Following twice weekly applications for 12 weeks, the number of surviving mice (18/20) was the same as the control group (18/20) in which DMBA was applied without promoter. Of the surviving animals, the number with papillomas was increased over controls (11/20 compared to 0/20), but no carcinomas were present. The effect of octylphenol alone was not studied.

A C_9APE with EO number undefined was studied for cocarcinogenic activity with N-methyl-N'-nitro-N-nitroso guanidine (MNNG) (Takahashi et al., 1975). Both MNNG at 100

Table 5-6. Rabbit Eye Irritation

Surfactant/ Trade Name	Concentration[a,b]	Irritation Category	Reference
p-t-C_8APE_1	15%	"irritant threshold"	Finnegan and Dienna, 1953
p-t-C_8APE_3	15%	"irritant threshold"	Finnegan and Dienna, 1953
p-t-C_8APE_5 (Triton X-45)	5%	"irritant threshold"	Finnegan and Dienna, 1953
p-t-C_8APE_9 (Triton X-100)	0.5%	"irritant threshold"	Finnegan and Dienna, 1953
p-t-$C_8APE_{12.5}$ (Triton X-102)	1%	"irritant threshold"	Finnegan and Dienna, 1953
NPE_2	undiluted	minimal	CIR Expert Panel, 1983
NPE_4 (Sterox ND)	undiluted 10%	severe (corrosive) slight	Monsanto Chemical Company, 1975
NPE_4	undiluted	moderate	CIR Expert Panel, 1983
NPE_4 (Surfonic N-40)	undiluted	minimal (13.0/110)	Texaco Chemical Company, 1992f
NPE_5 (Sterox NE)	undiluted 10%	severe (corrosive) severe (corrosive)	Monsanto Chemical Company, 1975
NPE_6	undiluted	severe	CIR Expert Panel, 1983
NPE_7 (Sterox NG)	undiluted 10%	severe (corrosive) 27.3/110	Monsanto Chemical Company, 1975
NPE_7	undiluted	moderate	CIR Expert Panel, 1983
$NPE_{9.2}$ (Sterox NJ)	undiluted 10%	severe (corrosive) severe (corrosive)	Monsanto Chemical Company, 1975
NPE_9 (Neutronyx 600)	20%	34.4/100	Gershbein and McDonald, 1977
NPE_9	undiluted	moderate	CIR Expert Panel, 1983
$NPE_{9.5}$	1% (2 drops) 5% (2 drops) 25% (2 drops)	very slight slight - moderate moderate - severe	Olson et al., 1962
$NPE_{9.5}$ (Surfonic N-95)	undiluted	minimal (14.4/110)	Texaco Chemical Company, 1992g

Table 5-6. (cont.)

Surfactant/ Trade Name	Concentration[a,b]	Irritation Category	Reference
$NPE_{10.3}$ (Sterox NK)	undiluted 10%	severe (corrosive) 27.6/110	Monsanto Chemical Company, 1975
NPE_{10}	undiluted	severe	CIR Expert Panel, 1983
NPE_{12} (Sterox NL)	undiluted	moderate	Monsanto Chemical Company, 1975
NPE_{12}	undiluted	severe	CIR Expert Panel, 1983
NPE_{13} (Sterox NM)	undilutd 10%	severe (corrosive) 22.1/110	Monsanto Chemical Company, 1975
NPE_{13}	undiluted	severe	CIR Expert Panel, 1983
NPE_{15}	undiluted	moderate	CIR Expert Panel, 1983
NPE_{15}	10, 15% (0.5 mL)	minimal	CIR Expert Panel, 1983
NPE_{15}	20%	slight to moderate	CIR Expert Panel, 1983
NPE_{30}	25% (0.5 mL)	non-irritating	CIR Expert Panel, 1983
NPE_{40}	undiluted	non-irritating	CIR Expert Panel, 1983
$C_{12}APE_{6.2}$ (Sterox DF)	undiluted 10%	severe (corrosive) 20.5/110	Monsanto Chemical Company, 1975
$C_{12}APE_{11}$ (Sterox DJ)	undiluted 10%	severe (corrosive) 28.5/110	Monsanto Chemical Company, 1975
nonylphenol	undiluted	57/110	Texaco Chemical Company, 1985a
nonylphenol	undiluted	58/110	Monsanto Chemical Company, 1985
nonylphenol	undiluted	10/10	Smyth et al., 1969

[a]Volume of 0.1 mL unless otherwise stated.
[b]Dilutions are aqueous solutions.

mg/L and the surfactant at 2000 mg/L were supplied in drinking water to 15 rats for 36 weeks. The surfactant plus carcinogen resulted in stomach adenocarcinomas in 12/15 rats (80%) compared to 8/13 (52%) in the group treated with MNNG alone. The incidence of tumors of the small intestine was 7/15 in the first group compared to 1/13 in the MNNG group. The authors suggested that the surfactant may help the carcinogen penetrate the mucosal cells.

7. Genotoxicity

APE surfactants and NP are non-genotoxic when tested for mutations or genetic damage in a variety of *in vitro* and *in vivo* test systems. These short-term tests include forward and reverse mutations in bacterial (*Salmonella typhimurium* and *Escherichia coli*) and mammalian cell systems; clastogenic effects such as aneuploidy in fungi (*Aspergillus nidulans*) and chromosome aberrations and breaks both *in vivo* and *in vitro* in bacterial and mammalian cells; unscheduled DNA synthesis in mammalian cells *in vitro*; and mammalian cell transformations (Table 5-7). In the *in vitro* studies, the use of a microsomal enzyme activating system is indicated by the presence (+) or absence (-) of S-9. In these studies APE surfactants were applied as aqueous or solvent diluted products. Unpublished and published studies performed through 1982 have been reviewed by Yam et al. (1984).

All of the bacterial and mammalian cell mutational assay results reported in Table 5-7 were negative as were the *in vivo* tests for genetic damage in somatic and germ cells. Although the results of many of the other studies were negative, a few of the tests, such as the cell transformation and unscheduled DNA synthesis assays, produced conflicting results. A weakly positive result was observed in the only viral transposition induction assay. These latter assays are of tentative importance in genotoxicity and in predicting carcinogenicity or cocarcinogenicity, but probably need further study to resolve conflicting results. Positive results were seen with two surfactants: C_8APE_9 (Nonidet P40) in an unscheduled DNA synthesis inhibition assay and the viral transposition assay and C_9APE_9 (NP-9) in one of two cell transformation assays.

In one study, wastewater effluents were tested for mutagenic activity. Using *Salmonella typhimurium* TA100, mutagenic activity was detected in wastewater effluents by Reinhard et al. (1982). The activity was present following chlorination, was removed after activated carbon treatment, but reappeared after final chlorination. The mutagenic fraction was isolated by Amberlite XAD resin, separated by liquid-solid chromatography on silica gel, taken up in a methanol fraction, and characterized by GC/MS. The compounds present were primarily carboxylic acids including brominated and nonbrominated alkylphenol polyethoxy carboxylic acids. However, mutagenicity testing of chemically-produced brominated alkylphenol ethoxylate carboxylic acids failed to confirm the mutagenicity of these compounds.

8. Developmental/Reproductive Toxicity

APE surfactants such as C_9APE_9 (nonoxynol-9) have been used in contraceptive preparations and thus have been studied for potential irritant and teratogenic effects. In humans, the amount applied ranges from 50 to 140 mg/vaginal application (0.7 to 2.0 mg/kg, assuming a 70-kg individual). In the following studies, effects were noted on the reproductive

Table 5-7. Genotoxicity Assays

Chemical	Assay	Indicator organism	Application/ Activating system	Concentration/Dose	Response[a]	Reference
Bacterial Assays						
p-Nonylphenol	Reverse mutation histidine locus	Salmonella typhimurium TA100, TA1535, TA98, TA1537, TA1538	Plate incorporation/ ±S-9	0-100 µg/plate	–	Shimizu et al., 1985
p-Nonylphenol	Reverse mutation tryptophan locus	Escherichia coli WP2uvrA	Plate incorporation/ ±S-9	0-100 µg/plate	–	Shimizu et al., 1985
Nonylphenol	Reverse mutation histidine locus	Salmonella typhimurium TA97, TA98, TA100, TA102, TA104	Preincubation spot test/±S-9	200 µg/plate	–	Varma et al., 1986
C_9APE_9 (NP-9)	Reverse mutation histidine locus	Salmonella typhimurium TA100, TA98	Preincubation plate incorporation/±S-9	100-10,000 µg/plate	–	Shibuya et al., 1985
C_9APE_9 (Nonoxynol 9)	Reverse mutation histidine locus	Salmonella typhimurium TA1535, TA100, TA1537, TA98	Plate incorporation/ ±S-9	40-2500 µg/plate	–	Meyer et al., 1988
C_9APE_5 (NP5)	Not reported	Salmonella typhimurium all strains	Plate incorporation/ ±S-9	Not reported	–	Imperial Chemical Industries, 1981
C_9APE_8 (NP8)	Not reported	Salmonella typhimurium all strains	Plate incorporation/ ±S-9	Not reported	–	Imperial Chemical Industries, 1981
C_9APE_{20} (NP20)	Not reported	Salmonella typhimurium all strains	Plate incorporation/ ±S-9	Not reported	–	Imperial Chemical Industries, 1981
C_8APE_1 (Triton X-15)	Not reported	Salmonella typhimurium all strains	Plate incorporation/ ±S-9	Not reported	–	Procter & Gamble Company, 1979
C_8APE_4 (Tergitol NP-14)	Not reported	Salmonella typhimurium all strains	Plate incorporation/ ±S-9	Not reported	–	Procter & Gamble Company, 1979
C_8APE_{20} (Tergitol NP-40)	Not reported	Salmonella typhimurium all strains	Plate incorporation/ ±S-9	Not reported	–	Procter & Gamble Company, 1979

Table 5-7. (cont.)

Chemical	Assay	Indicator organism	Application/ Activating system	Concentration/Dose	Response[a]	Reference
C$_9$APE$_4$ (Surfonic N-40)	Reverse mutation histidine locus	*Salmonella typhimurium* TA98, TA100, TA1535, TA1537, TA1538	Plate incorporation/ ±S-9	167-10,000 µg/plate	–	Texaco Chemical Company, 1991e
C$_9$APE$_{9.5}$ (Surfonic N-95)	Reverse mutation histidine locus	*Salmonella typhimurium* TA98, TA100, TA1535, TA1537, TA1538	Plate incorporation/ ±S-9	100-10,000 µg/plate	–	Texaco Chemical Company, 1983a
Mammalian Cells						
C$_8$APE$_1$ (Triton X-15)	Forward mutation thymidine kinase locus (TK+/-)	Mouse lymphoma L5178Y cells	Cells in culture/±S-9	Not reported	–	Coppinger et al., 1981
C$_9$APE$_9$ (Nonoxynol-9)	Forward mutation 8-azaguanine resistance	Rat liver T51B cells	Cells in culture/ not reported	0-25 µg/mL	–	Buttar et al., 1986
C$_8$APE$_9$ (Octoxynol-9)	Forward mutation 8-azaguanine resistance	Rat liver T51B cells	Cells in culture/ not reported	0-40 µg/mL	–	Buttar et al., 1986
C$_8$APE$_9$ (Triton X-100)	Forward mutation thymidine kinase locus (TK+/-)	Mouse lymphoma L5178Y cells	Cells in culture/-S-9	1.0-45 µg/L	–	Wangenheim and Bolcsfoldi, 1986, 1988
Chromosome Number						
C$_8$APE$_9$ (Triton X-100)	Aneuploidy	*Aspergillus nidulans*	Cultures/-S-9	0.005-0.03%	–	Assinder and Upshall, 1985
Chromosome Aberrations						
C$_8$APE$_9$ (Triton X-100)	Chromosome aberrations	Chinese hamster lung cells	*in vitro*/±S-9	Not reported	–	Matsuoka et al., 1986
C$_8$APE$_1$ (Triton X-15)	Chromosome aberrations in bone marrow	Sprague-Dawley rats, ♂ and ♀	ip and po/*in vivo*	0.58-0.96 mg/kg (ip) 2.2-11.0 mg/kg (po)	–	Thompson and Gibson, 1984

Table 5-7. (cont.)

Chemical	Assay	Indicator organism	Application/ Activating system	Concentration/Dose	Response[a]	Reference
DNA Breaks						
C_8APE_1 (Triton X-15)	Testicular cell DNA breaks detected by alkaline elution	Sprague-Dawley rats, ♂	ip/*in vivo*	0.15-0.88 mg/kg, single injection or daily injections for 5 days	–	Skare and Schrotel, 1984
C_8APE_9 (Triton X-100)	DNA breaks detected by alkaline elution	Mouse lymphoma L5178Y cells	Cells in culture/S-9	0-100 μL/L for 3 hours	–	Garberg and Bolcsfoldi, 1985; Garberg et al., 1988
C_9APE_4 (Surfonic N-40)	Micronucleus test	Mouse bone marrow cells	intraperitoneal injection	200 mg/kg	–	Texaco Chemical Company, 1992j
$C_9APE_{9.5}$ (Surfonic N-95)	Micronucleus test	Mouse bone marrow cells	intraperitoneal injection	75 mg/kg	–	Texaco Chemical Company, 1992k
C_9APE_{12} (Surfonic N-120)	Micronucleus test	Mouse bone marrow cells	intraperitoneal injection	40 mg/kg	–	Texaco Chemical Company, 1991f
Genetic Rearrangements						
C_9APE_9 (Nonidet-P40)	Viral transposition induction	*Escherichia coli* K12, phage λ::Tn9-infected cells	Spot test/S-9	10 mg/mL	+	Datta et al., 1983
Unscheduled DNA Synthesis						
C_9APE_9 (Nonidet P40)	Semi-conservative DNA synthesis inhibition	Mouse spleen leukocytes	Cell suspensions/ not reported	0.0002-0.02%	+	Tuschl et al., 1975
C_9APE_9 (Nonidet P40)	Unscheduled DNA synthesis inhibition	Irradiated mouse spleen leukocytes	Cell suspensions/ not reported	0.0005-0.001%	+	Tuschl et al., 1975
C_9APE_9 (Nonidet P40)	Unscheduled DNA synthesis inhibition	Irradiated human lymphocytes	Cell suspensions/ not reported	0.0005%	+	Tuschl et al., 1975

Table 5-7. (cont.)

Chemical	Assay	Indicator organism	Application/ Activating system	Concentration/Dose	Response[a]	Reference
C_8APE_9 (Octoxynol-9)	Unscheduled DNA synthesis	Rat primary hepatocytes	Cells in culture/ not reported	0-25 µg/mL	−	Buttar et al., 1986
C_9APE_9 (Nonoxynol-9)	Unscheduled DNA synthesis	Rat primary hepatocytes	Cells in culture/ not reported	0-50 µg/mL	−	Buttar et al., 1986
C_9APE_4 (Surfonic N-40)	Unscheduled DNA repair	Rat primary hepatocytes	Cell cultures	0.1-5000 µg/mL	−	Texaco Chemical Company, 1992h
$C_9APE_{9.5}$ (Surfonic N-90)	Unscheduled DNA repair	Rat primary hepatocytes	Cell cultures	0-0.001%	−	Texaco Chemical Company, 1984
C_9APE_{12} (Surfonic N-120)	Unscheduled DNA repair	Rat primary hepatocytes	Cell cultures	0.005-1.0 µg/mL	−	Texaco Chemical Company, 1992i
DNA Synthesis						
C_9APE_a	Inhibition of DNA synthesis	Mouse testis cells	in vivo	Not reported	−	Zhang et al., 1989
Germinal Cell Abnormalities						
C_9APE_9 (Nonoxynol-9)	Sperm abnormalities	Mouse germ cells	ip/in vivo	0-60 mg/kg/day for five days	−	Buttar et al., 1986
Cell Transformation						
C_8APE_9 (Triton X-100)	Cell transformation	BALB/3T3/A31-11 cells	in vitro/not reported	0.00001-0.001% 48 hours to 3 weeks	−	Long et al., 1982
C_9APE_9 (Nonidet P40)	Cell transformation	BALB/3T3/A31-11 cells	in vitro/not reported	0.00001-0.001% 48 hours to 3 weeks	−	Long et al., 1982
C_9APE_9 (Nonoxynol-9)	Cell transformation	BALB/3T3/A31-11 cells	in vitro/not reported	0.00001-0.001% 11 days to 3 weeks	+	Long et al., 1982
C_9APE_9 (Nonoxynol-9)	Cell transformation	Mouse fibroblast 10T½ cells	in vitro/not reported	0.00001% once per week for 5 weeks	+	Long et al., 1982
$C_9APE_{9.5}$ (Surfonic N-95)	Cell transformation	Balb/3T3 cells	in vitro	0.9-21.0 nL/mL	−	Texaco Chemical Company, 1983b

Table 5-7. (cont.)

Chemical	Assay	Indicator organism	Application/ Activating system	Concentration/Dose	Response[a]	Reference
C$_8$APE$_9$ (Octoxynol-9)	Cell transformation	Rat liver T51B cells	Cells in culture/ not reported	50 µg/mL	−	Buttar et al., 1986
C$_9$APE$_9$ (Nonoxynol-9)	Cell transformation	Rat liver T51B cells	Cells in culture/ not reported	30 µg/mL	−	Buttar et al., 1986
C$_9$APE$_9$ (Nonoxynol-9)	Cell transformation	BALB/3T3-A31-1-1 cells	in vitro/-S-9	0.01-1000 µg/mL 48 hours or 13 days	−	Sheu et al., 1988

[a]Results are indicated as negative (−) or positive (+).

system and genital tract of female test animals administered doses several times greater than the clinical dose. There was no evidence of teratogenic effects in offspring of rats treated orally or vaginally with doses that resulted in maternal or fetal toxicity.

Fourteen mice fed 0.3% v/w of $C_8APE_{9.5}$ (Triton X-100) in the diet for 26 weeks had 10 ovarian cysts compared to one cyst in 39 mice in the control group (Goldhammer et al., 1970). The mechanism of action for cyst formation appeared to be fluid distension of the ovarian subcapsular space associated with obstruction of the ostium. No further details were given.

In some studies APE surfactants were found to have an inflammatory effect on the vaginal epithelium. C_9APE_9 (nonoxynol-9) was administered intravaginally to rabbits and rats and the morphologic and chemical changes in the vaginal tissue were reported. Collagen sponges containing 2.5, 5, 20, or 50 mg of the chemical were inserted into the vaginas of New Zealand female rabbits, three per group, for 10 days (Chvapil et al., 1980a). Moderate inflammatory changes consisting of edema, small erosions, and infiltration of the submucosal layer by polymorphonuclear leukocytes were present in rabbits exposed to 2.5 mg. Inflammatory cells had also infiltrated the sponges. Inflammatory changes became more pronounced with increasing dosages, i.e., at 50 mg the mucosal epithelium was absent and the submucosal layer was inflamed.

Young adult Sprague-Dawley rats injected intravaginally with a dose of 50 mg/kg body weight for 5, 10, 15, or 20 days were studied for changes in vaginal cell DNA, collagen, and non-collagenous protein (Chvapil et al., 1980a). The DNA content was increased only on day 5 and the collagen on days 15 and 20, indicating little inflammatory effect. Noncollagen protein was not affected. The vagina of the rabbit is considered more sensitive to irritating substances than the vagina of the rat.

Ten virgin female Wistar rats were administered 50 mg/kg body weight of nonoxynol-9 as a 5% w/v solution in water by the intravaginal route (Tryphonas and Buttar, 1982). Groups of nine rats were sacrificed at 1.5, 3, 6, 12, and 24 hours and at 1, 2, 3, 4, and 6 weeks post-treatment. Gross findings included distension of the vagina by a greenish, opaque, flocculent fluid and a moderate enlargement of the iliac lymph nodes that were first observed at 6 hours post-treatment and were absent by one week post-treatment. Beginning one week after treatment and persisting for six weeks, distension, stenosis, and occlusion of the uterus, cervix and vagina were present in 50-85% of the treated females. The incidence and severity varied among groups. Histologically, degenerative changes and acute inflammation appeared in the vagina, cervix, and uterus by 1.5 hours post-treatment. These changes reached a maximum severity at 24 hours, abated by the end of the first week, and were entirely absent by the end of the second week. It was concluded that nonoxynol-9 is capable of inducing mucosal damage and inflammation in the rat vagina; abnormal healing can lead to stenosis and occlusion which can give rise to distension of the organ in the area above the injury.

Although nonoxynol-9 is considered a potent spermicide *in vitro*, it did not have potent *in vivo* contraceptive activity. Two mL of a 1 or 10 mg/mL preparation were applied vaginally to four female rabbits. The rabbits were mated and the percent fertilization of ova present in the oviducts was noted. The percent fertilizations at the two concentrations ranged from 0 to 95.5% and 0 to 62.2%, respectively, indicating an inconsistent antifertility effect (Kaminski et al., 1985).

The embryotoxic potential of nonoxynol-9 was investigated in a series of experiments. Nonoxynol-9 at doses of 50, 100, 250, or 500 μg injected into one of two of the uterine horns of six Sprague-Dawley rats on day 1 of pregnancy reduced the number of pregnancies and the mean number of viable embryos at the highest dose only (Stolzenberg et al., 1976). A single vaginal application of an aqueous solution of nonoxynol-9 (25 mg/kg) given to rats between gestational days 3 and 9 produced embryo- and fetocidal effects, but did not result in teratogenic effects (Buttar, 1982a).

Gross and histopathologic lesions were described in a later study following a single vaginal application of 50 mg/kg on days 3 or 7 of gestation (Tryphonas and Buttar, 1986). Dams treated on day 3 of pregnancy had 1 or less normal implantations/uterus and 11.5 resorptions/uterus. For dams treated on day 7 of pregnancy, the figures were 9.2 normal implantations and 4.8 resorptions/uterus. Control dams administered physiological saline on day 3 or 7 of gestation had 12.5 and 13.8 normal implantation sites and 0.72 and 0.32 resorption sites/uterus, respectively. The differences between treated and control implantations were statistically significant. Acute, transient vaginitis was present in all treated females. Acute endometritis in some treated females indicated nonoxynol-9 may have passed through the cervix into the uterus.

Pregnant Long-Evans Hooded rats were treated intravaginally on days 6 through 15 of gestation with nonoxynol-9 at dosages of 4 or 40 mg/kg/day (2 and 20 times the clinical usage) (Abrutyn et al., 1982). No meaningful differences were observed between the control and treated groups in maternal toxicity, gross and microscopic appearance of the vagina, maternal reproductive performance, fetal toxicity or fetal malformations. Results were also negative in a similar experiment in which Sprague-Dawley rats were administered a formulation containing C_8APE_9 (octoxynol-9) (Saad et al., 1984).

Rats were administered nonoxynol-9 according to several dosing regimes: 50, 250, or 500 mg/kg by gavage on days 6-15 of pregnancy, 500 mg/kg by gavage on days 1-20 of pregnancy, or 50 or 500 mg/kg dermally on days 6-15 of pregnancy (Meyer et al., 1988). The dams administered the two highest oral doses exhibited a statistically significant decrease in weight gain. A slight, but statistically significant smaller litter size and pre-implantation loss, not dose-related, was observed in rats treated orally. There was also a statistically significant dose-related increase in fetuses with both extra ribs and slightly dilated pelvic cavities. These reproductive and teratogenic effects may be attributable to the high toxic doses used in this study, as evidenced by the decrease in weight gain of the dams. No effects were present in rats treated dermally. No toxic, reproductive, or teratogenic effects were present in dams treated in a similar manner with nonoxynol-30 (C_9APE_{30}).

In a short-term preliminary *in vivo* assay, C_8APE_9 (Triton X-100) and C_9APE_{10} (Tergitol NP-10) administered by gavage at a dose level corresponding to the LD_{10} (800 and 600 mg/kg, respectively) to pregnant CD-1 mice on days 6-13 of pregnancy were nonteratogenic as measured by litter size, birth weight, and neonatal growth and survival to postnatal day 3 (Hardin et al., 1987).

Buttar et al. (1985) studied the effects of nonoxynol-9 and C_8APE_9 (octoxynol-9) at doses of 0.25 to 10 ug/mL on mouse embryos in culture. At 10 ug/mL, both spermicides were lethal within 24 hours. Control embryos were viable up to 72 hours.

C_8APE_9 (Triton X-100) was tested in a teratogenic screening assay based on the ability of chemicals to interfere with the normal growth and differentiation of murine neuroblastoma cells (Mummery et al., 1984). At a concentration of 0.001%, Triton X-100 tested positive as a modifier of induction of differentiation and was classified as a teratogen. The application of the results of this *in vitro* test are questionable since this compound was not teratogenic in the above *in vivo* tests.

Soto et al. (1991) reported estrogenic activity of nonylphenol. An *in vitro* cell proliferation assay using MCF_7 human breast cancer cultures indicated a low estrogenic potency for NP; relative potency of the hormone estradiol-17β compared to NP was in the range of 1×10^5 to 3×10^6. Ovariectomized rats injected subcutaneously with NP extracted from polystyrene tubes showed a weak *in vivo* endometrial proliferative response, 1×10^5 to 3×10^5 times less than estradiol. Similar conclusions were reached in subsequent work by Soto et al. (1992).

Xenobiotic estrogens are of concern because they may cause developmental and reproductive system effects and induce cell proliferation. Although Soto's results suggest that NP activates estrogen receptors, they are not conclusive evidence. Further, their relevance to environmental effects is unclear because NP potency is very weak.

9. Other Effects.

The cardiotoxicity of APE to dogs, but not to rats was discussed in Sections 2 and 3 (Smyth and Calandra, 1969). The action appears to be a direct pharmacologic effect on the heart muscle and is present after an administration period as brief as five days. Regardless of the alkyl group - octyl, nonyl, or dodecyl - focal myocardial lesions were present for APE of 15, 17.5, or 20 EO units.

There appeared to be a relationship between emesis and myocardial degeneration, suggesting a secondary effect on the heart from electrolyte imbalance. Electrolyte measurements, however, did not support this hypothesis. In addition, studies on the isolated, perfused cat heart showed a direct effect on contraction amplitude, with contractility declining most severely for NPE_{20} (Smyth and Calandra, 1969).

Denk et al. (1971) studied the interaction of several Triton APE (C_8APE_1, C_8APE_9, C_8APE_{30}, and C_9APE_{10}) with microsomal cytochrome P-450. They described a series of actions including blocking of a P-450-dependent enzyme and culminating in solubilization and degradation of P-450. In another study, C_8APE_9 at a concentration up to 1.6% in the incubation medium did not inhibit pancreatic amylase (Furuichi, 1974). A concentration of 6 μM NP inhibited the ATPase activity of the skeletal muscle sarcoplasmic reticulum of rabbits by 50% (Michelangeli et al., 1990).

10. Fate and Disposition

The metabolic fate of tritium-labeled p-tert-C_8APE_{40} (labeled in the octylphenol moiety) following a single oral dose to four rats and four dogs was studied by Larson et al. (1963). Within 24 hours, approximately 90% of the material was excreted in the feces, probably indicating low absorption. About 1% of the radioactivity appeared in the urine.

The fate of orally administered NP, uniformly labeled in the phenol ring, nonylphenol ethoxylate (Tergitol TP-9 or C_9APE_9) labeled with ^{14}C uniformly in either the ethoxylate moiety or in the phenol ring and of several pure nonylphenol ^{14}C-ethoxylates with 7, 10, 12, or 15 EO units (isolated by silica gel column chromatography) were studied in 150 g male rats (Knaak et al., 1966). For all labeled compounds, 90-95% of the radioactivity was excreted in either the urine or the feces by the seventh day. Rats fed 10 mg of the ethylene oxide-labeled C_9APE_9 excreted approximately 52% of the dose in the feces, 40% in the urine, and 1.2% as $^{14}CO_2$. Rats fed 10 mg of the NP ring-labeled surfactant excreted 78% of the dose in the feces and 20% in the urine. For NP labeled in the phenol ring, the excretion pattern was similar to that of the ring-labeled nonylphenol ethoxylates, with 19% excreted in the urine and 70% in the feces by day four. No $^{14}CO_2$ was detected in the expired air when the label was attached to the phenol ring.

Rats fed 10 mg (67 mg/kg) of ^{14}C-ethylene oxide-labeled pure C_9APE_7, C_9APE_{10}, C_9APE_{12}, or C_9APE_{15} homologs showed an excretion pattern similar to that of the ethoxylate-labeled mixture above. With increasing EO chain length, urinary and pulmonary excretion of radioactivity decreased and fecal excretion of the label increased. Excretion of $^{14}CO_2$ was inversely proportional to the number of ethylene oxide units. The study provided evidence that absorption of APE from the intestinal tract decreases with increasing ethoxylate chain length (Knaak et al., 1966).

As part of the same study, urinary metabolites were characterized. The principal urinary metabolites of C_9APE_9 were neutrals followed by acids: the latter were mono- and dicarboxylic acids of polyethylene glycol and the glucuronic acid conjugates of NP. About 1% of the dose appeared in the urine as free NPE, with the 14- and 15-mole adducts present in the highest concentration (Knaak et al., 1966).

Paulson et al. (1980) and Gardner et al. (1980) investigated the metabolism of pure t-C_8APE_6 in which the β-carbon of the ethylene oxide unit next to the ring was labeled. Excretion was primarily in the feces via the bile. By day four, 77% and 89% of the label appeared in the feces of the goat and rat, respectively. Collection of bile via a cannula showed that most of the dose appearing in the feces had been absorbed and underwent enterohepatic circulation. In addition to recovery of 89% of the dose in the feces by 96 hours for the rat, 6% was eliminated in the urine and 2% as carbon dioxide (Gardner et al., 1980). Organ analysis showed that only about 3% remained in the rat at that time (Table 5-8). Numerous metabolites, many of which could not be characterized by MS, were isolated from the bile and urine. None of the metabolites indicated central fission to octylphenol and PEG. Oxidation occurred at both the alkyl and ethylene moiety, with the ethoxylated chain being shortened one unit at a time. The alkyl chain was not shortened. Some of the oxidized metabolites appeared as conjugated metabolites.

^{14}C-ethylene oxide labeled C_9APE_9 in aqueous solution applied to the vaginal walls of nonpregnant rats and rabbits (Chvapil et al., 1980b) and pregnant rats (Buttar, 1982b) was rapidly absorbed into the blood stream. After six hours the level of ^{14}C in the uterus and placenta was in equilibrium with that of the maternal plasma (1.3 ug/mL). The amounts in the amniotic fluid and fetus were one-third of that observed in the mother's plasma.

In rats, intravaginally-administered C_9APE_9 (in saline) was excreted primarily in the feces (70%) while 23% was detected in the urine (Chvapil et al., 1980b). Approximately 95% of the administered dose was recovered within three days. In contrast to the rat, rabbits administered the chemical via intravaginal sponge excreted most of the dose in the urine (40%) with fecal excretion only 10%.

Table 5-8. ^{14}C in Tissues of the Rat 96 Hours after Oral Administration of Ethylene Oxide-Labeled C_8APE_6

Organ	Percent of total ^{14}C
Stomach	—
Small intestine	1.670
Large intestine	—
Lungs	0.150
Heart	0.008
Liver	0.300
Kidneys	0.020
Rest of carcass	0.900

Source: Gardner et al., 1980

B. HUMAN STUDIES

The APE surfactants as well as octylphenol and nonylphenol appear on the EPA Toxic Substances Control Act (TSCA) Chemical Inventory prepared in 1986 (RTECs, 1987). (See Table 1-1).

1. Dermal Irritation and Sensitization

Finnegan and Dienna (1953) tested five OPE (octylphenol with 1, 3, 4, 9.7, or 12.5 EO units) on humans in 48-hour patch tests. No primary irritation was present. Two of 50 subjects exhibited a mild sensitization when challenged on the other arm with C_8APE_1 two weeks later.

Undiluted C_9APE_4 was tested on the backs of 25 male and 25 female subjects in a repeated insult patch test (CIR Expert Panel, 1983). The primary application was left in place for 48 hours; 14 induction patches were applied for 24 hours each. A challenge patch, applied two weeks later, resulted in neither immediate nor delayed reactions. A similar test with C_9APE_9 and using twice the number of subjects also resulted in no responses.

Another repeated insult patch test using 50% solutions of C_9APE_{15} or C_9APE_{50} was performed on 53 male and 115 female subjects (CIR Expert Panel, 1983). The test material was applied at 48 hour intervals, three times per week for three weeks. After a three-week interval, the test area and an untreated site were challenged. Although there were transient reactions during the test period, neither compound was an irritant or sensitizer.

Twenty-eight of the above subjects were exposed to C_9APE_{15} or C_9APE_{50} using the same test protocol as above, except that exposure was on the forearm (CIR Expert Panel, 1983). The subjects were evaluated for phototoxic or photosensitization reactions following exposure to ultraviolet light. Neither ultraviolet A nor ultraviolet B light induced phototoxicity or photosensitization.

Cosmetic formulations containing C_9APE_4, C_9APE_9, or C_9APE_{12} were tested for cumulative skin irritation (CIR Expert Panel, 1983). Ten applications were applied for 24 hours each. There was a range of effects, from slight to mild irritation.

2. Pharmacokinetics

No studies on the absorption, metabolism, or excretion of APE in humans were found.

3. Therapeutic/Contraceptive Uses

Nonoxynol-9 (NPE_9) is reported to be a potent spermicide. When mixed with spermatozoa *in vitro* at concentrations of <1 mg/mL, it was completely spermicidal within 30 seconds (Vickery et al., 1983; Kaminski et al., 1985). The surfactant disrupted the lipid membrane and rendered the spermatozoa immobile (Schill and Wolf, 1981; Wilborn et al., 1983).

Lichtman et al. (1973) reported on a spermicidal formulation containing 10% C_9APE_4 that was tested as a contraceptive in 30 women. Two women experienced a mild local vaginal irritation and one reported pruritus.

Following a review of animal toxicity studies and assessment of the safety of intravaginal use in humans at daily doses of 124 mg, the Food and Drug Administration Panel on Review of Contraceptives has concluded that both octoxynol 9 (C_8APE_9) and nonoxynol 9

(C_9APE_9) are safe and effective for over-the-counter use as vaginal contraceptives (Fed. Reg. 45:241, December 12, 1980).

Some epidemiology studies have tentatively linked the high occurrence of spontaneous abortions and a number of congenital defects in children with the use of nonoxynol-9 type spermicides prior to conception, especially among infants conceived as a result of contraceptive failure (Smith et al., 1977; Warburton et al., 1980; Jick et al., 1981; Strobino et al., 1980). Other epidemiology studies with more carefully controlled factors, however, suggest that the number of congenital anomalies or malformations does not exceed that observed with other methods of contraception (Codero and Layde, 1981; Huggins et al., 1982; Mills et al., 1982; Polednak et al., 1982; Shapiro et al., 1982; Linn et al., 1983; Louik et al., 1987; Warburton et al., 1987).

4. Cosmetic Uses

The CIR Expert Panel (1983) assessed the safety of nonoxynols (C_9APE) which are used in human cosmetic products. Results of published and unpublished data submitted by the Cosmetic, Toiletry and Fragrance Association indicated that the nonoxynols ranging in EO units from 2 to 50 are safe as cosmetic ingredients at the concentrations presently used ($\leq 0.1\%$ to $>50\%$).

5. Epidemiology

No reports on human exposure to APE during manufacture were located.

Epidemiology studies on contraceptive use are cited above. The contraceptives, which contain 2 to 12.5% C_9APE_9 as the active ingredient (50-140 mg/vaginal application), have been found safe for human contraceptive use.

REFERENCES

Abrutyn, D., B.E. McKenzie and N. Nadaskay. 1982. Teratology study of intravaginally administered nonoxynol-9-containing contraceptive cream in rats. Fertil. Steril. 37:113-117.

Ariyoshi, T., H. Hasegawa, Y. Nanri and K. Arizono. 1990. Profile of hemoproteins and heme-metabolizing enzymes in rats treated with surfactants. Bull. Environ. Contam. Toxicol. 44:369-376.

Arthur D. Little. 1977. Human Safety and Environmental Aspects of Major Surfactants. A Report to the Soap and Detergent Association. Arthur D. Little, Inc., Cambridge, MA.

Assinder, S.J. and A. Upshall. 1985. Paramorphogenic and genotoxic activity of Triton X-100 and sodium dodecyl sulphate in *Aspergillus nidulans*. Mutat. Res. 142:179-181.

Boutwell, R.K. and D.K. Bosch. 1959. The tumor-promoting action of phenol and related compounds for mouse skin. Cancer Res. 19:413-427.

Brown, V.K.H. 1969. The influence of some detergents on the uptake of oxygen by "hairless" mouse skin. J. Soc. Cosmet. Chem. 20:413-420.

Brown, V.K.H. 1971. A comparison of predictive irritation tests with surfactants on human and animal skin. J. Soc. Cosmet. Chem. 22:411-420.

Buttar, H.S. 1982a. Assessment of the embryotoxic and teratogenic potential of nonoxynol-9 in rats upon vaginal administration. The Toxicologist 2:39. (Abstract).

Buttar, H.S. 1982b. Transvaginal absorption and disposition of nonoxynol-9 in gravid rats. Toxicol. Lett. 13:211.

Buttar, H.S., J.H. Moffatt and C. Bura. 1985. Effects of the vaginal spermicides, nonoxynol-9 and octoxynol-9, on the development of mouse embryos *in vitro*. Teratology 31:51A. (Abstract).

Buttar, H.S., S.H.H. Swierenga and T.I. Matula. 1986. Evaluation of the cytotoxicity and genotoxicity of the spermicides nonoxynol-9 and octoxynol-9. Toxicol. Lett. 31:65-73.

Chvapil, M., W. Droegemueller, J.A. Owen, C.D. Eskelson and K. Betts. 1980a. Studies of nonoxynol-9. I. The effect on the vaginas of rabbits and rats. Fertil. Steril. 33:445-450.

Chvapil, M., C.D. Eskelson, V. Stiffel and W. Droegemueller. 1980b. Studies on nonoxynol-9. II. Intravaginal absorption, distribution, metabolism, and excretion in rats and rabbits. Contraception 22:325-339.

CIR (Cosmetic Ingredient Review) Expert Panel. 1983. Final report on the safety assessment of nonoxynols -2, -4, -8, -9, -10, -12, -14, -15, -30, -40, and -50. J. Am. Coll. Toxicol. 2:35-60.

Codero, J.F. and P.M. Layde. 1981. Vaginal spermicides, chromosome abnormalities and limb reduction defects. Am. J. Hum. Genet. 33:74A. (Abstract).

Coppinger, W.J., S.A. Berman and E.D. Thompson. 1981. Testing of surfactants for mutagenic potential in the L5178Y TK+/- mouse lymphoma assay. Environ. Mutagen. 3:320. (Abstract).

Datta, A.R., B.W. Randolph and J.L. Rosner. 1983. Detection of chemicals that stimulate Tn9 transposition in *Escherichia coli* K12. Mol. Gen. Genet. 189:245-250.

Denk, H., J.B. Schenkman, P.G. Bacchin, F. Hutterer, F. Schaffner and H. Popper. 1971. Mechanism of cholestasis. III. Interaction of synthetic detergents with the microsomal cytochrome P-450 dependent biotransformation system *in vitro*. Exp. Mol. Pathol. 14:263-276.

Draize, J.H. 1959. Dermal toxicity. In: Appraisal of the Safety of Chemicals in Foods, Drugs and Cosmetics. Association of Food and Drug Officials of the U.S., Baltimore, MD. The Editorial Committee Publication, pp. 48-59.

Draize, J.H., G. Woodward and H.O. Calvery. 1944. Methods for the study of irritation and toxicity of substances applied topically to the skin and mucous membrane. J. Pharmacol. Exptl. Therap. 82:377-390.

Etnier, E. 1985. Chemical Hazard Information Profile for nonylphenol. Office of Toxic Substances, U.S. Environmental Protection Agency, Washington, DC.

Federal Register 45(241):82014-82049. 1980. Proposed rulemaking for vaginal contraceptive drug products for OTC use. December 12, 1980.

Finnegan, J.K. and J.B. Dienna. 1953. Toxicological observations on certain surface-active agents. Proc. Sci. Sect. Toilet Goods Assoc. 20:16-19.

Fitzhugh, O.G. and A.A. Nelson. 1948. Chronic oral toxicities of surface-active agents. J. Am. Pharm. Assoc. 37:29-32.

Furuichi, Y. 1974. Studies on the effects of surface active agents on amylase activity. I. Effects of nonionic surface active agents. Yakagaku 23:255-256.

Garberg, P. and G. Bolcsfoldi. 1985. Evaluation of a genotoxicity test measuring DNA strandbreaks in mouse lymphoma cells by alkaline unwinding and hydroxylapatite chromatography. Environ. Mutagen. 7 (Suppl. 3):73. (Abstract).

Garberg, P., E.L. Akerblom and G. Bolcsfoldi. 1988. Evaluation of a genotoxicity test measuring DNA-strand breaks in mouse lymphoma cells by alkaline unwinding and hydroxyapatite elution. Mutat. Res. 203:155-176.

Gardner, K.D., G.D. Paulson and G.L. Larsen. 1980. Metabolism of the nonionic surfactant ^{14}C-labeled α[(p-1,1,3,3-tetramethylbutyl)phenol]-w-hydroxylhexa(oxyethylene) in rats. Pest. Biochem. Physiol. 14:129-138.

Gerarde, H.W. 1963. Toxicology studies on hydrocarbons. XI. Aspiration hazard and toxicity of hydrocarbons and hydrocarbon mixtures. Arch. Environ. Health 6:35-47.

Gershbein, L.L. and J.E. McDonald. 1977. Evaluation of the corneal irritancy of test shampoos and detergents in various animal species. Food Cosmet. Toxicol. 15:131-134.

Goldhammer, H., W.R. McManus and R.A. Osburn. 1970. The effect of a range of Triton non-ionic surfactants on rodent ovaries.

Gosselin, R.E., R.A. Smith and H.C. Hodge. 1984. *Clinical Toxicology of Commercial Products*, 5th ed. Williams & Wilkins Co., Baltimore.

Goyer, M.M. J.H. Perwak, A. Sivak and P.S. Thayer. 1981. Human safety and environmental aspects of major surfactants (Supplement). A report to the Soap and Detergent Association by Arthur D. Little, Cambridge, MA.

Hardin, B.D., R.L. Schuler, J.R. Berg, G.M. Booth, K.P. Hazelden, K.M. MacKenzie, V.J. Piccirillo and K.N. Smith. 1987. Evaluation of 60 chemicals in a preliminary developmental toxicity test. Terat. Carc. Mut. 7:29-48.

Huggins, G., M. Vessey, R. Flavel, D. Yeates, and K. McPherson. 1982. Vaginal spermicides and outcome of pregnancy: findings in a large cohort study. Contraception 25:219-230.

Imperial Chemical Industries. 1981. Unpublished data cited in Yam et al., 1984.

Jick, H., A.M. Walker, K.J. Rothman, J.R. Hunter, L.B. Holmes, R.N. Watkins, D.C. D'Ewart, A. Danford and S. Madsen. 1981. Vaginal spermicides and congenital disorders. J. Am. Med. Assoc. 245:1329.

Kaminski, J.M., N.A. Nuzzo, L. Bauer, D.P. Waller and L.J.D. Zaneveld. 1985. Vaginal contraceptive activity of aryl 4-guanidinobenzoates (acrosin inhibitors) in rabbits. Contraception 32:183-189.

Knaak, J.B., J.M. Eldridge and L.J. Sullivan. 1966. Excretion of certain polyethylene glycol ether adducts of nonylphenol by the rat. Toxicol. Appl. Pharmacol. 9:331-340.

Larson, P.S., J.F. Borzelleca, E.R. Bowman, E.M. Crawford, R.B. Smith, Jr., and G.R. Hennigar. 1963. Toxicologic studies on a preparation of p-tertiary octylphenoxy-polyethoxy ethanols (Triton X-405). Toxicol. Appl. Pharmacol. 5:782-798.

Lichtman, A.S., V. Davajan and D. Tucker. 1973. C-film: a new vaginal contraceptive. Contraception 8:291-297.

Linn, S., S.C. Schoenbaum, R.R. Monson, B. Rosner, P.G. Stubblefield and K.J. Ryan. 1983. Lack of association between contraceptive usage and congenital malformations in offspring. Am. J. Obstet. Gynecol. 147:923-928.

Long, S.D., A.J. Warren and J.B. Little. 1982. Effect of nonoxynol-9, a detergent with spermicidal activity, on malignant transformation *in vitro*. 1982. Carcinogenesis 3:553-557.

Louik, C., A.A. Mitchell, M.M. Werler, J.W. Hanson and S. Shapiro. 1987. Maternal exposure to spermicides in relation to certain birth defects. N. Engl. J. Med. 317:474-478.

Matsuoka, A., T. Sofuni and M. Ishidate, Jr. 1986. Effect of surfactants on the induction of chromosomal aberrations in Chinese hamster cells in culture. Mutat. Res. 164:273-274.

Meyer, O., P. Haubro Anderson, E.V. Hansen and J.C. Larsen. 1988. Teratogenicity and *in vitro* mutagenicity studies on nonoxynol-9 and -30. Pharmacol. Toxicol. 62:235-238.

Michelangeli, F., S. Orlowski, P. Champeil, J.M. East and A.G. Lee. 1990. Mechanism of inhibition of the $(Ca^{2+}-Mg^{2+})$-ATPase by nonylphenol. Biochemistry 29:3091-3101.

Mills, J.L., E.E. Harley, G.F. Reed and H.W. Berendes. 1982. Are spermicides teratogenic? J. Am. Med. Assoc. 248:2148-2151.

Monsanto Chemical Company. 1959. Unpublished data.
Monsanto Chemical Company. 1964. Unpublished data.
Monsanto Chemical Company. 1972. Unpublished data.

Monsanto Chemical Company. 1975. Unpublished data.

Monsanto Chemical Company. 1985. Material Safety Data Sheet for nonylphenol. (Cited in Etnier, 1985).

Mummery, C.L., C.E. Van den Brink, P.T. Van der Saag and S.W. DeLaat. 1984. A short-term screening test for teratogens using differentiating neuroblastoma cells in vitro. Teratology 29:271-279.

Myers, R.C., S.M. Christopher and E.H. Fowler. 1988. Tergitol and carbowax samples assessment of toxicity and pulmonary effects in the rat following single endotracheal injection. Union Carbide Corporation, Bushy Run Research Center Project Report 50-128.

Olson, K.J., R.W. Dupree, E.T. Plomer and V.K. Rowe. 1962. Toxicological properties of several commercially available surfactants. J. Soc. Cosmet. Chem. 13:469-479.

Osterberg, R.E., S.P. Bayard and A.G. Ulsamer. 1976. Appraisal of existing methodology in aspiration toxicity testing. J. Assoc. Off. Anal. Chem. 59:516-525.

Paulson, G.D., E.R. Mansager and G.L. Larsen. 1980. The metabolism of the nonionic surfactant^{14}C-labeledα[p-(1,1,3,3-tetramethylbutyl)phenyl]-w-hydroxyhexa(oxyethylene) in the goat. Pest. Biochem. Physiol. 14:111-128.

Polednak, A.P., D.T. Janerich and D.M. Glebatis. 1982. Birth weight and birth defects in relation to maternal spermicide use. Teratology 26:27.

Procter & Gamble Company. 1979. Unpublished data cited in Yam et al., 1984.

Rantuccio, F., D. Sinisi, C. Corviello, A. Conte and A. Scardigno. 1984. Histological changes in rabbits after application of medicaments and cosmetic bases (III). Contact Dermat. 10:212-219.

Reinhard, M., N. Goodman and K.E. Mortelmans. 1982. Occurrence of brominated alkylphenol polyethoxy carboxylates in mutagenic wastewater concentrates. Environ. Sci. Technol. 16:351-362.

RTECs (Registry of Toxic Effects of Chemicals). 1987. U.S. Government Printing Office, Washington, DC.

Saad, D.J.C., R.M. Kirsch, L.L. Kaplan and D.E. Rodwell. 1984. Teratology of intravaginally administered contraceptive jelly containing octoxynol-9 in rats. Teratology 30:25-30.

Schick, M.J. (Ed.). 1967. *Nonionic Surfactants*, Volume 1, Surfactant Science Series. Marcel Dekker, Inc., NY.

Schill, W.B. and H.H. Wolf. 1981. Ultrastructure of human spermatozoa in the presence of the spermicide nonoxynol-9 and a vaginal contraceptive containing nonoxynol-9. Andrologia 13:42.

Shapiro, S., D. Slone, O.P. Heinonen, D.W. Kaufman, L. Rosenberg, A.A. Mitchell and S.P. Helmrich. 1982. Birth defects and vaginal spermicides. J. Am. Med. Assoc. 247:2381.

Sheu, C.W., F.M. Morland, J.K. Lee and V.C. Dunkel. 1988. *In vitro* BALB/3T3 cell transformation assay of nonoxynol-9 and 1,4-dioxane. Environ. Mol. Mutagen. 11:41-48.

Shibuya, T. N. Tanaka, M. Katoh, Y.T. Matsuda and K. Morita. 1985. Mutagenicity testing of ST-film with the Ames test, chromosome test *in vitro* and micronucleus test in female mice.

Shimizu, H., Y. Suzuki, N. Takemura, S. Goto and H. Matsushita. 1985. The results of microbial mutation test for forty-three industrial chemicals. Jpn. J. Ind. Health 27:400-419.

Skare, J.A. and K.R. Schrotel. 1984. Alkaline elution of rat testicular DNA: detection of DNA strand breaks after *in vivo* treatment with chemical mutagens. Mut. Res. 130:283-294.

Smith, E.S.O., C.S. Dafve and J.R. Miller. 1977. An epidemiological study of congenital reduction deformities of the limbs. Br. J. Prev. Soc. Med. 31:39-41.

Smyth, H.F., Jr. and J.C. Calandra. 1969. Toxicologic studies of alkylphenol polyoxyethylene surfactants. Toxicol. Appl. Pharmacol. 14:315-334.

Smyth, H.F., C.P. Carpenter, C.S. Weil, U.C. Pozzani and J.A. Striegel. 1962. Range-finding toxicity data: List VI. Am. Indust. Hyg. Assoc. J. 23:95-107.

Smyth, H.F., C.P. Carpenter and C.S. Weil. 1969. Range-finding toxicity data: List VII. Am. Indust. Hyg. Assoc. J. 30:470-476.

Soto, A.M., H. Justica, J.W. Wray and C. Sonnenschein. 1991. *p*-Nonylphenol: An estrogenic xenobiotic released from "modified" polystyrene. Environ. Health Perspect. 92:167-173.

Soto, A.M., T.-M. Lin, H. Justica, R.M. Silvia and C. Sonnenschein. 1992. In: T. Colborn and C. Clement, eds., *Chemically-Induced Alterations in Sexual and Functional Development*, Princeton Scientific Publishing Co., Princton, NJ, pp. 295-309.

Stolzenberg, S.J., R.M. Parkhurst, and E.J. Reist. 1976. Blastocidal and contraceptive actions by an extract and compounds from endod (*Phytolacca dodecandra*). Contraception 14:39-51.

Strobino, B., J. Klein, J. Stein, et al. 1980. Exposure to contraceptive creams, jellies and douches and their effect on the zygote. Am. J. Epidemiol. 112:434. (Cited in Mills et al., 1980).

Takahashi, M., S. Fukushima and M. Hananouchi. 1975. Induction of undifferentiated adenocarcinoma in the stomach of rats by N-methyl-N'-nitro-N-nitrosoguanidine with various kinds of surfactant. Gann 17:255-267.

Texaco Chemical Company. 1983a. Ames *Salmonella*/microsome plate test. Unpublished data.

Texaco Chemical Company. 1983b. *In vitro* transformation of BALB/3T3 cells assay. Unpublished data.

Texaco Chemical Company. 1984. The hepatocyte primary culture/DNA repair assay on compound 5601-29-20 using rat hepatocytes in culture. Unpublished data.

Texaco Chemical Company. 1985a. Material Safety Data Sheet (Cited in Etnier, 1985).

Texaco Chemical Company. 1985b. Unpublished data (Cited in Etnier, 1985).

Texaco Chemical Company. 1985c. Unpublished data (Cited in Etnier, 1985).

Texaco Chemical Company. 1991a. Acute exposure oral toxicity: Surfonic N-40. Unpublished data.

Texaco Chemical Company. 1991b. Acute exposure oral toxicity: Surfonic N-95. Unpublished data.

Texaco Chemical Company. 1991c. Acute exposure oral toxicity: Surfonic N-120. Unpublished data.

Texaco Chemical Company. 1991d. Primary dermal irritation study: Surfonic N-95. Unpublished data.

Texaco Chemical Company. 1991e. Ames/*Salmonella* plate incorporation assay: Surfonic N-40. Unpublished data.

Texaco Chemical Company. 1991f. *In vivo* micronucleus test: Surfonic N-120. Unpublished data.

Texaco Chemical Company. 1992a. Acute exposure dermal toxicity: Surfonic N-40. Unpublished data.

Texaco Chemical Company. 1992b. Acute exposure dermal toxicity: Surfonic N-95. Unpublished data.

Texaco Chemical Company. 1992c. Acute exposure dermal toxicity: Surfonic N-120. Unpublished data.

Texaco Chemical Company. 1992d. Primary dermal irritation study: Surfonic N-40. Unpublished data.

Texaco Chemical Company. 1992e. Primary dermal irritation study: Surfonic N-120. Unpublished data

Texaco Chemical Company. 1992f. Primary eye irritation: Surfonic N-40. Unpublished data.

Texaco Chemical Company. 1992g. Primary eye irritation: Surfonic N-95. Unpublished data.

Texaco Chemical Company. 1992h. Rat hepatocyte primary culture/DNA repair test: Surfonic N-40. Unpublished data.

Texaco Chemical Company. 1992i. Rat hepatocyte primary culture/DNA repair test: Surfonic N-120. Unpublished data.

Texaco Chemical Company. 1992j. *In vivo* micronucleus test: Surfonic N-40. Unpublished data.

Texaco Chemical Company. 1992k. *In vivo* micronucleus test: Surfonic N-95. Unpublished data.

Thompson, E.D. and D.P. Gibson. 1984. A method for determining the maximum tolerated dose for acute *in vivo* cytogenetic studies. Fd. Chem. Toxic. 22:665-676.

Tryphonas, L. and H.S. Buttar. 1982. Genital tract toxicity of nonoxynol-9 in female rats: temporal development, reversibility and sequelae of the induced lesions. Fund. Appl. Toxicol. 2:211-219.

Tryphonas, L. and H.S. Buttar. 1986. Effects of the spermicide nonoxynol-9 on the pregnant uterus and the conceptus of rat. Toxicology 39:177-186.

Tuschl, H., W. Klein, F. Kocsis, E. Bernat and H. Altmann. 1975. Investigations into the inhibition of DNA repair processes by detergents. Environ. Physiol. Biochem. 5:84-91.

Tyler, T.R., R.C. Myers, S.M. Christopher and E.H. Fowler. 1988. Pulmonary toxicity of ethoxylates given endotracheally to rats. Toxicologist 8:147. (Abstract).

U.S. EPA. 1985. Reference Values for Risk Assessment. SRC TR 85-300. U.S. Environmental Protection Agency, Cincinnati, OH.

Varma, M.M., L. Wan and J.H. Johnson, Jr. 1986. Mutagenesis of the metabolite of nonionic detergents in water. University of the District of Columbia Water Resources Research Report No. 73. (PB87-172649).

Vickery, B.H., J.C. Goodpasture, K. Bergstrom, K.A.M. Walker, J.W. Overstreet and D.F. Katz. 1983. Assessment of a new spermidical agent against ejaculated dog and human spermatozoa *in vitro*. Fertil. Steril. 40:231-236.

Wangenheim, J. and G. Bolcsfoldi. 1986. Mouse lymphoma TK+/- assay of 30 compounds. Environ. Mutagen. 8 (Suppl 6):90. (Abstract).

Wangenheim, J. and G. Bolcsfoldi. 1988. Mouse lymphoma L5178Y thymidine kinase locus assay of 50 compounds. Mutagen. 3:193-205.

Warburton, D., Z. Stein and J. Kline. 1980. Environmental influences on rates of chromosome anomalies in spontaneous abortions. Am. J. Hum. Genet. 32:92. (Abstract).

Warburton, D., R.H. Neugut, A. Lustenberger, A.G. Nicholas and J. Kline. 1987. Lack of association between spermicide use and trisomy. N. Engl. J. Med. 317:478-482.

Wilborn, W.H., D.W. Hahn and J.J. McGuire. 1983. Scanning electron microscopy of human spermatozoa after incubation with the spermicide nonoxynol-9. Fertil. Steril. 39:717-719.

Yam, J., K.A. Booman, W. Broddle, L. Geiger, J.E. Heinze, Y.J. Lin, K. McCarthy, S. Reiss, V. Sawin, R.I. Sedlak, R.S. Slesinski and G.A. Wright. 1984. Surfactants: a survey of short-term genotoxicity testing. Fd. Chem. Toxicol. 22:761-769.

Zerkle, T.B., J.F. Ross and B.E. Domeyer. 1987. Alkyl ethoxylates: an assessment of their oral safety alone and in mixtures. J. Anal. Org. Chem. Soc. 64:269-272.

Zhang, H.-J., Z.-W. Wei and Y.-Z. Zhu. 1989. Inhibition of testicular DNA synthesis by chemical mutagens. Environ. Mol. Mutagen. 14:228. (Abstract).

APPENDIX

Laboratory Studies of Biodegradation

Appendix. Laboratory Studies of Biodegradation

Chemical[a]	Test System/Protocol	Extent of Biodegradation[b]	Measurement Method	Reference
C_8APE_5	Warburg respirator Oxygen demand	7% in 6 hours 16% in 5 days	Oxygen consumption Oxygen consumption	Bogan and Sawyer, 1955; Sawyer et al., 1956
$C_{12}APE_{12}$	Oxygen demand	0% in 5 days	Oxygen consumption	Sheets and Malaney, 1956a; 1956b
$t-C_9APE_{16}$	River-water dieaway	30% in 34 days	Phosphomolybdate	Huyser, 1960
C_8APE_8	Warburg respirator	17% in 1 day	Oxygen consumption	Winter, 1962
n-pri-ortho-C_8APE_9 n-pri-para-C_8APE_9 n-sec-C_8APE_{10}	River-water dieaway River-water dieaway River-water dieaway	95% in 10 days 98% in 8 days 50% in 17 days	Surface tension Surface tension Surface tension	Blankenship and Piccolini, 1963
C_9APE_n	Continuous-flow activated sludge	60% in 6 hours	Foaming	Eldib, 1963
C_9APE_{10}	Batch activated sludge	100% in 3 days 96% in 3 days	UV absorption Phosphomolybdate	Sato et al., 1963
C_9APE_n	River-water dieaway	75% in 7 days	Surface tension	Wayman and Robertson, 1963
C_9APE_4 C_9APE_{10}	River-water dieaway River-water dieaway	58% in 34 days 83% in 34 days 65% in 34 days	Infrared spectrometry Infrared spectrometry UV absorption	Frazee et al., 1964
n-sec-C_9APE_9	Inoculated medium Inoculated medium Inoculated medium Shake flask Shake flask Shake flask Warburg respirator	57% in 9 days 66% in 9 days 75% in 9 days 62% in 7 days 60% in 7 days 0–50% in 7 days 0.1 g/g sample	Cobalt thiocyanate Surface tension Foaming Cobalt thiocyanate Surface tension Foaming Oxygen consumption	Garrison and Matson, 1964

Appendix (cont.)

Chemical[a]	Test System/Protocol	Extent of Biodegradation[b]	Measurement Method	Reference
br-C_9APE_9	Inoculated medium	33% in 9 days	Cobalt thiocyanate	Garrison and Matson, 1964 (cont.)
	Shake flask	10% in 7 days	Cobalt thiocyanate	
	Inoculated medium	32% in 9 days	Surface tension	
	Shake flask	18% in 7 days	Surface tension	
	Inoculated medium	0% in 9 days	Foaming	
	Shake flask	0% in 7 days	Foaming	
	Warburg respirator	0% in 5 days	Oxygen consumption	
n-sec-C_8APE_7	Shake flask	40% in 5 days	Cobalt thiocyanate	Huddleston and Allred, 1964a; 1964b
	River-water dieaway	54% in 26 days	Cobalt thiocyanate	
	Continuous-flow activated sludge	75% in 4 hours	Cobalt thiocyanate	
n-sec-C_9APE_9	Shake flask	65% in 5 days	Cobalt thiocyanate	
	River-water dieaway	65% in 26 days	Cobalt thiocyanate	
	Continuous-flow activated sludge	88% in 4 hours	Cobalt thiocyanate	
n-sec-$C_{10}APE_{10}$	Shake flask	92% in 5 days	Cobalt thiocyanate	
	River-water dieaway	96% in 26 days	Cobalt thiocyanate	
	Continuous-flow activated sludge	100% in 4 hours	Cobalt thiocyanate	
n-sec-$C_{12}APE_{17}$	Shake flask	90% in 7 days	Cobalt thiocyanate	
	River-water dieaway	91% in 26 days	Cobalt thiocyanate	
	Continuous-flow activated sludge	100% in 4 hours	Cobalt thiocyanate	
n-sec-$C_{14}APE_{15}$	River-water dieaway	93% in 26 days	Cobalt thiocyanate	
tp-C_9APE_9	Shake flask	30% in 5 days	Cobalt thiocyanate	
	River-water dieaway	54% in 26 days	Cobalt thiocyanate	
	Continuous-flow activated sludge	55% in 4 hours	Cobalt thiocyanate	
br-C_9APE_9	Warburg respirator	10% in 3 days	Oxygen consumption	Myerly et al., 1964

Appendix (cont.)

Chemical[a]	Test System/Protocol	Extent of Biodegradation[b]	Measurement Method	Reference
n-pri-C_nAPE_{10}	Warburg respirator	5%	Oxygen consumption	Steinle et al., 1964
n-pri-C_nAPE_{12}	Warburg respirator	5%	Oxygen consumption	
n-sec-C_nAPE_{12}	River-water dieaway	90% in 21 days	Cobalt thiocyanate	
		90% in 16 days	Surface tension	
tp-$C_{12}APE_{12}$	Warburg respirator	15%	Oxygen consumption	
	Warburg respirator	0% in 2 days	Oxygen consumption	
n-pri-$C_{12}APE_{12}$	Warburg respirator	7% in 20 hours	Oxygen consumption	Vath, 1964
n-sec-$C_{12}APE_{12}$	River-water dieaway	100% in 28 days	Cobalt thiocyanate	
	River-water dieaway	60:40% in 28 days	Surface tension; foam	
tp-C_9APE_{10}	Warburg respirator	15% in 2 days	Oxygen consumption	
	Warburg respirator	7% in 1 day	Oxygen consumption	
br-C_nAPE_n	River-water dieaway	80% in 35 days	Surface tension	Weil and Stirton, 1964
tert-C_8APE_9	Warburg respirator	3-4% in 10 days	Oxygen consumption	Barbaro and Hunter, 1965
n-sec-$C_{7,9}APE_8$	Shake flask	95-100% in 7 days	Cobalt thiocyanate	Booman et al., 1965
tert-$C_8APE_{7.5}$	Shake flask	100% in 5-7 days	Cobalt thiocyanate	
tert-C_8APE_{10}	Shake flask	97-100% in 4-5 days	Cobalt thiocyanate	
	Continuous-flow activated sludge	92-97% in 6 hours	Cobalt thiocyanate	
tert-$C_9APE_{12.5}$	Septic tank	50% in 2 days	Cobalt thiocyanate	
	Shake flask	100% in 4 days	Cobalt thiocyanate	
n-sec-C_9APE_9	Shake flask	42% in 7 days	Cobalt thiocyanate	Huddleston and Allred, 1965
		45% in 7 days	Surface tension	
		20% in 7 days	Foaming	
	River-water dieaway	94% in 20 days	Cobalt thiocyanate	
		80% in 20 days	Surface tension	
		90% in 20 days	Foaming	
		46% in 20 days	Organic carbon	

Appendix (cont.)

Chemical[a]	Test System/ Protocol	Extent of Biodegradation[b]	Measurement Method	Reference
n-sec-C_9APE_9 (cont.)	Activated sludge	99% in 1 day 85% in 1 day 75% in 1 day 60% in 1 day	Cobalt thiocyanate Surface tension Foaming Organic carbon	Huddleston and Allred, 1965 (cont.)
n-sec-$C_{12}APE_{12}$	Shake flask	58% in 7 days 60% in 7 days 30% in 7 days	Cobalt thiocyanate Surface tension Foaming	
	River-water dieaway	93% in 20 days 80% in 20 days 70% in 20 days 25% in 20 days	Cobalt thiocyanate Surface tension Foaming Organic carbon	
	Activated sludge	90% in 1 day 65% in 1 day 50% in 1 day 5% in 1 day	Cobalt thiocyanate Surface tension Foaming Organic carbon	
C_8APE_{10}	Shake flask	8% in 7 days 10% in 7 days 25% in 7 days	Cobalt thiocyanate Surface tension Foaming	
	River-water dieaway	95% in 20 days 20% in 20 days 80% in 20 days 53% in 20 days	Cobalt thiocyanate Surface tension Foaming Organic carbon	
	Activated sludge	97% in 1 day 50% in 1 day 75% in 1 day 40% in 1 day	Cobalt thiocyanate Surface tension Foaming Organic carbon	
C_9APE_a (Lissapol SNX)	Batch digester (anaerobic)	25-50% in 13 days	Thin-layer chromatography	Bruce et al., 1966

Appendix (cont.)

Chemical[a]	Test System/ Protocol	Extent of Biodegradation[b]	Measurement Method	Reference
n-sec-$C_{11}APE_{11}$	River-water dieaway	97% in 28 days 50% in 28 days	Cobalt thiocyanate Surface tension	Conway and Waggy, 1966
	Continuous-flow activated sludge	77% in 8 hours	Cobalt thiocyanate, foaming	
	Shake flask	75% in 8 days 48% in 8 days 0% in 8 days	Cobalt thiocyanate Surface tension Foaming	
	Bottle procedure	+	Biological oxygen demand	
	Inoculated medium	+	Carbon dioxide formation	
t-C_8APE_{10}	River-water dieaway	62% in 28 days 25% in 28 days	Cobalt thiocyanate Surface tension	
	Continuous-flow activated sludge	55% in 8 hours	Cobalt thiocyanate, foaming	
	Shake flask	15% in 7 days 11% in 7 days 0% in 7 days	Cobalt thiocyanate Surface tension Foaming	
	Bottle procedure	+	Oxygen consumption	
	Inoculated medium	+	Carbon dioxide formation	
br-C_9APE_{11}	Continuous-flow activated sludge	65% in 8 hours	Cobalt thiocyanate, foaming	
	Inoculated medium	+	Carbon dioxide formation	
	Shake flask	75% in 7 days 35% in 7 days 0% in 7 days	Cobalt thiocyanate Surface tension Foaming	
n-sec-C_9APE_n (Igepal LO-630)	Continuous-flow activated sludge	88% in 4 hours	Cobalt thiocyanate	Huddleston, 1966
	Shake flask	44% in 8 days	Cobalt thiocyanate	
	River-water dieaway	40% in 11 days 35% in 11 days 40% in 11 days	Cobalt thiocyanate Surface tension Foaming	
	Bottle procedure	0% in 5 days	Oxygen consumption	

Appendix (cont.)

Chemical[a]	Test System/ Protocol	Extent of Biodegradation[b]	Measurement Method	Reference
t-C_8APE_{10} (Triton X-100)	River-water dieaway	30% in 11 days 0% in 11 days 0% in 11 days 10% in 8 days	Cobalt thiocyanate Surface tension Foaming Cobalt thiocyanate	Huddleston, 1966 (cont.)
	Shake flask Bottle procedure Continuous-flow activated sludge	0% in 5 days 55% in 4 hours	Oxygen consumption Cobalt thiocyanate	
br-C_9APE_n (Igepal CO-630)	Shake flask	30% in 8 days	Cobalt thiocyanate	
p-t-C_8APE_{10} (Triton X-100)	Batch activated sludge	98% in 1 day 65% in 1 day 90-100% in 7 days	Cobalt thiocyanate ^{14}Carbon Cobalt thiocyanate	Lashen et al., 1966
	Shake flask Continuous-flow activated sludge	90% in 3 hours 95% in 6 hours 65% in 6 hours	Cobalt thiocyanate Foaming ^{14}Carbon	
(^{14}C-EO chain) (^{3}H-ring labeled)	Continuous-flow activated sludge Bottle procedure Septic tank simulation Septic tank + soil	0% in 6 hours 23-29% in 7 days 58% in 3 days 63% in 3 days 7% in 3 days 93% in 3+ days 84% in 3+ days 46% in 3+ days	^{3}H Oxygen consumption Cobalt thiocyanate Foaming ^{14}Carbon Cobalt thiocyanate Foaming ^{14}Carbon	
n-sec-C_9APE_{10} n-sec-C_9APE_{11} n-sec-$C_{12}APE_{11}$ br-C_8APE_9	River-water dieaway River-water dieaway River-water dieaway River-water dieaway	90% in 36 days 93% in 36 days 60% in 36 days 87% in 35 days 56% in 35 days	Infrared spectroscopy Infrared spectroscopy Infrared spectroscopy Cobalt thiocyanate Infrared spectroscopy	Osburn and Benedict, 1966
br-C_9APE_{10}	River-water dieaway	87% in 34 days 84% in 35 days 97% in 34 days 69% in 34 days	Infrared spectroscopy Ultraviolet absorption Cobalt thiocyanate Infrared spectroscopy	
br-$C_{12}APE_{11}$	River-water dieaway	0% in 36 days	Infrared spectroscopy	

Appendix (cont.)

Chemical[a]	Test System/Protocol	Extent of Biodegradation[b]	Measurement Method	Reference
br-C_nAPE_9 (Lissapol NX)	Sewage dieaway	55% in 42 days	Thin-layer chromatography	Patterson et al., 1966
n-sec-C_9APE_{15} (Tergitol NP-35)	Trickling filter	26;66%	Mercuric iodide	Schönborn, 1966
	Continuous-flow activated sludge	77% in 3 hours	Mercuric iodide	
		75% in 3 hours	Paper chromatography	
br-C_9APE_6 (Merpoxen NO-60)	Trickling filter	44%	Mercuric iodide	
	Continuous-flow activated sludge	70% in 3 hours	Mercuric iodide	
		83% in 3 hours	Mercuric iodide	
		90% in 3 hours	Paper chromatography	
ortho-2-$C_{10}APE_{9.5}$	Inoculated medium	95% in 7 days	Cobalt thiocyanate	Smithson, 1966
	Continuous-flow activated sludge	100% in 6 hours	Cobalt thiocyanate	
n-sec-$C_{10}APE_{8.5}$	Inoculated medium	95% in 7 days	Cobalt thiocyanate	
	Continuous-flow activated sludge	91% in 6 hours	Cobalt thiocyanate	
n-sec-C_8APE_8	Inoculated medium	71% in 28 days	Organic carbon	Borstlap and Kortland, 1967
n-sec-C_8APE_{15}	Inoculated medium	51% in 28 days	Organic carbon	
br-C_8APE_9	Inoculated medium	46% in 28 days	Organic carbon	
br-C_8APE_{15}	Inoculated medium	49% in 28 days	Organic carbon	
br-C_9APE_9	Inoculated medium	0% in 7 days	Cobalt thiocyanate	Bunch and Chambers, 1967
br-C_9APE_{15}	Inoculated medium	0–60% in 7 days	Cobalt thiocyanate	
n-pri-$C_{10}APE_{10}$	Continuous-flow activated sludge	~100% in 3 hours	Thin-layer chromatography	Bürger, 1967
$C_{10}APE_{10}$	Continuous-flow activated sludge	0% in 3 hours	Organic carbon	
br-C_8APE_{6-7}	Warburg respirometer	10–58% in 8 hours	Oxygen consumption	Hartmann et al., 1967

Appendix (cont.)

Chemical[a]	Test System/ Protocol	Extent of Biodegradation[b]	Measurement Method	Reference
C_9APE_4 C_9APE_{10} C_9APE_{20} C_9APE_{30}	Batch activated sludge Batch activated sludge Batch activated sludge Batch activated sludge	44% in 1 day 29% in 1 day 8% in 1 day 6; 40% in 1 day	Sulfation methylene blue Sulfation methylene blue Sulfation methylene blue Sulfation methylene blue	Han, 1967
br-C_9APE_9	Recirculating trickling filter	100% in 6 days 80% in 9 days	Cobalt thiocyanate Thin-layer chromatography	Jenkins et al., 1967
t-C_8APE_{10}	River-water dieaway Continuous flow activated sludge	90-95% in 4-24 days 95-100% in 2-3 days 90-95% in 2-3 days	Foaming Cobalt thiocyanate Thin-layer chromatography Surface tension Foaming	Lashen et al., 1967; Lashen and Booman, 1967; Lashen and Lamb, 1967
tp-C_9APE_7	Continuous-flow activated sludge Biological oxygen demand Continuous-flow activated sludge	24% in 3 hours 14% in 3 hours 2% in 20 days 28% in 3 hours 30-50% in 3 hours	Organic carbon Chemical oxygen demand Oxygen consumption Mercuric iodide Chemical oxygen demand	Janicke, 1968a; Janicke, 1968b
n-sec-C_nAPE_9 br-C_9APE_4 br-C_9APE_9 br-C_9APE_{16}	Inoculated medium Inoculated medium Inoculated medium Inoculated medium	60-80% in 28 days 50-80% in 42 days 50-70% in 42 days 30-75% in 63 days	Thin-layer chromatography Thin-layer chromatography Thin-layer chromatography Thin-layer chromatography	Patterson et al., 1968
br-C_9APE_4 br-C_9APE_8	Inoculated medium Inoculated medium	53% in 20 days 31% in 20 days 14% in 20 days 29% in 20 days 16% in 20 days 11% in 20 days 0% in 20 days	Phosphotungstate Chemical oxygen demand Oxygen consumption Phosphotungstate Chemical oxygen demand Oxygen consumption UV absorption	Pitter, 1968

Appendix (cont.)

Chemical[a]	Test System/ Protocol	Extent of Biodegradation[b]	Measurement Method	Reference
br-C_9APE_{10}	Inoculated medium	11% in 20 days	Chemical oxygen demand	Pitter, 1968 (cont.)
		6% in 20 days	Oxygen consumption	
			Phosphotungstate	
br-C_9APE_{15}	Inoculated medium	13% in 20 days	Chemical oxygen demand	
		6% in 20 days	Oxygen consumption	
		3% in 20 days	Phosphotungstate	
br-C_9APE_{30}	Inoculated medium	0% in 20 days	Chemical oxygen demand	
		2% in 20 days	Oxygen consumption	
		0% in 20 days	Phosphotungstate	
		0% in 20 days	UV absorption	
n-sec-C_9APE_a	Batch activated sludge	65% in 1 day	Foaming	Mausner et al., 1969
	River-water dieaway	94% in 1 day	Cobalt thiocyanate	
		92% in 28 days	Cobalt thiocyanate	
		100% in 28 days	Foaming	
	Shake flask	0% in 14 days	Cobalt thiocyanate	
		68% in 14 days	Foaming	
n-sec-$C_{9-10}APE_{10.4}$	Batch activated sludge	85% in 1 day	Cobalt thiocyanate	
	River-water dieaway	99% in 1 day	Cobalt thiocyanate	
		99% in 28 days	Foaming	
		100% in 28 days	Cobalt thiocyanate	
	Shake flask	6% in 14 days	Foaming	
		87% in 14 days	Cobalt thiocyanate	
	Inoculated medium	75% in 7 days	Foaming	
t-C_8APE_{10}	Batch activated sludge	<60% in 1 day	Cobalt thiocyanate	
	River-water dieaway	95% in 1 day	Foaming	
		80% in 28 days	Cobalt thiocyanate	
		100% in 28 days	Foaming	
	Shake flask	0% in 14 days	Cobalt thiocyanate	
		44% in 14 days	Foaming	
	Inoculated medium	<60% in 7 days	Cobalt thiocyanate	

Appendix (cont.)

Chemical[a]	Test System/ Protocol	Extent of Biodegradation[b]	Measurement Method	Reference
tp-C_9APE_n	Batch activated sludge	50% in 1 day 95% in 1 day	Foaming Cobalt thiocyanate	Mausner et al., 1969 (cont.)
	River-water dieaway	92% in 28 days 98% in 28 days	Foaming Cobalt thiocyanate	
	Shake flask	0% in 14 days 19% in 14 days	Foaming Cobalt thiocyanate	
tp-$C_{12}APE_{10}$	Batch activated sludge	20% in 1 day 96% in 1 day	Foaming Cobalt thiocyanate	
	River-water dieaway	87% in 28 days 98% in 28 days	Foaming Cobalt thiocyanate	
	Shake flask	15% in 14 days 0% in 14 days	Foaming Cobalt thiocyanate	
t-C_8APE_9	Continuous-flow activated sludge	80-90% in 3 hours	Thin-layer chromatography	Mann and Reid, 1971
	Inoculated medium	10%	Thin-layer chromatography	
tp-C_9APE_{10}	Inoculated medium	0.06G	Oxygen consumption	Zika, 1971
n-pri-C_8APE_8	Inoculated medium	28% in 30 days	Oxygen consumption	Fischer, 1972
n-pri-C_8APE_8	Inoculated medium	5% in 30 days	Oxygen consumption	
n-pri-$C_{12}APE_{10}$	Inoculated medium	14% in 30 days	Oxygen consumption	
C_9APE_7	Inoculated medium	4% in 30 days	Oxygen consumption	
C_9APE_{10}	Inoculated medium	4-6% in 30 days	Oxygen consumption	
tp-C_9APE_{10}	Inoculated medium	50% 90%	Oxygen consumption Gas chromatography	Rudling, 1972
	Shake flask	0-70% in 8 days	Thin-layer chromatography	
	Continuous-flow activated sludge	>95% 3 hours	Thin-layer chromatography	

Appendix (cont.)

Chemical[a]	Test System/Protocol	Extent of Biodegradation[b]	Measurement Method	Reference
n-sec-C_9APE_9	Inoculated medium	64% in 21 days 61% in 21 days	Thin-layer chromatography Foaming	Stead et al., 1972
n-sec-$C_{9-10}APE_9$	Inoculated medium	100% in 21 days 98% in 21 days	Thin-layer chromatography Foaming	
tp-C_9APE_8	Inoculated medium	22% in 21 days 14% in 21 days	Thin-layer chromatography Foaming	
tp-C_9APE_9	Continuous-flow activated sludge	34% in 3 hours	Bismuth iodide	Gerike and Schmid, 1973
	Inoculated medium	10% in 30 days 1% in 30 days	Bismuth iodide Oxygen consumption	
tp-$C_{12}APE_{11}$	Continuous-flow activated sludge	26% in 3 hours	Bismuth iodide	
	Inoculated medium	0% in 30 days 2% in 30 days	Bismuth iodide Oxygen consumption	
C_9APE_9	Seawater microcosm	95-99% in 4-14 days	Cobalt thiocyanate	Lacaze, 1973
n-sec-C_9APE_9	Continuous-flow activated sludge	93-96% in 6 hours	Thin-layer chromatography	Stiff et al., 1973a; 1973b
t-C_8APE_9	Continuous-flow activated sludge	92-95% in 6 hours	Thin-layer chromatography	
tp-C_9APE_9	Continuous-flow activated sludge	95% in 6 hours	Thin-layer chromatography	
n-sec-C_9APE_9	Inoculated medium	43% in 28 days 40% in 26 days 92% in 25 days 100% in 3 days	Carbon dioxide formation Carbon dioxide formation Cobalt thiocyanate Foaming	Sturm, 1973
n-sec-$C_{10}APE_9$	Inoculated medium	37% in 28 days	Carbon dioxide formation	
br-C_9APE_8	Inoculated medium	3% in 28 days	Carbon dioxide formation	
t-C_8APE_a	Inoculated medium	20% in 8 days	Surface tension	Throckmorton et al., 1973; 1974

Appendix (cont.)

Chemical[a]	Test System/Protocol	Extent of Biodegradation[b]	Measurement Method	Reference
C_9APE_{10}	Inoculated medium	<40% in 2 days	Bismuth iodide	Treccani et al., 1973
n-sec-$C_{8-9}APE_9$	Inoculated medium	66% in 8 days 36% in 8 days 64% in 8 days	Surface tension Foaming Thin-layer chromatography	Albanese and Capuci, 1974
n-sec-$C_{10}APE_9$	Inoculated medium	82% in 8 days 73% in 8 days 77% in 8 days	Surface tension Foaming Thin-layer chromatography	
br-C_8APE_{10}	Inoculated medium	69% in 8 days 77% in 8 days	Surface tension Surface tension	
br-C_9APE_{10}	Inoculated medium	48% in 8 days 73% in 7 days	Foaming Bismuth iodide	
C_8APE_7	Sewage Recycle trickling filter	100% in 3 days 100% in 3 days	Iodine method	Baleux and Caumette, 1974
t-C_8APE_{10}	Sewage Recycle trickling filter	95% in 5 days 100% in 5 days	Iodine method	
n-sec-$C_{8-10}APE_9$	Inoculated medium	96% in 8 days 85% in 8 days	Bismuth iodide Surface tension	Brüschweiler, 1974
C_9APE_{10}	Inoculated medium	5% in 30 days 2% in 30 days	Bismuth iodide Oxygen consumption	Fischer et al., 1974
tp-C_9APE_2	Inoculated medium	50% in 28 days	Gas chromatography	Rudling and Solyom, 1974
tp-C_9APE_8	Inoculated medium	90% in 5 days 96% in 5 days	Bismuth iodide Bismuth iodide	
	Continuous-flow activated sludge	>91% in 3 hours	Thin-layer chromatography	
tp-C_9APE_{10}	Inoculated medium	90% in 5 days 96% in 30 days	Bismuth iodide Bismuth iodide	
	Continuous-flow activated sludge	>91% in 3 hours	Thin-layer chromatography	

Appendix (cont.)

Chemical[a]	Test System/ Protocol	Extent of Biodegradation[b]	Measurement Method	Reference
tp-C_9APE_{14}	Inoculated medium	90% in 12 days	Bismuth iodide	Rudling and Solyom, 1974 (cont.)
		96% in 30 days	Bismuth iodide	
	Continuous-flow activated sludge	>91% in 3 hours	Thin-layer chromatography	
tp-C_9APE_{16}	Continuous-flow activated sludge	95-96% in 3 hours	Bismuth iodide	
tp-C_9APE_{30}	Continuous-flow activated sludge	83-93% in 3 hours	Bismuth iodide	
C_9APE_6	Inoculated medium	40% in 10 days	Bismuth iodide	Wencker et al., 1974
C_9APE_9	Inoculated medium	26% in 10 days	Bismuth iodide	
n-sec-C_9APE_{10}	Continuous-flow activated sludge	75% in 3 hours	Bismuth iodide	Wickbold, 1974
tp-C_9APE_9	Continuous-flow activated sludge	30-40% in 3 hours	Bismuth iodide	
n-sec-$C_{8-10}APE_9$	Inoculated medium	95% in 8 days	Bismuth iodide	Brüschweiler, 1975
		82% in 8 days	Surface tension	
C_9APE_4	Inoculated medium	12% in 8 days	Surface tension	
C_9APE_{10}	Inoculated medium	90% in 8 days	Bismuth iodide	
		30% in 8 days	Surface tension	
C_9APE_{30}	Inoculated medium	31% in 8 days	Bismuth iodide	
		25% in 8 days	Surface tension	
o-C_9APE_9	Inoculated medium	80% in 14 days	Bismuth iodide	Fischer et al., 1975
		84% in 30 days	Bismuth iodide	
		29% in 30 days	Oxygen consumption	
	Continuous-flow activated sludge	90% in 3 hours	Bismuth iodide	
		61% in 3 hours	Chemical oxygen demand	

Appendix (cont.)

Chemical[a]	Test System/ Protocol	Extent of Biodegradation[b]	Measurement Method	Reference
C_9APE_9	Inoculated medium	10% in 30 days 1% in 30 days	Bismuth iodide Oxygen consumption	Fischer et al., 1975 (cont.)
	Continuous-flow activated sludge	97% in 3 hours 49% in 3 hours 48% in 3 hours	Bismuth iodide Oxygen consumption Organic carbon	
C_9APE_{10}	Inoculated medium	87% in 32 days 5% in 30 days 2% in 30 days	Bismuth iodide Bismuth iodide Oxygen consumption	
	Continuous-flow activated sludge	>90% in 3 hours 69% in 3 hours	Bismuth iodide Chemical oxygen demand	
br-C_9APE_9	River-water dieaway	97% in 3 days 60% in 30 days 24% in 30 days 12% in 30 days	Cobalt thiocyanate Chemical oxygen demand Cobalt thiocyanate Chemical oxygen demand	Kurata and Koshida, 1975
br-C_9APE_9	Inoculated medium	98% in 8 days 50% in 15 days	Surface tension Organic carbon	Sekiguchi et al., 1975
br-C_9APE_{10}	Continuous-flow activated sludge	85-95% in 36 hr 60-70% in 36 hr 40-50% in 36 hr	Cobalt thiocyanate Chemical oxygen demand Organic carbon	Kurata and Koshida, 1976
n-pri-$C_{10}APE_{6.5}$ n-pri-$C_{10}APE_{10}$ n-pri-$C_{10}APE_{15}$ n-pri-$C_{10}APE_{21}$ n-pri-$C_{10}APE_{30}$ n-pri-$C_{12}APE_{11}$ n-pri-$C_{16}APE_{14}$ n-pri-$C_{18}APE_{15}$	Shake flask Shake flask Shake flask Shake flask Shake flask Shake flask Shake flask Shake flask	76-90% in 4 days 66-77% in 4 days 59-71% in 4 days 46-58% in 4 days 34-49% in 4 days 61-68% in 4 days 49-56% in 4 days 43-51% in 4 days	Cobalt thiocyanate Cobalt thiocyanate Cobalt thiocyanate Cobalt thiocyanate Cobalt thiocyanate Cobalt thiocyanate Cobalt thiocyanate Cobalt thiocyanate	Marei et al., 1976
t-C_8APE_9	River-water dieaway	86% in 77 days 76% in 77 days	Bismuth iodide Surface tension	Reiff, 1976

Appendix (cont.)

Chemical[a]	Test System/Protocol	Extent of Biodegradation[b]	Measurement Method	Reference
C_8APE_{10}	River-water dieaway	78-95% in 11 days 79-90% in 11 days	Cobalt thiocyanate Surface tension	Ruiz Cruz and Dobarganes Garcia, 1976
$tp-C_9APE_9$	Seawater Pond water	94% in 23 days 35% in 23 days	Bismuth iodide Bismuth iodide	Schöberl and Mann, 1976
C_9APE_9	Inoculated medium Continuous-flow activated sludge	40% in 13 days 92% in 3 hours	Organic carbon Bismuth iodide	Stache, 1976
$iso-C_8APE_8$	Inoculated medium	21% in 7 days 0% in 7 days 0% in 7 days	Foaming Bismuth iodide Oxygen consumption	Vaicum et al., 1976
$br-C_9APE_n$	Inoculated medium	22% in 7 days 20% in 7 days 0% in 7 days	Foaming Bismuth iodide Oxygen consumption	
$br-C_9APE_{14}$	Inoculated medium	5% in 7 days 7% in 7 days 0% in 7 days	Foaming Bismuth iodide Oxygen consumption	
$t-C_9APE_{10}$ $br-C_9APE_{10}$	River-water dieaway River-water dieaway	94-95% in 5 days 88% in 10 days 92% in 8 days	Cobalt thiocyanate Cobalt thiocyanate	Dobarganes Garcia and Ruiz Cruz, 1977
$br-C_9APE_{20}$	River-water dieaway	75% in 25 days 96% in 12 days	Cobalt thiocyanate	
C_8APE_6 C_9APE_{10}	Inoculated medium Inoculated medium	32% in 10 days 19% in 5 days	Oxygen consumption Oxygen consumption	Inoue et al., 1977
C_9APE_{10}	Inoculated medium	9-27% in 2 days	Bismuth iodide	Janicke and Hilge, 1977
$br-C_9APE_{10}$	River-water dieaway	37-50% in 4 days	Cobalt thiocyanate	Kurata et al., 1977

Appendix (cont.)

Chemical[a]	Test System/ Protocol	Extent of Biodegradation[b]	Measurement Method	Reference
n-sec-C_9APE_9	River-water dieaway	90% in 12 days	Cobalt thiocyanate	Ruiz Cruz and Dobarganes Garcia, 1977
tp-C_9APE_5	River-water dieaway	93% in 7 days	Cobalt thiocyanate	
tp-C_9APE_9	River-water dieaway	75-90% in 10-20 days	Cobalt thiocyanate	
tp-C_9APE_{10}	River-water dieaway	90% in 9 days	Cobalt thiocyanate	
tp-C_9APE_{12}	River-water dieaway	88% in 11 days	Cobalt thiocyanate	
tp-C_9APE_{15}	River-water dieaway	77% in 17 days	Cobalt thiocyanate	
tp-C_9APE_{20}	River-water dieaway	60% in 17 days	Cobalt thiocyanate	
tp-C_9APE_{30}	River-water dieaway	4% in 17 days	Cobalt thiocyanate	
tp-C_9APE_{40}	River-water dieaway	1% in 17 days	Cobalt thiocyanate	
tp-$C_{12}APE_9$	River-water dieaway	50% in 20 days	Cobalt thiocyanate	
	River-water dieaway	75% in 28 days	Cobalt thiocyanate	
C_9APE_{10}	River-water dieaway	48-71% in 8 days	Cobalt thiocyanate	Boyer et al., 1977
		83-95% in 8 days	Cobalt thiocyanate	
		51-73% in 8 days	Bismuth iodide	
		81-91% in 8 days	Bismuth iodide	
C_9APE_5	Inoculated medium	71% in 7 days	Chemical oxygen demand	Fuka and Pitter, 1978; Pitter and Fuka, 1979
		80-85% in 7 days	Organic carbon	
C_9APE_6	Inoculated medium	62% in 7 days	Chemical oxygen demand	
		84% in 7 days	Organic carbon	
		85% in 7 days	UV spectroscopy	
C_9APE_9	Inoculated medium	70% in 7 days	Chemical oxygen demand	
		88% in 7 days	Organic carbon	
		89% in 7 days	UV spectroscopy	
br-C_8APE_{10}	Shake-flask	20% in 28 days	Cobalt thiocyanate	Kravetz et al., 1978
		20% in 28 days	Dissolved organic carbon	
		10% in 28 days	CO_2 formation	
		20% in 28 days	HBr/GC	
	Shake flask	53% in 28 days	Biological oxygen demand	

Appendix (cont.)

Chemical[a]	Test System/ Protocol	Extent of Biodegradation[b]	Measurement Method	Reference
n-sec-C_9APE_9	River-water dieaway	93% in 12 days	Cobalt thiocyanate	Ruiz Cruz and Dobarganes Garcia, 1978
	Shake flask	70% in 8 days	Bismuth iodide	
	Batch activated sludge	91% in 1 day	Bismuth iodide	
	Inoculated medium	68% in 19 days	Bismuth iodide	
	Continuous-flow activated sludge	88% in 3 hours	Bismuth iodide	
tp-C_8APE_9	River-water dieaway	68% in 11 days	Cobalt thiocyanate	
	Shake flask	80% in 8 days	Bismuth iodide	
	Batch activated sludge	92% in 1 day	Bismuth iodide	
	Inoculated medium	57% in 19 days	Bismuth iodide	
	Continuous-flow activated sludge	90% in 3 hours	Bismuth iodide	
tp-C_9APE_9	River-water dieaway	91% in 16 days	Cobalt thiocyanate	
	Shake flask	32% in 8 days	Bismuth iodide	
	Batch activated sludge	89% in 1 day	Bismuth iodide	
	Inoculated medium	45% in 19 days	Bismuth iodide	
	Continuous-flow activated sludge	85% in 3 hours	Bismuth iodide	
tp-$C_{12}APE_9$	River-water dieaway	81% in 25 days	Cobalt thiocyanate	
	Shake flask	18% in 8 days	Bismuth iodide	
	Batch activated sludge	83% in 1 day	Bismuth iodide	
	Inoculated medium	12% in 19 days	Bismuth iodide	
	Continuous-flow activated sludge	75% in 3 hours	Bismuth iodide	
C_9APE_{10}	Inoculated medium	40% in 7 days	Phosphomolybdate	Davis et al., 1979
	Recycle trickling filter	98% in 42 hours	Phosphomolybdate	
br-$C_9APE_{9.5}$	Inoculated medium	10% in 10 days	Bismuth iodide	Itoh et al., 1979
		0% in 10 days	CO_2 formation	
n-pri-$C_{12}APE_9$	Inoculated medium	24% in 10 days	CO_2 formation	

Appendix (cont.)

Chemical[a]	Test System/ Protocol	Extent of Biodegradation[b]	Measurement Method	Reference
C_9APE_{10}	Continuous-flow activated sludge	60% in 3 hours 45% in 3 hours 15% in 3 hours 95% in 3 hours	Bismuth iodide Chemical oxygen demand UV spectroscopy Bismuth iodide	Moreno Danvila, 1979
C_9APE_5	Inoculated medium	94% in 7 days 88% in 7 days 78% in 7 days 85% in 7 days	Bismuth iodide UV spectroscopy Chemical oxygen demand Organic carbon	Pitter and Fuka, 1979; Fuka and Pitter, 1980
C_9APE_{10}	Inoculated medium	99% in 7 days 81% in 7 days 78% in 7 days 83% in 7 days	Bismuth iodide UV spectroscopy Chemical oxygen demand Organic carbon	
C_9APE_{15}	Inoculated medium	98% in 7 days 86% in 7 days 78% in 7 days 79% in 7 days	Bismuth iodide UV spectroscopy Chemical oxygen demand Organic carbon	
C_9APE_{20}	Inoculated medium	79% in 7 days 85% in 7 days 60% in 7 days 73% in 7 days	Bismuth iodide UV spectroscopy Chemical oxygen demand Organic carbon	
C_9APE_{25}	Inoculated medium	74% in 7 days 84% in 7 days 53% in 7 days 68% in 7 days	Bismuth iodide UV spectroscopy Chemical oxygen demand Organic carbon	
C_9APE_{35}	Inoculated medium	84% in 7 days 72% in 7 days 47% in 7 days 38% in 7 days	Bismuth iodide UV spectroscopy Chemical oxygen demand Organic carbon	
C_9APE_2 C_9APE_{10}	Batch activated sludge Inoculated medium	70-80% in 7 days 58% in 5 days 26% in 5 days	Gas chromatography Organic carbon Oxygen consumption	Geiser, 1980

Appendix (cont.)

Chemical[a]	Test System/Protocol	Extent of Biodegradation[b]	Measurement Method	Reference
tp-C_9APE_{13}	Inoculated medium	8-26% in 20 days 78% in 20 days 23% in 20 days	Chemical oxygen demand Cobalt thiocyanate Organic carbon	Narkis and Schneider-Rotel, 1980
tp-C_9APE_9	Continuous-flow activated sludge	18-21% in 3 hours 68-76% in 3 hours	Bismuth iodide Organic carbon	Schöberl et al., 1981
$C_8APE_{8.5}$ (olefin-derived)	Semi-continuous activated sludge	40% in 24 hr (20 mg C/L)	Dissolved organic carbon removal	Saeger et al., 1980
C_9APE_4 (Sterox ND)	Semi-continuous activated sludge	58% in 24 hr (20 mg C/L)	Dissolved organic carbon removal	
C_9APE_9 (Sterox NJ)	Semi-continuous activated sludge	55% in 24 hr (20 mg C/L)	Dissolved organic carbon removal	
C_9APE_{10} (Sterox NK)	Semi-continuous activated sludge	68% in 24 hr (20 mg C/L)	Dissolved organic carbon removal	
$C_{12}APE_6$ (Sterox DF)	Semi-continuous activated sludge	80% in 24 hr (20 mg C/L)	Dissolved organic carbon removal	
$C_{12}APE_{10}$ (Sterox DJ)	Semi-continuous activated sludge	79% in 24 hr (20 mg C/L)	Dissolved organic carbon removal	
$C_{12}APE_{10}$ (olefin-derived)	Semi-continuous activated sludge	65% in 24 hr (20 mg C/L)	Dissolved organic carbon removal	
C_9APE_9 (Sterox NJ)	Semi-continuous activated sludge	25-45% (20 mg C/L) 20-50% (20 mg C/L)	Dissolved organic carbon removal	Unpublished data cited in Saeger et al., 1980
$C_{12}APE_{10}$ (Sterox DJ)	Semi-continuous activated sludge	20-80% (20 mg C/L) 34% (20 mgC/L)	Dissolved organic carbon removal	
APE_8	Continuous-flow activated sludge	>76%	UV spectroscopy	Pudo and Erndt, 1981

Appendix (cont.)

Chemical[a]	Test System/Protocol	Extent of Biodegradation[b]	Measurement Method	Reference
C_9APE_6	Inoculated medium	100% in 7 days (5, 20 mg/L)	Cobalt thiocyanate	Tabak and Bunch, 1981
C_9APE_8	Inoculated medium	100% in 7 days (5, 20 mg/L)	Cobalt thiocyanate	
C_9APE_{10}	Inoculated medium	100% in 7 days (5, 20 mg/L)	Cobalt thiocyanate	
C_9APE_{30}	Inoculated medium	79-95% in 7 days (5 mg/L)	Cobalt thiocyanate	
C_9APE_9	Recycling biofilter	98% in 2 days	Chemical oxygen demand	Ackermann and Frahne, 1982
C_9APE_{11}	Inoculated medium	98% in 30 days 80% in 30 days 64% in 30 days	Bismuth iodide Organic carbon UV spectroscopy	Brüschweiler and Gämperle, 1982; Brüschweiler et al., 1983
C_9APE_{23}	Inoculated medium	98% in 30 days 74% in 30 days 62% in 30 days	Bismuth iodide Organic carbon UV spectroscopy	
C_9APE_{30}	Continuous-flow activated sludge	91% in 3 hours	Bismuth iodide	IRCA, 1982
br-C_9APE_9 (^{3}H-ring label) (^{14}C-EO label)	Continuous-flow activated sludge	98-100% in 8 hours >95% in 8 hours 25 (10-42)% in 8 hours 47-59% in 8 hours	Cobalt thiocyanate Foam, surface tension ^3H_2O $^{14}CO_2$ formation	Kravetz et al., 1982
C_8APE_4	Continuous-flow activated sludge	53% in 20 hours 18% in 20 hours	Polarography	Kozarac et al., 1983
C_8APE_{10}	Continuous-flow activated sludge	45% in 20 hours 46% in 10 hours	Polarography Bismuth iodide Polarography	
C_8APE_{16}	Continuous-flow activated sludge	56% in 10 hours	Bismuth iodide	

Appendix (cont.)

Chemical[a]	Test System/Protocol	Extent of Biodegradation[b]	Measurement Method	Reference
br-C_9APE_9	Continuous-flow activated sludge	>98% (25°C) >97% (12°C) 84% (8°C)	Cobalt thiocyanate Cobalt thiocyanate Cobalt thiocyanate	Kravetz et al., 1983; 1984
(^{3}H-ring label)		29% (25°C) 10% (10°C) 2% (8°C)	^3H_2O formation	
(^{14}C-EO label)		58% (25°C) 50% (12°C) 10% (8°C)	$^{14}CO_2$ formation	
n-pri-$C_{9-10}APE_9$ n-sec-$C_{8-10}APE_9$	Continuous-flow activated sludge Inoculated medium	96% in 3 hours 68% in 3 hours 84% in 19 days 29% in 30 days	Bismuth iodide Organic carbon Bismuth iodide Oxygen consumption	Berth et al., 1984
br-C_9APE_9	Continuous-flow activated sludge	97% in 3 hours 48% in 3 hours	Bismuth iodide Organic carbon	
C_8APE_9	Continuous-flow activated sludge	95% in 6 hours	Bismuth iodide	Birch, 1984
C_9APE_{10}	Inoculated medium Batch activated sludge Continuous-flow activated sludge	10% in 28 days 10% in 28 days 99.6% 59% in 3 hours	Oxygen consumption Organic carbon Bismuth iodide Organic carbon	Gerike, 1984; Gerike and Jasiak, 1984
C_9APE_9	River sediments standing (water phase) stirring (water phase) River-water dieaway	98% in 10 days (20 mg/L) 98% in 5 days (20 mg/L) 94% in 10-16 days 100% in 10 days	Cobalt thiocyanate, HPLC Cobalt thiocyanate, HPLC Cobalt thiocyanate HPLC	Yoshimura, 1986
br-C_9APE_{10}	Continuous-flow activated sludge	96% 87-89%	Bismuth iodide Chemical oxygen demand	Neufahrt et al., 1987

Appendix (cont.)

Chemical[a]	Test System/ Protocol	Extent of Biodegradation[b]	Measurement Method	Reference
$C_{10-12}APE_9$	Anaerobic reactor	45-50%	Methane; Phosphomolybdic acid	Wagener and Schink, 1987
br-NPE_9	Continuous-flow activated sludge	25-30% (50 mg/L)	CO_2 formation	Salanitro et al., 1988
C_8APE_{1-5} C_8 carboxylic acids br-C_8APE_{1-5} br-C_8 carboxylic acids	Activated sludge inoculum Primary sewage inoculum (batch study)	~100% in 127 days ~100% in 127 days Incomplete in 127 days Incomplete in 127 days	Biological oxygen demand Biological oxygen demand Biological oxygen demand Biological oxygen demand	Ball et al., 1989
NPE_{12}	Closed-bottle test Sturm (sludge) Gledhill (sludge) Gledhill (accl. sludge)	30% in 28 days 65% in 28 days 45% in 28 days 42% in 28 days	Oxygen consumption CO_2 evolution CO_2 evolution CO_2 evolution	Hughes et al., 1989
NP C_9APE_1 C_9APE_2	Sludge-treated soil	80% in 3 weeks 80% in 3 weeks 80% in 3 weeks	HPLC/UV HPLC/UV HPLC/UV	Marcomini et al., 1989
NPE_7	Closed-bottle test EPA Gledhill (no acclimation)	60% in 28 days (2 mg/L) 40% in 28 days (10 mg/L)	Oxygen consumption CO_2 evolution	Markarian et al., 1989
C_9APE_3 + AE (385 mg/L, 241 mg/L)	Continuous-flow activated sludge	99.9% 98.8% 74.5% 70.6%	Cobalt thiocyanate Biological oxygen demand Chemical oxygen demand Total organic carbon	Patoczka and Pulliam, 1990
br-NPE_9	Inoculated medium Inoculated medium Continuous-flow activated sludge	31% 30% 14-34% (10-50 mg/L)	Biological oxygen demand CO_2 formation CO_2 formation	Kravetz et al., 1991

[a] In the absence of specific information, the octyl, nonyl, and dodecyl APE probably represent the branched commercial products derived from diisobutylene, tripropylene, and tetrapropylene, respectively.
[b] Numbers in parenthesis are initial concentrations.

REFERENCES

Ackermann, K. and D. Frahne. 1982. Biofilter: Can it decolorize effluents? Melliand Textilber. (1):66-69.

Albanese, P., R. Capuci. 1974. Biodegradation of nonionic surfactants. Note 2: Measuring biodegradation. Riv. Ital. Sost. Grasse. 51:70-81.

Baleux, B. and P. Caumette. 1974. Biodegradation of nonionic surfactants. Study of a new experimental method and screening test. Rev. Inst. Pasteur Lyon. 7:278-297.

Ball, H.A., M. Reinhard and P.L. McCarty. 1989. Biotransformation of halogenated and nonhalogenated octylphenol polyethoxylate residues under aerobic and anaerobic conditions. Environ. Sci. Technol. 23:951-961.

Barbaro, R.D. and J.V. Hunter. 1965. Effect of clay minerals on surfactant biodegradability. Purdue Conf. 20:189-196.

Berth, P., P. Gerike, P. Gode and J. Steber. 1984. Ecological evaluation of important surfactants. Proceedings of the Eighth International Congress on Surface-Active Substances, Munich, West Germany, 1:227-236.

Birch, R.R. 1984. Biodegradation of nonionic surfactants. J. Am. Oil. Chem. Soc. 61:340-343.

Blankenship, F.A. and V.M. Piccolini. 1963. Biodegradation of nonionics. Soap Chem. Specialties. 39(12):75-78,181.

Bogan, R.H., C.N. Sawyer. 1955. Studies on the relation between chemical structure and biochemical oxidation. Sewage Ind. Waters 27:917-928.

Booman, K.A., D.E. Daugherty, J. Duprè and A.T. Hagler. 1965. Degradation studies on branched-chain EO surfactants. Soap Chem. Specialties. 41(1):60-63, 116, 118-119.

Borstlap, C. and C. Kortland. 1967. Biodegradability of nonionic surfactants under aerobic conditions. Fette-Seifen-Anstrichmittel 69:736-738.

Boyer, S.L., K.F. Guin, R.M. Kelley, M.L. Mausner, H.F. Robinson, T.M. Schmitt, C.R. Stahl and E.A. Setzkorn. 1977. Analytical method for nonionic surfactants in laboratory biodegradation and environmental studies. Environ. Sci. Technol. 11:1167-1171.

Bruce, A.M., J.D. Swanwick, R.A. Ownsworth. 1966. Synthetic detergents and sludge digestion: Some plant observations. JISP. 1966:427-447.

Brüschweiler, H. 1974. Remarks on primary biodegradation on nonionic surfactants. Communication to OECD expert group on biodegradability of nonionic surfactants, January 16, 1974. (Cited in Swisher, 1987).

Brüschweiler, H. 1975. Properties and biodegradation characteristics of surfactants. Chimia 29:31-42.

Brüschweiler, H., H. Gämperle. 1982. Primary and ultimate biodegradation of alkylphenol ethoxylates. XIII Jornadas Com. Español Deterg. 55-71.

Brüschweiler, H., H. Gämperle, F. Schwager. 1983. Primary biodegradation, ultimate biodegradation, and biodegradation intermediates of alkylphenol ethoxylates. Tenside 20:317-324.

Bunch, R.L. and C.W. Chambers. 1967. A biodegradability test for organic compounds. J. Water Pollut. Control Fed. 39:181-187.

Bürger, K. 1967. Mechanism of aerobic biodegradation of nonionic surfactants and its analytical characterization. Münchner Beitr. Abwasser-, Fisch.- Flussbiol. 9:56-63.

Conway, R.A. and G. T. Waggy. 1966. Biodegradation testing of typical surfactants in industrial usage. Am. Dyestuff Reptr. 55(16):P607-P614.

Davis, L., J. Blair, C. Randall. 1979. Mixed bacterial cultures break :non-biodegradable" detergent. Ind. Wastes. 25:26-28.

Dobarganes Garcia, M.C. and J. Ruiz Cruz. 1977. Pollution of natural waters by synthetic detergents. XI. Influence of experimental variables in the biodegradation of nonionic surfactants in river water. Grasas y Aceitas. 28:161-172.

Eldib, I.A. 1963. Testing biodegradability of detergents. Soap Chem. Specialties. 39:59-63.

Fischer, W.K. 1972. Correlation between chemical constitution and biodegradability for nonionic surfactants. Proceedings of the Sixth International Congress on Surface-Active Substances, Zurich, Switzerland, 3:735-752.

Fischer, W.K., P. Gerike and R. Schmid. 1974. Method combination for sequential testing and evaluation of biodegradability of synthetic substances, e.g., organic chelants, through generally applicable gross parameters (BOD, CO_2, COD, TOC). Z. Wass.-Abwass.-Forsch. 4:99-118.

Fischer, W.K., P. Gerike and W. Holtmann. 1975. Biodegradation determinations via unspecific analyses (COD, DOC) in coupled units of the OECD confirmatory test. II. Results. Water Res. 9:1137-1141.

Frazee, C.D., Q.W. Osburn and R.O. Crisler. 1964. Application of infrared spectroscopy to surfactant degradation studies. J. Am. Oil Chem. Soc. 41:808-812.

Fuka, T. and P. Pitter. 1978. Relation between molecular structure and biodegradability of organic compounds. VII. Biodegradability of nonionic sulfated surfactants. Sb. VSChT. F22:51-73.

Fuka, T. and P. Pitter. 1980. Relation between molecular structure and biodegradability of organic compounds. IX. Biodegradability of alkylphenolpolyglycol ethers. Sb. VSChT. F23:5-45.

Garrison, L.J. and R.D. Matson. 1964. A comparison by Warburg respirometry and die-away studies of the degradability of select nonionic surfactants. J. Am. Oil Chem. Soc. 41:799-804.

Geiser, R. 1980. Mikrobieller Abbau Eines Nichtionogenen Tensids in Gegenwart von Aktivkohle. Sc. D. Thesis, ETH, Zurich. (Cited in Swisher, 1987).

Gerike, P. and R. Schmid. 1973. Determination of nonionic surfactants with the Wickbold method in biodegradation research and in river waters. Tenside 10:186-189.

Gerike, P. 1984. The biodegradability testing of poorly water soluble compounds. Chemosphere 13:169-190.

Gerike, P. and W. Jasiak. 1984. Surfactants in the recalcitrant metabolite test. Proceedings of the Eighth International Congress on Surface-Active Substances, Munich, West Germany, 1:195-208.

Han, K.W. 1967. Determination of biodegradability of nonionic surfactants by sulfation and methylene blue extraction. Tenside 4:43-45.

Hartmann, L., P. Wilderer and W. Staub. 1967. Reaction-Kinetic investigations into the biodegradation of modern surfactants through trickle filter organisms. Tenside 4:138-143.

Huddleston, R.L. and R.C. Allred. 1964a. Evaluation of detergents by using activated sludge. J. Am. Oil Chem. Soc. 41:732-735.

Huddleston, R.L. and R.C. Allred. 1964b. Effect of structure on biodegradation of nonionic surfactants. Proceedings of the Fourth International Congress on Surface-Active Substances, 3:871-882.

Huddleston, R.L. and R.C. Allred. 1965. Biodegradability of ethoxylated alkyl phenol surfactants. J. Am. Oil Chem. Soc. 42:983-986.

Huddleston R.L. 1966. Biodegradable detergents for the textile industry. Am Dyestuff Reporter 55(2):P52-P54.

Hughes, A.I., D.R. Peterson and R.K. Markarian. 1989. Comparative biodegradability of linear and branched alcohol ethoxylates. Presented at the American Oil Chemists' Society Annual Meeting, May 3-7, Cincinnati, OH.

Huyser, H.W. 1960. Relation between the structure of detergents and their biodegradation. Proceedings of the Third International Congress on Surface-Active Substances, Cologne, Germany, University of Mainz Press, 3:295-301.

Inoue, Z., J. Fukuyama and A. Honda. 1977. Toxicity and biodegradation of surfactants. Mizu Shori Gijutsu 18:119-132.

IRCA (Institute Récherche Chimique Appliquée). 1982. Nonylphenol ethoxylate surfactants are over 90% degradable. Tenside. 19:55.

Itoh, S., S. Setsuda, A. Utsunomiya and S. Naito. 1979. Studies on the biodegradation test method of chemical substances. II. Ultimate biodegradabilities of some anionic, nonionic and cationic surfactants estimated by CO_2 production. Yukagaku 28:199-204.

Janicke, W. 1968a. Indirect determination of biodegradability of nonionic surfactants by organic carbon determination. Gas-Wasserfach. 109:240-249.

Janicke, W. 1968b. Indirect determination of biodegradation of organic compounds through COD analysis. Gesundh.-Ing. 89:309-314.

Janicke, W. and G. Hilge. 1977. Determination of the elimination-degree of water-endangering substances. (General water-elimination-test). Z. Wass.-Abwass.-Forsch. 10:4-9.

Jenkins, S.H., N. Harkness, A. Lennon and K. James. 1967. The biological oxidation of synthetic detergents in recirculating filters. Water Res. 1:51-53.

Kozarac, Z.D., D. Hrsak, B. Cosovic and J. Vrzina. 1983. Electroanalytical determination of the biodegradation of nonionic surfactants. Environ. Sci. Technol. 17:268-272.

Kravetz, L., H. Chung, J.C. Rapean, K.F. Guin and W.T. Shebs. 1978. Ultimate biodegradability of detergent range alcohol ethoxylates. Presented at American Oil Chemists' Society 69th Annual Meeting, St. Louis, May 1978.

Kravetz, L., H. Chung, K.F. Guin, W.T. Shebs and L.S. Smith. 1982. Ultimate biodegradation of an alcohol ethoxylate and a nonylphenol ethoxylate under realistic conditions. Household Pers. Prod. Ind. 19(3;4):46-52, 72:62-70.

Kravetz, L., H. Chung, K.F. Guin, W.T. Shebs and L.S. Smith. 1983. Primary and ultimate biodegradation of an alcohol ethoxylate and an alkylphenol ethoxylate under average winter conditions in the USA. Presented at the American Oil Chemists' Society 74th Annual Meeting, May 1983, Chicago, IL.

Kravetz, L., H. Chung, K.F. Guin, W.T. Shebs and L.S. Smith. 1984. Primary and ultimate biodegradation of an alcohol ethoxylate and an alkylphenol ethoxylate under average winter conditions in the USA. Tenside 21:1-6.

Kravetz, L., J.P. Salanitro, P.B. Dorn and K.F. Guin. 1991. Influence of hydrophobe type and extent of branching on environmental response factors of nonionic surfactants. J. Am. Oil Chem. Soc. 68:610-618.

Kurata, N. and K. Koshida. 1975. Biodegradability of nonionic surfactants in river die away test using Tama river water. Yukagaku 24:879-881.

Kurata, N. and K. Koshida. 1976. Biodegradability of sec-alcohol ethoxylate: Experimental results in continuous-flow activated sludge tests. Yukagaku 25:499-500.

Kurata, N., K. Koshida and T. Fujii. 1977. Biodegradation of surfactants in river water and their toxicity to fish. Yukagaku 26:115-118.

Lacaze, J.C. 1973. Influence of illumination on biodegradation of nonionic surfactant used for dispersion of marine oil spills. C.r. Acad. Sci. Paris. D277:409-412.

Lashen, E.S., F.A. Blankenship, K.A. Booman and J. Dupre. 1966. Biodegradation studies on a p-t-octylphenoxypolyethoxyethanol. J. Am. Oil Chem. Soc. 43:371-376.

Lashen, E.S., G.F. Trebbi, K.A. Booman and J. Duprè. 1967. Biodegradability of nonionic detergents. Soap Chem. Specialties. 43:55-58.

Lashen, E.S. and K.A. Booman. 1967. Biodegradability and treatability of alkylphenol ethoxylates: A class of nonionic surfactants. Water Sewage Works 114:R155-R163.

Lashen, E.S. and J.C. Lamb. 1967. III. Biodegradation of a nonionic detergent. Water Wastes Eng. 4:56-59.

Mann, A.H. and V.W. Reid. 1971. Biodegradation of synthetic detergents: evaluation by community trials. II. Alcohol and alkylphenol ethoxylates. J. Am. Oil. Chem. Soc. 48:588-594.

Marcomini, A., P.D. Capel, T. Lichtensteiger, P.H. Brunner and W. Giger. 1989. Behavior of aromatic surfactants and PCBs in sludge-treated soil and landfills. J. Environ. Qual. 18:523-528.

Marei, A., T.M. Kassen and B.A. Gebril. 1976. Alkylphenol ethoxylate surfactants. Indian J. Technol. 14:447-452.

Markarian, R.K., K.W. Pontasch, D.R. Peterson and A.I. Hughes. 1989. Review and analysis of environmental data on Exxon surfactants and related compounds. Technical Report, Exxon Biomedical Sciences, Inc., East Millstone, NJ.

Mausner, M., J.H. Benedict, K.A. Booman, T.E. Brenner, et al. 1969. The status of biodegradability testing of nonionic surfactants. J. Am. Oil Chem. Soc. 46:432-444.

Moreno Danvila, A. 1979. Toxicity of surfactants during biodegradation. Behavior of *Daphnia magna* in activated sludge effluents. X. Jornadas Com. Espanol Deterg. pp. 59-80.

Myerly, R.C., J.M. Rector, E.C. Steinle, C.A. Vath and H.T. Zika. 1964. Secondary alcohol ethoxylates as degradable detergent materials. Soap Chem Spec. 40(5):78-82.

Narkis, N., M. Schneider-Rotel. 1980. Ozone-induced biodegradability of a nonionic surfactant. Water Res. 14:1225-1232.

Neufahrt, A., K. Hofmann, G. Täuber and Z. Dano. 1987. Biodegradation of nonylphenol ethoxylate (NPE) and certain environmental effects of its catabolites. Unpublished paper sent to Monsanto Europe, Brussels, Belgium.

Osburn, Q.W. and J.H. Benedict. 1966. Polyethoxylated alkyl phenols: relationship of structure to biodegradation mechanism. J. Am. Oil Chem. Soc. 43:141-146.

Patoczka, J. and G.W. Pulliam. 1990. Biodegradation and secondary effluent toxicity of ethoxylated surfactants. Water Res. 24:965-972.

Patterson, S.J., K.B.E. Tucker and C.C. Scott. 1966. Nonionic detergents and related substances in British waters. WPR Conf. #3 2:103-116.

Patterson, S.J., C.C. Scott, K.B.E. Tucker. 1968. Nonionic detergent degradation. II. TLC and foaming properties of alkylphenol ethoxylates. J. Am. Oil Chem. Soc. 45:528-532.

Pitter, P. 1968. Relation between degradability and chemical structure of nonionic polyethylene oxide compounds. Proceedings of the Fifth International Congress on Surface-Active Substances, 1:115-123.

Pitter, P. and T. Fuka. 1979. Biodegradation of non-sulfated and sulfated nonylphenol ethoxylate surfactants. Env. Protect. Eng. 5:47-56.

Pudo, J. and E. Erndt. 1981. The influence of nonionic detergents on the organism community in activated sludge. Verh. Internat. Verein. Limnol. 21:1083-1087.

Reiff, B. 1976. The effect of biodegradation of three nonionic surfactants on their toxicity to rainbow trout. Proceedings of the Seventh International Congress on Surface-Active Substances, Moscow, USSR, 4:163-176.

Rudling, L. 1972. Biodegradability of Nonionic Surfactants: A Progress Report. Swedish Water and Air Pollution Research Laboratory (IVL), Report No. B134. (Cited in Swisher, 1987).

Rudling, L. and P. Solyom. 1974. The investigation of branched NPE's. Water Res. 8:115-119.

Ruiz Cruz, J. and M.C. Dobarganes Garcia. 1976. Pollution of natural waters by synthetic detergents. X. Biodegradation of nonionic surfactants in river water. Grasas Aceites 27:309-322.

Ruiz Cruz, J. and M.C. Dobarganes Garcia. 1977. Pollution of natural waters by synthetic detergents. XII. Relation between structure and biodegradation of nonionic surfactants in river water. Grasas Aceites 28:325-331.

Ruiz Cruz, J. and M.C. Dobarganes Garcia. 1978. Pollution of natural waters by synthetic detergents. XIII. Biodegradation of nonionic surfactants in river water and determination of their biodegradability by different test methods. Grasas Aceites 29:1-8.

Saeger, V.W., R.G. Kuehnel, C. Linck and W.E. Gledhill. 1980. Biodegradation of Sterox and Dimersol olefin-derived alkylphenol ethoxylates. Report No. ES-80-SS-46, Monsanto Industrial Chemicals Company, St. Louis, MO.

Salanitro, , J.P., G.C. Langston, P.B. Dorn and L. Kravetz. 1988. Activated sludge treatment of ethoxylate surfactants at high industrial use concentrations. Presented at the International Conference on Water and Wastewater Microbiology, Newport Beach, California, February 8-11.

Sato, M., K. Hashimoto and M. Kobayashi. 1963. Microbial degradation of nonionic synthetic detergent, poly EO nonylphenol ethers. Water Treat. Eng. 4:31-36.

Sawyer, C., R.H. Bogan and J.R. Simpson. 1956. Biochemical behavior of synthetic detergents. Ind. Eng. Chem. 48:236-240.

Schöberl, P. and H. Mann. 1976. Temperature on the biodegradation of nonionic surfactants in sea-and freshwater. Arch. Fischereiwiss 27:149-158.

Schöberl, P., E. Kunkel and K. Espeter. 1981. Comparative investigations on the microbial metabolism of a nonylphenol and an oxoalcohol ethoxylate. Tenside 18:64-72.

Schönborn, W. 1966. *Analytical Determination and Biodegradation of Nonionic Surfactants in Wastewaters*. No. 12, Erich Schmidt, Berlin. (Cited in Swisher, 1987).

Sekiguchi, H., K. Miura, K. Oba and A. Mori. 1975. Biodegradation of α-olefin sulfonates and other surfactants. Yukagaku 24:145-148.

Sheets, W.D. and G.W. Malaney. 1956a. Synthetic detergents and the BOD test. Sewage Ind. Wastes. 28:10-17.

Sheets, W.D. and G.W. Malaney. 1956b. The COD values of syndets, surfactants and builders. Purdue Conf. 11:185-196. (Cited in Swisher, 1987).

Smithson, L.H. 1966. Properties of ethoxylate derivatives of nonrandom alkylphenols. J. Am. Oil Chen. Soc. 43:568-571.

Stache, H. 1976. Properties of cycloaliphatic alcohol derivatives. Proceedings of the Seventh International Congress on Surface-Active Substances, Moscow, USSR, 1:378-391.

Stead, J.B., A.T. Pugh, I.I. Kaduji and R.A. Morland. 1972. A comparison of biodegradability of some alkylphenol ethoxylates using three methods of detection. Proceedings of the Sixth International Congress on Surface-Active Substances, Zurich, Switzerland, Carl Hanser Verlag, 3:721-734.

Steinle, E.C., R.C. Myerly and C.A. Vath. 1964. Surfactants containing ethylene oxide: Relationship of structure to biodegradability. J. Am. Oil Chem. Soc. 41:804-807.

Stiff, M.J., R.C. Rootham and G.E. Culley. 1973a. The effect of temperature on the removal of nonionic surfactants during small scale activated sludge sewage treatment. I. Comparison of alcohol ethoxylates with a branched chain alkylphenol ethoxylate. Water Res. 7:1003-1010.

Stiff, M.J., R.C. Rootham and G.E. Culley. 1973b. II. Comparison of a linear alkylphenol ethoxylate with branched chain alkylphenol ethoxylates. Water Res. 7:1407-1415.

Sturm, R.N. 1973. Biodegradability of nonionic surfactants: screening test for predicting rate and ultimate biodegradation. J. Am. Oil Chem. Soc. 50:159-167.

Swisher, R.D. 1987. *Surfactant Biodegradation*, 2nd. Ed., Surfactant Science Series, Vol. 18. Marcel Dekker, Inc., New York.

Tabak, H.H. and R.L. Bunch. 1981. Measurement of nonionic surfactants in aqueous environments. Proceedings of the Purdue Industrial Waste Conference, 36:888-907.

Throckmorton, P.E., D. Aelony, R.R. Egan, F.H. Otey. 1973. New biodegradable surfactants derived from starch: preparation and properties. Tenside. 10:1-7.

Throckmorton, P.E., R.R. Egan, D. Aelony, G.K. Mulberry, F.H. Otey. 1974. Biodegradable surfactants derived from corn starch. J. Am. Oil Chem. Sci. 51:486-494.

Treccani, V., G. Braggi, E. Galli, G. Pensotti and V. Andreoni. 1973. The determination of biodegradability of surfactants: Elective culture test. Riv. Ital. Sost. Grasse. 50:418-422.

Vaicum, L., M. Cicei and L. Stefanescu. 1976. Research on biodegradability and toxicity of some polyethoxylate nonionic surfactants. Studii Epurarea Apelor 17:27-43.

Vath, C.A. 1964. A sanitary engineer's approach to biodegradation of nonionics. Soap Chem. Spec. 40(2):56-58.

Wagener, S. and B. Schink. 1987. Anaerobic degradation of nonionic and anionic surfactants in enrichment cultures and fixed-bed reactors. Water Res. 21:615-622.

Wayman, C.H., J.B. Robertson. 1963. Biodegradation of anionic and nonionic surfactants under aerobic and anaerobic conditions. Biotechnol. Bioeng. 5:367-384.

Weil, J.K. and A.J. Stirton. 1964. Biodegradation of some tallow-based surfactants in river water. J. Am. Oil Chem. Soc. 41:355-358.

Wencker, D.E., E. Allenbach and P. Laugel. 1974. Some observations on the biodegradation of surfactants. Bull. Soc. Pharm. Strasbourg 17:135-145.

Wickbold, R. 1974. Analytical comments on surfactant biodegradation. Tenside 11:137-144.

Winter, W. 1962. Biodegradation of detergents in sewage treatment. Wasserwirtsch-Wassertech. 12:265-271.

Yoshimura, K. 1986. Biodegradation and fish toxicity of nonionic surfactants. J. Am. Oil Chem. Soc. 63:1590-1596.

Zika, H.T. 1971. The use of biodegradable linear alcohol surfactants in textile wet processing. J. Am. Oil Chem. Soc. 48:273-278.

INDEX

INDEX

A

Abbreviations
   for alcohol ethoxylates, 7–12
   for alkylphenol ethoxylates, 195–198
Absorption
   of alcohol ethoxylates, 130–131, 133
   of alkylphenol ethoxylates, 194, 325
Acetaldehyde, 16
Acetate, 50
Acid-catalyzed reactions, 15
Acrylic processes, 200
Activated sludge
   alcohol ethoxylates and, 35, 37, 44, 83
   alkylphenol ethoxylates and, 236, 237, 240, 244, 250
Acute toxicity
   of alcohol ethoxylates
     in algae, 68–77
     in animal studies, 95–132
       dermal, 103–110
       inhalation, 110, 111
       oral, 95–102, 299–305
     dermal, 103–110
     in fish, 60–68
     inhalation, 110, 111
     intraperitoneal, 110, 112
     intrapleural, 110
     intravenous, 110
     in invertebrates, 68
     oral, 95–102, 299–305
     subcutaneous, 110
   of alkylphenol ethoxylates, 192, 263–270
     in algae, 276–280
     in animal studies, 299–305
     in fish, 263–270
     in invertebrates, 270–275
     oral, 299, 300–302
Adsorption
   of alcohol ethoxylates, 20, 25, 81, 83
   of alkylphenol ethoxylates, 235, 237, 247
AE, see Alcohol ethoxylates
Aerobic biodegradation, 48–50
AES, see Alcohol ethoxysulfates
Agricultural uses
   of alcohol ethoxylates, 3, 13, 86
   of alkylphenol ethoxylates, 191, 200
Air emission standards, 217
Alcohol ethoxylates (AE), see also specific types
   abbreviations for, 7–12
   absorption of, 130–131, 133
   acute toxicity of, see under Acute toxicity
   adsorption of, 20, 25, 81, 83
   advantages of, 3, 13
   aerobic biodegradation of, 48–50
   in algae, see under Algae
   alkyl chain structure of, 42–44
   anaerobic biodegradation of, 50

analgesic effects of, 5, 131–132
anesthetic effects of, 5, 87, 131–132
animal studies of, see under Animal studies
applications of, 3, 5, 13, 86, 133–134
aquatic toxicity of, see under Aquatic toxicity
behavioral responses to, 5, 78
bioaccumulation of, 59, 84
bioconcentration of, 84
biodegradation of, see under Biodegradation
carcinogenicity of, 5, 124
CAS identification for, 16–17
chemical structure of, see under Chemical structure
chemistry of, 14–16
chronic toxicity of, 75, 78–83, 114–115
clearance of, 84
defined, 7, 13
desorption of, 20
detection of, 3, see also specific methods
developmental toxicity of, 125–129
disposition of, 130–131, 133
distribution of, 4, 15, 84
elimination of, 4
environmental levels of, see under Environmental levels
extraction of, 20–21, 25, 26
field studies of biodegradation of, 37–40, 46
in fish, see under Fish
foaming ability of, 3, 21
genotoxicity of, 5, 125–128
half-lives of, 59, 84
in higher plants, 85–86
history of, 13–14
human studies of, 132–134
hydrolysis of, 4, 35, 47, 49, 83
hydrophobicity of, 40, 86
inhalation toxicity of, 5, 110, 111
in situ monitoring of, 75, 83
in invertebrates, 59, 68, 78, 81, 86
irritation from
   in animals, 115–120, 121–124
   in humans, 132–133
laboratory studies of biodegradation of, see under
   Laboratory studies of biodegradation
lipophilicity of, 87
liposolubility of, 81, 82
manufacture of, 3, 8–10, 14–16
metabolic pathways of biodegradation of, 47–50
metabolism of, 130–131, 133
mineralization of, 40, 85
modification of properties of, 15
nomenclature for, 7–12, 16–17
oral toxicity of, see under Oral toxicity
persistence of, 19
phytotoxicity of, 86
in plants, 85–86
production of, 13
reproductive toxicity of, 5, 125–129
sampling of, 3

separation of, 3, 19, 20–21
in soil, 37
structure-activity relationships of, 81–83
sublethal toxicity of, 5, 78
surface tension of, 3, 21, 43, 81
therapeutic uses of, 133–134
toxicity of, see under Toxicity
uptake of, 4, 84
Alcohol ethoxysulfates (AES), 3, 13, see also specific types
Algae
   alcohol ethoxylates in, 59, 68–77
      chronic toxicity of, 82
      modes of action of, 87
   alkylphenol ethoxylates in, 193, 276–280
Alkoxylation, 15
Alkylbenzene sulfonates, 36, 43, 44, 213
Alkyl chain structure
   of alcohol ethoxylates, 42–44
   of alkylphenol ethoxylates, 251–252
   biodegradation and, 42–44, 251–252
Alkylphenol carboxylic acids (APEC), 191, see also specific types
Alkylphenol ethoxylates (APE), 3, 26, see also specific types
   abbreviations for, 195–198
   absorption of, 194, 325
   acute toxicity of, 192, 263–270
   adsorption of, 235, 237, 247
   in algae, 193, 195, 276–280
   alkyl chain structure of, 251–252
   animal studies of, see under Animal studies
   applications of, 200–201, 325
   aquatic toxicity of, see under Aquatic toxicity
   bioaccumulation of, 193, 285–287
   biodegradation of, see under Biodegradation
   in birds, 291
   carcinogenicity of, 194, 311–314
   CAS identification for, 199, 201–202
   chemical structure of, 251–253, 299
   chemistry of, 199–200
   chronic toxicity of, 280–283, 308
   define, 191
   dermal toxicity of, 299–304, 307–311
   developmental toxicity of, 320–322
   disposition of, 322–324
   dissipation of, 239
   environmental levels of, see under Environmental levels
   excretion of, 193, 285, 325
   fate of, 239, 322–324
   field studies of biodegradation of, 240–251
   in fish, see under Fish
   genotoxicity of, 194, 314–319
   in higher plants, 288–291
   human studies of, 324–326
   inhalation toxicity of, 303
   in invertebrates, 270–275
   irritation from, 193, 194, 324–325
   lab studies of biodegradation of, see under Laboratory studies of biodegradation
   manufacture of, 199–200
   metabolic pathways of biodegradation of, 253–255
   metabolism of, 325
   mutagenicity of, 194
   nomenclature for, 195–198, 201–202
   oral toxicity of, 299–302, 305–307
   in plants, 288–291
   in soil, 250–251
   structure-activity relationships of, 281–284
   sublethal toxicity of, 280
   teratogenicity of, 194
   therapeutic uses of, 325
   uptake of, 193, 285
   in wildlife, 291
Anaerobic biodegradation, 50
Anaerobic sludge digesters, 50, 239
Analgesic effects, 5, 131–132
Analytical methods, see also specific types
   for alcohol ethoxylate determination, 3, 19–26
   for alkylphenol ethoxylate determination, 205–216
Anesthetic effects, 5, 87, 131–132
Anesthetics, see also specific types
Animal studies
   of alcohol ethoxylates, 5, 95–132
      absorption, 130–131
      acute toxicity, 95–112
         dermal, 103–110
         inhalation, 110, 111
         oral, 95–102
      analgesic, 131–132
      anesthetic, 131–132
      carcinogenicity, 124
      chronic toxicity, 114–115
      developmental toxicity, 125–129
      disposition, 130–131
      genotoxicity, 125–128
      irritation, 115–124
      metabolism, 130–131
      reproductive toxicity, 125–129
      subchronic toxicity, 112–114
   of alkylphenol ethoxylates, 299–324
      acute toxicity, 299–305
      carcinogenicity, 311–314
      chronic toxicity, 308
      dermal, 304, 308–311
      developmental toxicity, 314, 320–322
      disposition, 322–324
      fate, 322–324
      genotoxicity, 314–319
      reproductive, 314, 320–322
      subchronic toxicity, 305–308
Anion exchange resins, 20–21, 27
APE, see Alkylphenol ethoxylates
APEC, see Alkylphenol carboxylic acids
Applications

of alcohol ethoxylates, 3, 5, 13, 86, 133–134
of alkylphenol ethoxylates, 191, 200–201, 325
Aquatic toxicity
    of alcohol ethoxylates, 59–84
        acute, 60–77
            in algae, 68–77
            in fish, 60–68
            in invertebrates, 68, 78
        bioaccumulation and, 83
        chronic, 75, 78–81
        environmental variables and, 83
        structure-activity relationships and, 81–83
        sublethal, 78
    of alkylphenol ethoxylates, 235, 263–287
        acute, 263–270
            in algae, 276–280
            in fish, 263–270
        bioaccumulation and, 285–287
        chronic, 280–283
        environmental variables and, 285
        structure-activity relationships and, 281–284
        sublethal, 280
Autoradiography, 84

B

Barium salts, 23, 208
BAS, see Batch activated sludge
Batch activated sludge (BAS) system, 236
Batch jars, 46
BCF, see Bioconcentration factors
Behavioral responses, 5, 78
Bench-scale biotreaters, 46
BIAS, see Bismuth iodide active substances
Bioaccumulation
    of alcohol ethoxylates, 59, 84
    of alkylphenol ethoxylates, 193, 285–287
    toxicity and, 84, 285–287
Bioconcentration, 84
Bioconcentration factors (BCF), 193, 285, 286
Biodegradation
    aerobic, 48–50
    of alcohol ethoxylates, 3, 13, 35–50
        anaerobic, 50
        chemical structure and, 40–46
        environmental variables and, 46–47
        field studies of, 37–40, 46
        laboratory studies of, 35–37, 46, 143–179
        metabolic pathways of, 47–50
        resistance to, 49
        retardation of, 4
    alkyl chain structure and, 42–44, 251–252
    of alkylphenol ethoxylates, 192, 235–255
        chemical structure and, 251–253
        field studies of, 240–251
        metabolic pathways of, 253–255
        in soil, 250–251

        in wastewater, 240–250
        anaerobic, 50
        chain length and, 41–42, 252–253
        chemical structure and, 40–46, 251–253
        defined, 35
        environmental variables and, 46–47
        ethoxylate chain length and, 41–42
        field studies of
            for alcohol ethoxylates, 37–40, 46
            for alkylphenol ethoxylates, 240–251
        of hydrocarbons, 85
        laboratory studies of
            for alcohol ethoxylates, 35–37, 46, 143–179
            for alkylphenol ethoxylates, 235–239, 335–357
        metabolic pathways of, 47–50, 253–255
        primary, 35, 43, 45, 191, 235, 237–239, 254
        resistance to, 49
        retardation of, 4
        in soil, 37, 250–251
        speed of, 40
        temperature and, 46–47, 237
        ultimate, 35, 43, 235, 237–239, 255
Biodegradation intermediates, 27, 28
Biodegradation products, 19–21, 48, 192, see also specific types
Biological oxygen demand (BOD)
    alcohol ethoxylates and, 35–37, 41, 42, 44, 45
    alkylphenol ethoxylates and, 235, 236, 250
Biotransformation products, 244
Biphenyls, 85
Birds, 291
Bismuth iodide, 3
Bismuth iodide active substances (BIAS) method
    for alcohol ethoxylate detection, 22–23, 27, 36, 37, 40, 44
    for alkylphenol ethoxylate detection, 191, 207, 245
BOD, see Biological oxygen demand
Borate-modified oxidation, 15
Brackish pond water, 40

C

CAD, see Collisionally activated decomposition
Cadmium chloride, 23, 208
Calcium salts, 23, 208
Carbon dioxide, 47, 50, 194, 235
    evolution of, 43, 46, 48, 236
    formation of, 35, 44, 235, 238
    production of, 36
Carcinogenicity
    of alcohol ethoxylates, 5, 124
    of alkylphenol ethoxylates, 194, 311–314
Cardiotoxicity, 194, 299, 307, 322
CAS, see Continuous activated-sludge
CAS identification

of alcohol ethoxylates, 16–17
of alkylphenol ethoxylates, 199, 201–202
Cation exchange resins, 20–21, 27
Cell permeability, 86
Central fission, 48–49
Cetylpolyethylene glycol, 86
Chain branching, 42–44
Chain length, 41–42, 81, 84, 86, 252–253
Chemical ionization (CI), 191, 214, 215
Chemical methods, see also specific types
    for alcohol ethoxylate detection, 3, 21–24
    for alkylphenol ethoxylate detection, 206–208
Chemical non-specific methods, see also specific types
    for alcohol ethoxylate detection, 19, 35, 37, 59
    for alkylphenol ethoxylate detection, 191
Chemical oxygen demand (COD), 35, 235, 253
Chemical-specific methods, 19, see also specific types
Chemical structure
    of alcohol ethoxylates, 4, 5, 7, 14
        biodegradation and, 40–46
        determination of, 25
        toxicity and, 59, 75
    of alkylphenol ethoxylates, 251–253, 299
        biodegradation and, 40–46, 251–253
        determination of, 25
        toxicity and, 59, 75, 299
Chemistry
    of alcohol ethoxylates, 14–16
    of alkylphenol ethoxylates, 199–200
Chromatography, 24, 314, see also specific types
    gas, see Gas chromatography (GC)
    high performance liquid, see High performance liquid chromatography (HPLC)
    thin-layer, see Thin-layer chromatography (TLC)
Chronic toxicity
    of alcohol ethoxylates, 75, 78–83, 114–115
    of alkylphenol ethoxylates, 280–283, 308
CI, see Chemical ionization
Cleaners, 3, 191, 200, 201
Clearance, 84
Cleavage, 40, 48, 50, 192
CMCs, see Critical micelle concentrations
Cobalt thiocyanate active substances (CTAS) method
    for alcohol ethoxylate detection, 3, 22, 26–28, 37, 40, 43, 44, 47, 81
    for alkylphenol ethoxylate detection, 191, 205–207, 226, 245, 250, 252, 253
Cobalt thiocyanate spectrophotometry, 21
COD, see Chemical oxygen demand
Collisionally activated decomposition (CAD), 216
Colorimetry, 21
Continuous activated-sludge (CAS) test
    for alcohol ethoxylate detection, 37
    for alcohol ethoxylate detection, 35, 36, 49, 50
Continuous bench-scale biotreaters, 46
Continuous flow-through activated sludge units, 60

Contraceptive uses
    of alcohol ethoxylates, 5, 133–134
    of alkylphenol ethoxylates, 325
Cosmetics, 3, 5, 13, 134, 326
Critical micelle concentrations (CMCs), 21
Crustaceans, 59, 193
CTAS, see Cobalt thiocyanate active substances

D

DCI, see Desorption chemical ionization
Depolymerization, 50
Dermal sensitization
    alcohol ethoxylates and, 120–124, 132–133
    alkylphenol ethoxylates and, 311, 324–325
Dermal toxicity
    of alcohol ethoxylates
        acute, 103–110
        chronic, 115
        in humans, 132–133
        subchronic, 114
    of alkylphenol ethoxylates, 299–304, 307–311
Desorption
    of alcohol ethoxylates, 20
    of alkylphenol ethoxylates, 212, 216
Desorption chemical ionization (DCI), 216
Detection, see also specific methods
    of alcohol ethoxylates, 3
    of alkylphenol ethoxylates, 191
Detergents, 3, 13
Developmental toxicity
    of alcohol ethoxylates, 125–129
    of alkylphenol ethoxylates, 314, 320–322
Die-away screening test, 44
1,4–Dioxane, 16
Disc-flame ionization detection, 25
Disposition
    of alcohol ethoxylates, 130–131, 133
    of alkylphenol ethoxylates, 322–324
Dissipation, 239
Dissolved organic carbon (DOC), 36, 44, 45, 236
Distribution, 4, 15, 84
DOC, see Dissolved organic carbon

E

ECD, see Electron capture detection
Effluent
    alcohol ethoxylates in, 25, 37, 40, 60, 81, 83
    alkylphenol ethoxylates in, 192, 226, 235, 240, 245, 246
EI, see Electron impact
Elastomers, 191, 200
Electron capture detection (ECD), 215
Electron impact (EI), 25, 191, 214, 215, 226
Elimination, 4

ELS, see Evaporative light scattering
Environmental levels
   of alcohol ethoxylates, 19–28
      analytical methods for determination of, 19–26
      in water, 27–28
   of alkylphenol ethoxylates, 205–226
      analytical methods for determination of, 205–216
      toxicity and, 287
      in water, 217–226
Environmental variables, 46–47, 83, 285, see also specific types
Enzymes, 86, see also specific types
EO (ethylene oxide), see Alkylphenol ethoxylates (APE)
Epidemiology, 134, 326
Ethoxylate chain length, 41–42
Ethoxylation, 14, 24, 84
Ethylene oxide (EO), see Alkylphenol ethoxylates (APE)
Evaporative light scattering (ELS), 3, 214
Excretion, 193, 285, 325
Extraction, see also specific methods
   of alcohol ethoxylates, 20–21, 25, 26
   of alkylphenol ethoxylates, 191, 211, 214
   liquid-liquid, 26
   solid-phase, 20
   toluene, 21, 25
Eye irritation
   from alcohol ethoxylates, 119–124
   from alkylphenol ethoxylates, 194, 308–313

F

FAB, see Fast atom bombardment
Fast atom bombardment (FAB), 26, 216, 222
Fate, 239, 322–324
FD, see Field desorption
FID, see Flame ionization detection
Field desorption (FD), 212, 216
Field studies of biodegradation
   of alcohol ethoxylates, 37–40, 46
   of alkylphenol ethoxylates, 240–251
Fish
   alcohol ethoxylates in, 4, 5, 60–68
      bioaccumulation of, 59, 84
      chronic toxicity of, 81–83
      mode of action of, 86, 87
      sublethal effects of, 78
   alkylphenol ethoxylates in, 192, 193, 285
      acute toxicity of, 263–270
Fission, 47–49
Fixed-bed reactors, 50, 239
Flame ionization detection (FID), 25, 212, 214, 222
Flow-through activated sludge units, 60, 81
Fluorescence detection
   of alcohol ethoxylates, 3, 19, 24
   of alkylphenol ethoxylates, 213, 214, 223, 244, 245
Foaming ability, 3, 21, 205

Foam stripping (solvent sublation, gas stripping), 20, 211, 213
Formaldehyde, 16

G

Gas chromatography (GC)
   for alcohol ethoxylates, 3, 19, 21, 24–25, 28, 43, 48
   for alkylphenol ethoxylates, 208, 212, 214–215, 222, 244, 252, 255, 314
Gas exchange, 86
Gas stripping (solvent sublation, foam stripping), 20, 211, 213
GC, see Gas chromatography
Genotoxicity
   of alcohol ethoxylates, 5, 125–128
   of alkylphenol ethoxylates, 194, 314–319
Glucuronic acid, 194
Glycol ethers, 200
Gravimetry, 23, 208
Groundwater, 226

H

Half-lives, 59, 84
Higher plants
   alcohol ethoxylate toxicity to, 85–86
   alkylphenol ethoxylate toxicity in, 288–291
High performance liquid chromatography (HPLC)
   for alcohol ethoxylates, 3, 19, 25–26
   for alkylphenol ethoxylates, 191, 208, 211–216, 223, 244, 245
Household laundry detergents, 3
HPLC, see High performance liquid chromatography
Human studies
   of alcohol ethoxylates, 132–134
   of alkylphenol ethoxylates, 324–326
Hydrocarbons, 85, 215, see also specific types
Hydrolysis, 4, 35, 47, 49, 83
Hydrophobicity, 40, 86

I

Industrial cleaners, 3, 13, 200, 201
Influent, 37, 40, 60, 240, 246
Infrared detection
   of alcohol ethoxylates, 24, 28
   of alkylphenol ethoxylates, 216, 253
Inhalation toxicity
   of alcohol ethoxylates, 5, 110, 111
   of alkylphenol ethoxylates, 303
In situ monitoring, 75, 83
Instrumental analyses, see also specific types
   of alcohol ethoxylates, 24–26
   of alkylphenol ethoxylates, 208–216
Intramolecular scission, 49

Invertebrates
    alcohol ethoxylates in, 59, 68, 78, 81, 86
    alkylphenol ethoxylates in, 270–275
Iodobismuthate titration, 21
Ion-exchange, 20, 25
Ion exchange, 25, 28, 86, 211
IR, see Infrared
Irritation
    from alcohol ethoxylates
        in animals, 115–124
        in humans, 132–133
    from alkylphenol ethoxylates, 193, 194, 324–325
    eye, see Eye irritation
    mucosal, 120

L

Laboratory studies of biodegradation
    of alcohol ethoxylates, 35–37, 46, 143–179
    of alkylphenol ethoxylates, 235–239, 335–357
LAS, see Linear alkylbenzene sulfonates
Laundry detergents, 3, 13
Life-cycle studies, 81
Light scattering, 3, 214
Linear alkylbenzene sulfonates (LAS), 43, 44, 213
Linear alkyl chains, 45
Lipophilicity, 87
Liposolubility, 81, 82
Liquid-liquid extraction, 26
Liquid scintillation counting, 84
Liquid-solid chromatography, 314
Lung injury, 5

M

Magnesium sulfate, 24
Manufacture
    of alcohol ethoxylates, 3, 8–10, 14–16
    of alkylphenol ethoxylates, 199–200
Mass spectrometry (MS)
    for alcohol ethoxylates, 3, 19, 24–26, 28
    for alkylphenol ethoxylates, 191, 212, 214–216, 222, 223, 226, 244, 252, 255, 314
Membrane permeability, 86
Mercuric chloride, 23, 208
Metabolic pathways of biodegradation
    of alcohol ethoxylates, 47–50
    of alkylphenol ethoxylates, 253–255
Metabolism, 130–131, 133, 325
Methane, 50
Methyl branching, 40–41
Microorganisms, 46;
    acclimation of, 46
    air-borne, 50
    alcohol ethoxylate effects on, 5
    biodegradation and, 35
    source of, 46

toxicity and, 84–85, 287–288
    unacclimated, 36
Mineralization, 40, 85
Mixed-bed ion exchange, 25
Modification of properties, 15
MS, see Mass spectrometry
Mucosal irritation, 120
Mutagenicity, 194

N

Nitrification, 85
NMR, see Nuclear magnetic resonance
Nomenclature
    for alcohol ethoxylates, 7–12, 16–17
    for alkylphenol ethoxylates, 195–198, 201–202
Nonylphenol ethoxylates (NPE), see Alkylphenol ethoxylates (APE)
NPE (nonylphenol ethoxylates), see Alkylphenol ethoxylates (APE)
Nuclear magnetic resonance (NMR)
    for alcohol ethoxylates, 26, 28
    for alkylphenol ethoxylates, 236, 237

O

Oil industry, 3, 13
Oral toxicity
    of alcohol ethoxylates
        acute, 95–102
        chronic, 114–115
        subchronic, 113
    of alkylphenol ethoxylates, 299, 300–302, 305–307
Oxidation
    of alcohol ethoxylates, 4, 15, 35, 47, 49, 50
    of alkylphenol ethoxylates, 192
    borate-modified, 15

P

Paper industry, 3, 13, 191
PEGs, see Polyethylene glycols
Persistence, 19, 205
Personal care products, 191, 201
Pesticide spraying programs, 251
Phenanthrene, 85
Phosphate, 13
Physical methods, see also specific types
    for alcohol ethoxylate detection, 3, 21
    for alkylphenol ethoxylate detection, 205–206
Physicochemical methods, see also specific types
    for alcohol ethoxylates, 24
    for alkylphenol ethoxylates, 208
Phytotoxicity, 86
Plants, 85–86, 288–291
Plastics, 191, 200

INDEX

Polyethylene glycols (PEGs), 21, 24, 27, 28, 40, 42, 47, 49
    alkylphenol ethoxylates and, 206, 239
Polymerization, 200
Pond water, 40
Potassium picrate active substances (PPAS) method
    for alcohol ethoxylate detection, 23, 28
    for alkylphenol ethoxylate detection, 205, 208
PPAS, see Potassium picrate active substances
Process industries, 3, 13
Production, 13
Propionate, 50
Propylene oxide, 15, 45–46
Proteins, 86, see also specific types

R

Reproductive toxicity
    of alcohol ethoxylates, 5, 125–129
    of alkylphenol ethoxylates, 314, 320–322
River water
    alcohol ethoxylates in, 26, 35, 37
    alkylphenol ethoxylates in, 208, 215, 218–219, 221
River-water dieaway method, 235

S

Sampling, 3
SARs, see Structure-activity relationships
SCAS, see Semi-continuous activated-sludge
Scission, 49, 50
Secondary alcohols, 44, see also specific types
Sediment
    alcohol ethoxylates in, 50, 81
    alkylphenol ethoxylates in, 223–226
    toxicity and, 81
Semi-continuous activated-sludge (SCAS) test
    for alcohol ethoxylate detection, 35, 36, 43–46, 60, 75
    for alkylphenol ethoxylate detection, 236
Separation, see also specific methods
    of alcohol ethoxylates, 3, 19, 20–21
    of alkylphenol ethoxylates, 191, 208
Sewage, 26, 36, 192, 215, 246, 250
Sewage retention time (SRT), 237
Sewage treatment plants, 37, 44, 83, 246
Shake-flask test
    of alcohol ethoxylates, 35, 36, 43, 44, 46
    of alkylphenol ethoxylates, 235, 236
Silica gel adsorption, 25
Sludge, 36
    activated, see Activated sludge
    alcohol ethoxylates in, 26, 35, 44, 50, 60
    alkylphenol ethoxylates in, 192, 237, 246–250
    unacclimated, 44
Sludge retention time (SRT), 37, 47
Soil, 37, 250–251

Solid-phase extraction, 20
Solvent sublation (foam stripping, gas stripping), 20, 211, 213
Spectrometry, 23, 208, see also specific types
    mass, see Mass spectrometry (MS)
SRT, see Sludge retention time
Standards, 27, 217
Structure-activity relationships (SARs)
    of alcohol ethoxylates, 81–83
    of alkylphenol ethoxylates, 281–284
Subchronic toxicity, 112–114, 193, 305–308
Sublethal toxicity
    of alcohol ethoxylates, 5, 78
    of alkylphenol ethoxylates, 280
Surface tension, 3, 21, 43, 81, 205

T

Tap water, 83
Temperature, 46–47, 83, 237
Teratogenicity, 194
Textile industry, 3, 13, 191, 200, 222
Therapeutic uses, 133–134, 325
Thermal conductivities, 214
Thin-layer chromatography (TLC)
    for alcohol ethoxylate determination, 24, 28, 37, 40, 46, 48, 84
    for alkylphenol ethoxylate determination, 207, 208, 217, 251
Titration, 21
TLC, see Thin-layer chromatography
TOC, see Total organic carbon
Toluene extraction method, 21, 25
Total organic carbon (TOC), 35, 36, 44, 235
Toxicity
    acute, see Acute toxicity
    of alcohol ethoxylates, 4, 5, 59–87
        acute, see Acute toxicity
        in algae, 68–77
        aquatic, see under Aquatic toxicity
        bioaccumulation and, 84
        chemical structure and, 59, 75
        chronic, 75, 78–83
        developmental, 125–129
        environmental variables and, 83
        in fish
            acute, 60–68
            chronic, 81–83
            mode of action of, 86, 87
        in higher plants, 85–86
        inhalation, 5, 110, 111
        in invertebrates, 59, 68, 78, 81, 86
        irritation, 115–120
        microorganisms and, 84–85
        mode of action of, 86–87
        in plants, 85–86
        reproductive, 5, 125–129

sublethal, 5, 78
of alkylphenol ethoxylates, 192, 193
  acute, see Acute toxicity
  in algae, 276–280
  aquatic, see under Aquatic toxicity
  bioaccumulation and, 285–287
  chronic, 280–283
  environmental variables and, 285
  in fish, 263–270
  in higher plants, 288–291
  inhalation, 303
  in invertebrates, 270–275
  microorganisms and, 287–288
  mode of action of, 291
  in plants, 288–291
  reproductive, 314, 320–322
  sublethal, 280
aquatic, see Aquatic toxicity
bioaccumulation and, 84, 285–287
cardio-, 194, 299, 307, 322
chemical structure and, 59, 75, 299
chronic, see Chronic toxicity
dermal, see Dermal toxicity
developmental, 125–129, 314, 320–322
environmental variables and, 83, 285
geno-, 5, 125, 126–128, see Genotoxicity
inhalation, 5, 110, 111, 302
microorganisms and, 84–85, 287–288
mode of action of, 86–87, 291
oral, see Oral toxicity
reproductive, 5, 125–129, 314, 320–322
sediment and, 81
subchronic, 112–114, 193, 305–308
sublethal, 5, 78, 280
temperature and, 83
Trickling filter systems, 37, 46, 246
Typical maximum, 16

U

Ultraviolet detection
  of alcohol ethoxylates, 3, 19, 24, 26, 28
  of alkylphenol ethoxylates, 192, 208, 212, 213, 216, 238, 244, 245
Upper bound concentration, 16
Uptake
  of alcohol ethoxylates, 4, 84
  of alkylphenol ethoxylates, 193, 285
Uses, see Applications
UV, see Ultraviolet

V

Vinyl acetate, 200
Volatilization, 237

W

Wastewater
  alcohol ethoxylates in, 4, 19, 25, 26, 37–40, 46
  alkylphenol ethoxylates in, 192, 208, 213, 215, 222, 235, 237, 240–250
Water, 27–28, see also specific types
  brackish pond, 40
  hardness of, 83
  natural bodies of, 27–28, 217–226, see also specific types
  pond, 40
  quality standards for, 27, 217
  river, 26, 35, 37, 208, 215, 218–219, 221
  tap, 83
  toxicity in, see Aquatic toxicity
  waste-, see Wastewater
Water fleas, 59, 81
Wetting agents, 86